AF807196

The Human Rights of Older Persons

Bridget Lewis · Kelly Purser · Kirsty Mackie

The Human Rights of Older Persons

A Human Rights-Based Approach to Elder Law

Springer

Bridget Lewis
Australian Centre for Health
Law Research
Faculty of Law
Queensland University of Technology
Brisbane, Qld, Australia

Kelly Purser
Australian Centre for Health
Law Research
Faculty of Law
Queensland University of Technology
Brisbane, Qld, Australia

Kirsty Mackie
Faculty of Law
University of the Sunshine Coast
Sunshine Coast, Qld, Australia

ISBN 978-981-15-6734-6 ISBN 978-981-15-6735-3 (eBook)
https://doi.org/10.1007/978-981-15-6735-3

© Springer Nature Singapore Pte Ltd. 2020
This work is subject to copyright. All rights are reserved by the Publisher, whether the whole or part of the material is concerned, specifically the rights of translation, reprinting, reuse of illustrations, recitation, broadcasting, reproduction on microfilms or in any other physical way, and transmission or information storage and retrieval, electronic adaptation, computer software, or by similar or dissimilar methodology now known or hereafter developed.
The use of general descriptive names, registered names, trademarks, service marks, etc. in this publication does not imply, even in the absence of a specific statement, that such names are exempt from the relevant protective laws and regulations and therefore free for general use.
The publisher, the authors and the editors are safe to assume that the advice and information in this book are believed to be true and accurate at the date of publication. Neither the publisher nor the authors or the editors give a warranty, expressed or implied, with respect to the material contained herein or for any errors or omissions that may have been made. The publisher remains neutral with regard to jurisdictional claims in published maps and institutional affiliations.

This Springer imprint is published by the registered company Springer Nature Singapore Pte Ltd.
The registered company address is: 152 Beach Road, #21-01/04 Gateway East, Singapore 189721, Singapore

In memory of Ian Mackie.

Preface

The ageing global population presents challenges not only for individuals themselves but also for a range of other stakeholders, notably governments and policy-makers, legal, financial and health professionals, and the aged care sector. The paid and unpaid workforces, financial markets, housing and transportation infrastructure, health and aged care, social security and welfare systems, for example, will all be affected. There are also significant implications for families, friends and unpaid carers of older people, which also highlight the importance of intergenerational bonds existing within society (discussed in Chap. 1).

The perception that 'ageing' is a process to be endured, not enjoyed, and that 'old age' is something to be feared, is persistent. The idea that older people are vulnerable, frail, dependent, inflexible and a drain on finite financial and health resources also endures. Such notions are representative of the ageist assumptions that pervade modern society, often reinforcing poor intergenerational relations and discriminatory conduct, particularly in the workforce (discussed in Chap. 5). Capacity (discussed in Chap. 6) also continues to raise challenges, especially in the light of mentally disabling conditions such as dementia. Elder abuse, although underreported, is unacceptably common (discussed in Chap. 7). Negative outcomes can also result such as financial insecurity (discussed in Chap. 8), lack of secure housing (Chap. 9), as well as a lack of access to quality health and aged care services (Chap. 10).

The human rights of older persons are violated where they experience abuse, neglect or maltreatment, or when they are unable to access justice, appropriate instead healthcare or an adequate standard of living including safe housing. When older persons are unable to participate in education or employment due to age-based discrimination or are precluded from fully participating in their communities because of physical, systemic or social barriers, their human rights are also infringed. Significant too is the fact that human rights harms can frequently result in further human rights violations, thus often reinforcing vulnerabilities and inequities experienced throughout the life course.

Human rights are guaranteed under international treaties such as the *Universal Declaration of Human Rights*, *International Covenant on Civil and Political Rights* and the *International Covenant on Economic, Social and Cultural Rights*, as well as regional instruments and domestic laws (discussed in Chaps. 2–4). Governments are therefore obliged to take steps to respect and protect human rights, including ensuring that both government and non-government actors adhere to appropriate standards. However, although older people are entitled to the same rights as all individuals, they have specific needs which often go unmet, a situation which is compounded by ageist attitudes. Absent an international convention on the rights of older persons (CROP), countries lack the specific guidance and obligations to adequately address older people's human rights, and the international community lacks the necessary oversight and enforcement which would come with a dedicated treaty.

The law, both internationally and nationally, therefore has a significant role to play. The multifaceted, intersecting and often complex areas of law which impact older persons, either directly or indirectly, comprise an amorphous grouping known as 'elder law'. They are frequently fragmented and inconsistent across jurisdictions. Crucially, the law is also the medium through which many human rights threats are perpetrated against older persons, whilst simultaneously offering a means through which to seek remedial relief if harms occur. The issue of access to justice is thus fundamental in attempting to seek any legally enforceable relief.

The need to dedicate more attention to the various issues associated with ageing is increasingly being recognized. This work of research contributes to that discourse through synthesizing and critically analysing the existing literature across a number of intersecting areas to generate a new understanding of 'elder law' within a proposed human rights framework (set out in Chap. 3). Whilst this book argues for more research in dedicated areas (discussed throughout), one of the fundamental premises advanced is that the issues arising in 'old age' are often the culmination of experiences occurring throughout the whole life course and, thus, 'ageing' is not something that happens to one's future self. Consequently, a paradigm shift must occur transitioning away from traditional medical and charity-based models of responding to 'old age' to instead acknowledge older persons as active holders of legally enforceable rights. It is argued here that a CROP is an essential tool in achieving this, but that even without a CROP there is much to be gained from a human rights-based approach.

The critical discussion undertaken exploring the role of the human rights principles of autonomy, dignity, equality, liberty, non-discrimination and participation, as well as the specific rights they underpin (Chaps. 2–4), assists in generating a new appreciation for the range of impacts, vulnerabilities and inequities that older persons experience in the identified realms of financial security, accommodation, health and aged care, particularly when considering the crosscutting issues of ageism, capacity and elder abuse. The concluding chapter (Chap. 11) therefore draws together a selection of the key themes emerging from this novel analysis before presenting the main recommendations designed to promote human rights-based legal and policy reform. Accordingly, this work is useful for people within a number of disciplines including those involved in academia, policy, as well as legal and health practice. However, this work does not, and indeed cannot, purport to delve into jurisdictional intricacies given that its focus is on raising conceptual issues relevant to developing a human rights framework for 'elder law'. It is the critical discussion of the literature combined with the authors' practical experience and theoretical knowledge of the broader and interconnected contextual settings that combine to contribute new knowledge to the existing discourse. No other work has argued for this specific approach in relation to the domains of finance, accommodation and health analysed in the context of the fundamental issues of ageism, capacity and abuse.

We do recognize, however, that authentically addressing the ageist assumptions and practices pervasive in the current legal and policy approaches to ageing will require detailed and considered input from a wide variety of stakeholders. To this end, this book also champions the necessity of understanding the lived experiences of older persons and hearing their voices in order to achieve meaningful outcomes. We have therefore undertaken this work with a view to contributing to the literature on what is a crucial measure of our society—the treatment of older people.

Brisbane, Australia Bridget Lewis
Brisbane, Australia Kelly Purser
Sunshine Coast, Australia Kirsty Mackie

Acknowledgements The authors acknowledge the financial assistance received from the Faculty of Law, Queensland University of Technology and the Australian Centre for Health Law Research, which supported the research and writing of this book.

We also wish to thank the following people for their invaluable research assistance: Helene Jacmon, Wyatt Cook-Revell and Jesse To. Any errors which remain are entirely ours.

Contents

Abbreviations

ADL	Activities of daily living
AHD	Advanced health directive
CCRC	Continuing Care Retirement Community
CEDAW	Convention on the Elimination of All Forms of Discrimination Against Women
CESCR	Committee on Economic, Social and Cultural Rights
CRC	Convention on the Rights of the Child
CROP	Convention on the Rights of Older Persons
CRPD	Convention on the Rights of Persons with Disabilities
ECH	Extra care housing
EPA	Enduring power of attorney
EPAU	Elder Abuse Prevention Unit
HRC	Human Rights Committee
IADL	Instrumental activities of daily living
ICCPR	International Covenant on Civil and Political Rights
ICESCR	International Covenant on Economic, Social and Cultural Rights
OECD	Organisation for Economic Cooperation and Development
OEWGA	Open-Ended Working Group on Ageing
UDHR	Universal Declaration of Human Rights
UN	United Nations
WHO	World Health Organization

Chapter 1
Introduction

The world is experiencing a demographic shift as life expectancies extend and birth rates drop. The associated economic and social transformations present numerous challenges for our societies but, to date, policy responses have often been inadequate or inappropriate and have failed to sufficiently address systemic problems such as ageism and elder abuse. In many cases, laws which are designed to support or protect older persons in fact perpetuate infringements of their human rights, either through poor design, insufficient resourcing, denial of access to justice or a combination of these factors. This book argues that an approach based on the laws and principles of human rights would provide a better foundation for the design and implementation of all laws which affect and interact with the lives of older persons. This body of 'elder law' includes the laws which regulate discrimination, legal capacity, elder abuse, financial and social security, estate planning, accommodation services, and health and aged care. It is argued that a shift to a human rights-based approach would help to alter community attitudes about older persons, moving away from ideas premised on protection and dependence, towards a greater valuing of each individual's inherent dignity, autonomy and capability, regardless of their age.

1.1 Introduction

The fact that populations are ageing at unprecedented rates is an inescapable global phenomenon. The United Nations (UN) has identified that 'population ageing is poised to become one of the most significant social transformations of the twenty-first century'.[1] This demographic change presents a number of challenges, only some of which can be fully appreciated at the present time, with novel issues expected to materialise into the future. These challenges are developing alongside, and in connection with, rapid advancements in technological innovation, which promises

[1] United Nations, Ageing.

© Springer Nature Singapore Pte Ltd. 2020
B. Lewis et al., *The Human Rights of Older Persons*,
https://doi.org/10.1007/978-981-15-6735-3_1

potential solutions but also raises new problems requiring careful consideration. They are also occurring in the age of climate change, which introduces new threats to health and well-being, and presents an enormous economic and social challenge of its own.[2]

The challenges presented by the globally ageing population confront not only the ageing individuals themselves but also a range of other stakeholders. The scale of the demographic shift is a huge economic challenge for governments, as more people move out of the paid workforce (thereby diminishing the tax-base) and come to require greater support through social security, as well as health and aged care. There will be implications across society involving the paid workforce, financial markets, housing and transportation infrastructure, health and aged care, and social security. As these existing systems, and the people who work within them, are placed under greater strain there are also significant impacts for the families, friends and carers of older people, as well as for intergenerational bonds within society generally. The consequences for older individuals can be particularly damaging and in recent years reports of neglect, maltreatment and abuse have become increasingly common.[3] As the number of older people within our communities increases, these problems become more challenging to address, while at the same time becoming more pressing.

This is not to say that ageing is a negative experience—it is not. Nor should the ageing population be viewed as a 'threat' or our thinking limited to problematisation. Nevertheless, ageist stereotypes continue to pervade many societies, represented by notions that older persons are vulnerable, dependent, frail, inflexible and a burden on the community.[4] Attention is increasingly being focused on 'rethinking' ageing, in particular reconsidering the idea that 'ageing' is tied to a specific chronological age, in an effort to combat these attitudes.[5] The age at which a person becomes 'older' is the subject of much debate, and there is growing realisation that negative attitudes can attach to a particular age if too much emphasis is placed on a specific number.[6] It is also essential to recognise that older persons are by no means a homogenous group. However we define 'older persons', the category will represent a diverse mix of backgrounds, experiences, preferences, interests and skills—as diverse as any age-defined group within society. Having said this, there is some utility to defining age and 'older persons' by reference to a chronological age, particularly in relation to population estimates and for research purposes such as understanding the prevalence of elder abuse.

Experiences of ageing, and the negative associations that currently attach to ageing, can have significant consequences for individuals as they grow older. For example, autonomy can be questioned with older people being viewed as vulnerable and therefore in need of protection. Further, financial pressures may result in more older people being required to remain in paid work longer before they are able to

---

[2]Lewis (2018a), 159–60.

[3]United Nations, Ageing.

[4]Lagacé et al. (1989); Butler (1989), 139.

[5]See, for example: World Health Organization (2015), 64–6; Beard et al. (2016), 163–6.

[6]See, for example: Australian Law Reform Commission (2016), 21–2.

retire, which can therefore mean that they are exposed to discriminatory practices, particularly if they are re-entering the workforce at an older age. Older people may also experience a heightened risk of vulnerability to abuse. Such abuse can include physical, psychological, sexual and financial harm.[7] Significantly, perpetrators of elder abuse are often the very individuals and/or groups tasked with caring for older persons, and can include family members, friends and carers, as well as strangers.

The various and interconnected social and economic challenges presented by the ageing population have significant implications for our legal systems. The areas of law which impact, either directly or indirectly, on older persons comprise a body known collectively as 'elder law'. These areas are multifaceted, intersecting and frequently complex. They are also fragmented and inconsistent across different jurisdictions. This has been recognised by national and international advocacy groups who have highlighted the need for changes to a variety of legal and policy frameworks.[8]

Elder law encompasses any laws, regulations or processes that older people come into contact with but most frequently tends to include laws related to age-based discrimination, employment, legal capacity, elder abuse, financial management, estate planning, accommodation services, as well as health and aged care. It must also be acknowledged that accessing justice can be difficult for older persons no matter what legal issue they are confronting. There can be many barriers, particularly for those with impaired or lost capacity as well as victims of abuse who can experience a lack of autonomy and/or participation in decision-making. Any assessment of elder law must therefore take into account not only the substantive content of the law but also its processes for access, implementation and enforcement, to ensure that barriers to justice are adequately addressed.

As will be seen throughout this book, much of the discussion around improving elder law adopts the objective of preserving individual autonomy for persons as they age wherever possible. This requires a nuanced understanding of autonomy, for example through the concept of relational autonomy, which recognises the value of relationships of care and support which facilitate the exercise of autonomy.[9] This more relational understanding is particularly important given the role that familial and/or carer support can play in retaining independence as people age. It is also influential in the paradigm shift which has been occurring, for example, from substitute to supported decision-making for people with impaired capacity.[10] Given the centrality of autonomy and decision-making in many areas of elder law and the implications a loss of autonomy can have for access to justice, it is a recognised touchstone throughout the book. We do not, however, propose to examine the particular theory of relational autonomy in detail given that our aim here is restricted to exploring a human rights framework for elder law.

---

[7] World Health Organization (2017a).

[8] See, for example: United Nations General Assembly (2013); AGE Platform Europe (2018); Doron and Apter (2010); Fredvang and Biggs (2012), 21; International Expert-Conference on the Human Rights of Older Persons (2018).

[9] See, for example: O'Connor (2010).

[10] Convention on the Rights of Persons with Disabilities (2006), art 12 ('*CRPD*').

Older people frequently report diminishing respect for their autonomy and dignity, and a loss of connectedness as they age.[11] These experiences are linked to pervasive ageist attitudes within society and can all be viewed as threats to older persons' human rights. Where older persons experience abuse, neglect or maltreatment, or when they are unable to access justice, appropriate healthcare or an adequate standard of living, their human rights are violated. When they are unable to participate in education or employment due to aged-based discrimination or are precluded from fully participating in their communities because of physical, systemic or social barriers, their rights are infringed in ways that can lead to further human rights harms.

These rights are guaranteed under international treaties, including the *International Covenant on Civil and Political Rights* (ICCPR) and the *International Covenant on Economic, Social and Cultural Rights* (ICESCR), as well as regional instruments and domestic laws. Under such laws, governments are obliged to take steps to respect, protect and fulfil the human rights of older persons. This includes ensuring that both government and non-government entities adhere to human rights standards.

With this in mind, this book argues that a human rights-based approach is an ideal and novel framework in which to analyse the various challenges associated with the ageing population and to develop appropriate legal responses. A human rights-based approach is one which emphasises the inherent dignity, autonomy and liberty of each individual, and which champions the economic, social, cultural, civil and political rights of all people. It is therefore uniquely placed to address the varying interconnected issues facing older persons and falling within the scope of elder law. No other work has applied such an approach to the domains of finance, accommodation and health while incorporating analysis of the fundamental cross-cutting issues of ageism, capacity and abuse.

The specific rights found within human rights law emphasise the importance of a number of key principles which are fundamental to human rights generally, most notably, respect for inherent human dignity. This is recognised as the source of all human rights and a foundational principle in the *Universal Declaration of Human Rights* (UDHR) as well as other human rights treaties.[12] It is the essential value underlying rights such as the rights to the highest attainable standard of health,[13] privacy,[14] freedom from cruel, inhuman and degrading treatment,[15] an adequate standard of living,[16] and social security.[17] Respect for the autonomy of an individual is another

---

[11] See, for example: Australian Institute of Health and Welfare (2019a); Barbosa Neves et al. (2019); Relationships (Australia 2018); Sutin et al. (2018).

[12] Universal Declaration of Human Rights (1948), preamble ('*UDHR*'). See also preambles of International Covenant on Civil and Political Rights (1966) ('*ICCPR*') and *International Covenant on Economic, Social and Cultural Rights* (1966) ('*ICESCR*'), and CRPD, art 3.

[13] ICESCR, art 12.

[14] ICCPR, art 17.

[15] CRPD, art 15.

[16] ICESCR, art 11.

[17] UDHR, art 22 and ICESCR, art 9.

fundamental principle of human rights, which can clearly be seen translated into specific rights such as freedom of movement[18] and freedom to choose one's employment.[19] Another key principle of human rights is non-discrimination.[20] International human rights treaties all emphasise that, because people possess human rights by virtue of their being human, it follows that there can be no grounds to discriminate against any person in the fulfilment of their human rights—human rights belong to all persons equally and all persons are entitled to the full range of human rights. Taken together, these principles and the specific rights they underpin provide a useful way of understanding the range of impacts and vulnerabilities that are experienced by older persons. They furnish a language for recognising and protecting the opportunities and contributions of older people and can be used as the basis for a human rights-based approach to elder law. Such an approach would require that, at all times, the basic dignity and autonomy of individuals must be respected, and that older persons' freedom to make decisions for themselves must be respected and supported wherever possible.

However, existing mechanisms for the protection of these rights are inadequate. Countries need to increase their efforts to strengthen the protection of the rights of older people, both on a national and an international scale. There is currently no international instrument which specifically addresses the rights of older people, despite over a decade's worth of work within the UN human rights bodies advocating for such protections. There was even a call in the 2016 report by the Independent Expert on the Enjoyment of all Human Rights by Older Persons for states to 'step up their efforts to determine the best way to strengthen the protection of the human rights of older persons and to consider the various proposals that have been made, notably the elaboration of a convention on the rights of older persons'.[21]

Although older people are entitled to the same rights as all other individuals and are protected by the major international human rights instruments like the ICCPR and ICESCR, they have particular needs which too often are not met, a situation which is compounded by the ageist attitudes pervasive in society. Older people also experience unique forms of abuse and exploitation which warrant protection under human rights law. Absent an international convention on the rights of older persons (CROP), responsibility falls back to individual countries to implement and safeguard the general rights found within international human rights law in ways which adequately address the needs and experiences of older persons, but without the specific guidance or commitment which would come with a dedicated treaty.

A human rights-based approach to elder law would significantly enhance existing domestic protections of older persons' rights. In Australia, for example, the international instruments and principles have implications for several key policy areas relating to older persons. They represent important standards applying to issues such as driving, housing, making and revoking enduring documents, the provision

---

[18] ICCPR, art 12.

[19] Respect for individual autonomy is a general principle of the CRPD, art 3.

[20] UDHR, art 2.

[21] Kornfeld-Matte (2016), para 125.

of health and aged care, end of life and palliative care, and financial management. However, although human rights principles provide an innovative normative framework in which to approach the issues in this area, human rights are not currently well-protected in Australian domestic law. This means that there is, currently, limited ability for older individuals to enforce their rights, particularly through legal action. Nevertheless, there has recently been increasing attention dedicated to protecting vulnerable older people in Australia. For instance, in 2017 the Australian Law Reform Commission (ALRC) conducted an inquiry into elder abuse and a Royal Commission into Aged Care Quality and Safety was established in 2019.[22] Both of these bodies adopted the language of human rights, with the ALRC recommending that responses to elder abuse adopt a human rights framework.[23] Without stronger human rights laws, however, relying on existing national frameworks of human rights protections and international treaties will provide limited benefits for older persons, as the existing protections are often incomplete and inadequately enforced. This is a problem faced not only by Australia but one which is present worldwide.

The ability of older persons to both participate in, and benefit equitably from, societal development needs to be protected and facilitated.[24] This book responds to this need, seeking not only to explore the international context but also to apply a human rights lens to the existing mechanisms within domestic laws to protect the rights of older people. We seek to demonstrate the significant potential for approaches to ageing to draw further on human rights, both with respect to protecting specific rights as well as acknowledging the fundamental human rights principles of autonomy, dignity and non-discrimination.

To this end, this book provides a detailed and comprehensive human rights analysis of some of the key areas of law affecting older persons. We have drawn on multidisciplinary scholarship and international advocacy to identify areas for examination. For example, the Stanford Centre on Longevity identifies three domains for well-being into older age, financial security, social engagement and healthy living, and these domains can be helpful in thinking about the wide range of issues which fall within the scope of elder law.[25] Variations on these three themes repeat throughout the literature, and are predictably expanded and developed within relevant disciplines such as public health, gerontology, social policy, psychology, health law and medical ethics. International instruments like the *United Nations Principles for Older Persons* also highlight the importance of independence, participation, care, self-fulfilment and dignity as framing principles, and these too can help to identify areas of the law which require attention.[26]

[22]There have been a number of phases to the ALRC Inquiry. An initial Issues Paper was released for comment in June 2016, with a Discussion Paper published in December 2016 which set out preliminary proposals for law reform and invited further public comment. The Final Report was handed down on 15 June 2017.

[23]Australian Law Reform Commission (2017).

[24]United Nations Department of Economic and Social Affairs, Population Division (2017), 2.

[25]Stanford Center on Longevity (n.d.).

[26]United Nations Principles for Older Persons (1991).

From this broad base of scholarship and advocacy this book identifies a number of areas for analysis. Several issues cut across various areas of law, for example, ageism (discussed in Chap. 5), capacity (Chap. 6), abuse and access to justice (Chap. 7), while others can be more easily categorised according to the specific legal questions that they raise such as financial security (Chap. 8), accommodation (Chap. 9), and health and aged care (Chap. 10). These particular topics form the focus of the remainder of this book. First, however, this chapter will provide more detail on the nature and extent of the challenges posed by the ageing population. This will include a discussion in the next section of some of the key terms in this area, such as what is meant by 'older' and 'elder law', and clarify the way that terminology will be used throughout the book. Later sections in this chapter will then outline the demographic changes occurring before introducing the main topics for analysis. The chapter will also identify two recurrent themes in the elder law context, access to justice and women's experiences of ageing.

1.2 Defining Elder Law

Before we can analyse 'elder law' through a human rights lens, it is important to acknowledge some of the terminological and conceptual complexities that exist in this area. At the outset, we need to recognise that the idea of speaking of the human rights specifically of older persons or of a dedicated body of elder law is itself contested. Some have questioned whether delineating older persons as a particular cohort in need of special attention (including through dedicated human rights instruments) is itself an ageist undertaking.[27] However, identifying and advocating for the rights of older persons is not intended to segregate older persons or cast them as different in any negative sense. Rather, it is intended to recognise that ageing is a continuing process everyone experiences throughout the life course, and that human rights do not diminish as people age. All people continue to be entitled to the same human rights, no matter their age or level of ability. This book argues in favour of a dedicated human rights instrument for older persons, not to give them different rights, but to proclaim strongly their entitlement to the same rights while recognising the particular ways in which those rights can be impacted as people age. In the same way that the *Convention on the Rights of Persons with Disabilities* articulated meaningful entitlements and obligations in the process of unpacking core human rights for persons with disabilities, a CROP would be able to respond to the lived experiences of older persons, including their experiences of ageism, and develop a roadmap for better implementation within domestic law. From that starting point, this section will now address other areas of definitional uncertainty and specify the terminology adopted throughout this book to lay the groundwork for later analysis.

---

[27] Williams (2003); Avers et al. (2011); Greengross (2019).

1.2.1 Older or Elder?

Language can be powerful, especially when considering the negative and discriminatory, that is, ageist, implications that can exist with some terms used in association with 'ageing'.[28] For example, the use of the word 'old' as a descriptor—the *old* man, the *old* woman, distinct from simply the man, the woman—is often more than just a factual adjective. It instead has a negative connotation that can affect an individual's interpretation of the situation. Take, for instance, the following two scenarios: first, the woman had a car accident; and second, the old woman had a car accident. The use of the word 'old' in the second scenario is more likely to give rise to a perception that it was the driver's fault as she was 'old'. It is important, then, to think critically around the language used and the effect that this language can have—either consciously or subconsciously.

As noted above, the question of the age at which a person should be considered 'older' is contested, as is the question of whether 'older' is even the appropriate term to be using given the ageist assumptions such language can engender. Significant too is the choice between 'older' or 'elder' as appropriate terms.[29] These terms are often used in substitution for one another. The term 'elder' can, however, attract particular cultural significance raising questions as to whether it is appropriate to use in a more general sense.[30] The use of the term 'elderly' has also been criticised as being ageist, for the reasons discussed above.[31] Nevertheless, the terms are so entrenched in the modern vernacular that it will be difficult to displace them. They also now attract 'brand recognition', for instance increasing recognition around 'elder law' and 'elder abuse'. The growing recognition of the terms can thus attract much needed attention, and potentially funding. Consequently, it may not be prudent to now try to change the language adopted when those terms are becoming more identifiable. The question also arises, if not 'elder' then what?

The UN Committee on Economic, Social and Cultural Rights previously adopted the term 'older person' in preference over 'elder' in 1995 but this debate is in no way settled.[32] What is important is appreciating the power of language and challenging the ageist assumptions that attach to the language currently used in connection with ageing although, as will be discussed, issues of ageism run deeper than linguistics alone. In this book, we tend towards the term 'older' but retain the phrases involving 'elder' such as elder law.

---

[28] See, for example: Sunlife (2019); Hill (2019).

[29] See, for example: Australian Law Reform Commission (2017).

[30] Australian Law Reform Commission (2016), 22.

[31] Avers et al. (2011), 153–5.

[32] Committee on Economic, Social and Cultural Rights (1995).

1.2.2 *What Age is 'Older'?*

At what age then does someone become 'older'? Different definitions of 'older' have been postulated globally, ranging from fifty-five years and older through to sixty-five years and over.[33] According to the United Nations, the term 'older' most commonly refers to people aged over sixty years.[34] However, in Australia, for example, a person is generally considered 'older' once they reach sixty-five years of age.[35] The World Health Organization (WHO) states that the categorisation differs amongst countries but is generally connected to the age at which a person can retire.[36]

In the ALRC Discussion Paper, a distinction was drawn between 'old' and 'old, old'.[37] 'Old, old' is generally understood as applying to any person over the age of eighty. There is indeed utility in the 'old'/'old, old' distinction.[38] Such a differentiation is useful given that the risk of neurodegenerative conditions, such as Alzheimer's disease, is age-related and often expressed for specific age cohorts. For example, the risk of dementia-related conditions, such as Alzheimer's disease, is substantially higher for people aged eighty years and over than it is for people aged sixty or sixty-five years and over (although this is not to ignore the risk of early on-set dementia).[39] Further, as people age, and especially as they enter the 'old, old' age range, there is an increased risk of geriatric syndromes including frailty, falls, pressure ulcers and incontinence.[40] Associated with this is the possibility of increased dependence or vulnerability and thus the threat of abuse may also escalate.

For our purposes, we define 'older' as anyone aged sixty-five years and over. We will indicate where any specific issues exist in relation to the 'old, old' category (that is, anyone aged eighty years and over). It should also be noted that the definition of 'older' for indigenous peoples can range from approximately fifty years and over given shorter life expectancies.[41] In Australia, for example, this is because Aboriginal and Torres Strait Islander people generally experience poorer health and have higher rates of disability than other Australians in a commensurate position.[42]

---

[33] See, for example: United Nations Department of Economic and Social Affairs, Population Division (2015a); Australian Institute of Health and Welfare (2017); World Health Organization (2004).

[34] United Nations Department of Economic and Social Affairs, Population Division (2015a).

[35] Australian Institute of Health and Welfare (2017).

[36] World Health Organization (2004).

[37] Australian Law Reform Commission (2016), 22.

[38] Ibid.

[39] See, for example: Australian Law Reform Commission (2017), para 2.8 (note however that the ALRC uses statistics based on eighty-five years of age).

[40] World Health Organization (2018).

[41] Australian Law Reform Commission (2017), 34.

[42] Australian Human Rights Commission (2015); Australian Institute of Health and Welfare (2019b).

1.2.3 Elder Law

Older people have diverse legal requirements that are influenced by an array of both conflicting and interconnected considerations. These factors can include a person's age, socio-economic situation, education level, health, cultural and linguistic background, as well as their familial and care-based networks. Geographical location can also be a significant consideration, particularly for those in regional, rural and remote areas. It is anticipated that as the population ages, the associated legal requirements will similarly become more diverse and complicated.[43]

The concept of 'elder law' is therefore a complex intersection of any and all substantive areas of law which apply to or interact with the challenges that can arise when a person is 'older'. These can present in any legal context but some of the most common include: succession and estate planning, especially wills and enduring powers of attorney; family law, with, for instance, the increasing prevalence of grandparents having responsibility for the care of grandchildren; equitable doctrines such as undue influence, unconscionable conduct and estoppel, all of which have prominent roles to play in addressing legal wrongs to older people; discrimination and employment law; as well as property and contract law in connection with accommodation and health needs such as retirement villages and aged care facilities. There is significant complexity not only in each discrete area of law but also in the way that these substantive areas interact with each other and with other socio-economic issues and frameworks, as well as with governmental and institutional responses. For example, the need to take on the care and responsibility for grandchildren, a family law issue, can force an older retired person back into the workforce thus potentially exposing them to discrimination and workforce issues. This can also have financial and estate planning, as well as accrued retirement benefit consequences, for the individual, not to mention the possible personal, familial and social impact.

Legal professionals may also not be equipped to adequately engage with older clients and their particular needs, such as accommodating hearing and sight impairments or early stage dementia. Such conditions can sometimes incorrectly be interpreted as signifying either a loss or lack of capacity. A legal professional's failure to recognise relational factors and/or individual attributes (such as language and speech) and personal abilities (such as visual, verbal and auditory functioning, and/or diminished or lost cognitive abilities) can also have a deleterious effect upon an older individual's ability to access justice and 'quality' legal representation. Significantly, in terms of a person's human rights, such an experience can, in turn, further infringe upon individual autonomy. An incorrect determination of a loss or lack of decision-making capacity can then attract a (potentially) unnecessary protectionist response, thus infringing further upon the person's own decision-making ability and other human rights.[44]

In addition to these difficulties, there are a number of practical challenges presented by age itself, especially in the 'old, old' category. While age alone is not

[43]Law Council of Australia (2018), 35.

[44]Purser and Sullivan (2019), 88–98.

indicative of a lack of capacity there are age-related mentally disabling conditions, particularly in the 'old, old' category, that can impede upon an individual's ability to engage with lawyers and the legal system.[45] Limited access to information can also infringe upon an older person's ability to effectively engage with the legal system, particularly when considering technological literacy and reduced mobility. While new technologies have the potential to both enhance and compromise the enjoyment of a wide range of human rights, it must be acknowledged that not all older people are conversive with, nor want to use, technology. For example, in Australia, older persons have the second lowest level of digital inclusion.[46] While new technologies offer the possibility of many benefits, they should not be adopted blindly without reference to the issues arising from such adoption, including the potential infringement on the right to privacy and their use as vehicles for abuse.[47]

With all this in mind, this book approaches elder law as not only a set of substantive legal fields, but also the associated and cross-cutting issues which determine the nature of older persons' interactions with the legal system. Ageism, legal capacity, abuse and access to justice are key considerations here. Dedicated chapters follow later on ageism, capacity and abuse, but these concepts, like access to justice, remain relevant to many substantive areas and connections will therefore be noted where relevant throughout.

1.2.4 Ageism

Ageism is 'stereotyping, prejudice, and discrimination against people on the basis of their age'.[48] The negative attitudes which are representative of ageism are rife. For example, consider the workforce where discrimination against older workers has been documented.[49] Consider also, the health and aged care settings where older adults are often at their most vulnerable.[50] Older people can erroneously be viewed as frail and dependent, as well as a burden on society. Ageism can also act as a powerful barrier to accessing justice, particularly for those who have experienced elder abuse.[51] It can therefore impede the development of sound policy and best practice.

Ageism can also represent a violation of specific human rights depending on the context in which it is present. Relevant rights include the right to work, and particularly the right to gain a living by work of one's own choosing.[52] The rights

---

[45] See, for example: Purser (2017); Purser and Lonie (2019).

[46] Australian Digital Inclusion Index (2020).

[47] See, for example: Lewis et al. (2018b).

[48] World Health Organization (2017b).

[49] Australian Human Rights Commission (2016).

[50] World Health Organization (2015).

[51] World Health Organization (2017b).

[52] ICESCR, art 6.

to education,[53] to equality before the law,[54] and to healthcare can all potentially be undermined where ageism is present. Human rights are also relevant to the steps which are taken by governments to protect and support older persons. The right to privacy,[55] for example, must be respected in all measures which relate to older persons, even when those measures are designed to act in the older person's best interests. The right to social security also imposes a requirement that support provided by the government to older persons must be sufficient to enable them to live a life of dignity, and must avoid ageist assumptions about the kind of lives that older people lead.[56]

Although ageism is garnering increasing recognition as a danger to the enjoyment of human rights by older persons, little is being done by way of active measures to identify, measure and/or combat its insidious effects.[57] Consequently, the promotion of stereotypes of older people as being incompetent, slow and/or an economic burden (amongst other negative imagery) can become a self-fulfilling prophecy, negatively affecting both the individual and society more generally.[58] By perpetuating harmful perceptions of older persons, and denying full and equal respect and participation, ageism therefore operates as a driver of elder abuse.

It is therefore clear that we must seek to improve understandings of the multiple benefits which flow from the full participation of older people in local communities and society more broadly.[59] Despite this, there has been limited discussion about the effective implementation of education measures, quantifiable outcomes and/or funding models to achieve these objectives. Instead, the discourse seems currently restricted to high level, aspirational statements about ageism which, although understandable, will need to be given practical effect if there is to be any success in combating the damaging effects of ageist attitudes. The harmful effects of ageism and the potential of a human rights-based approach to help combat age-based stereotypes will be discussed in greater detail in Chap. 5.

1.3 Rates of Demographic Change

A precondition of analysing and responding to the human rights implications of the ageing population is having an accurate understanding of the scale and nature of that demographic shift. The statistics around ageing are often said to be 'alarming', itself an example of how powerful and emotive the choice of language can be. However, they do demonstrate that populations worldwide are ageing. This is because of three

[53] Ibid., art 13.

[54] ICCPR, art 26.

[55] Ibid., art 17.

[56] ICESCR, art 9.

[57] For example: Recommendation 3–3(b) in Australian Law Reform Commission (2017), 9.

[58] Winick (1996), 21.

[59] See, for example: Australian Law Reform Commission (2017), Recommendation 3–3(b), 9.

main contributing factors: fertility, mortality and migration.[60] Globally, all regions have seen a significant growth in life expectancy since 1950.[61] Improvements to health and associated care in older age account for a significant proportion of the increased rates of longevity, especially when considered in light of slowing fertility rates.[62] Where countries experience significant rates of migration, this can also be a factor given that immigrants tend to be of working age.[63] The rates of ageing are also expected to increase as the 'baby boomer' generation retires.[64]

It is anticipated that the number of people aged sixty years and over worldwide is expected to reach approximately two billion by 2050.[65] This is just over double the number of older people in 2017 (962 million).[66] The year 2050 will also see 1 in 6 people being over sixty-five years of age (16%), up from 1 in 11 in 2019 (9%).[67] In fact, between 2015 and 2050 it is estimated that people aged over sixty years will almost double in number rising from 12 to 22% of the world's population.[68] Further, for the first time in history, 2018 saw the number of people aged sixty-five years and over outnumber children aged under five year globally.[69] By 2050 it is estimated that there will be more people aged over sixty than children and young adults aged 10–24 (2.1 billion as opposed to 2.0 billion).[70] The number of people in the 'old, old' category is anticipated to triple, rising from 143 million in 2019 to 425 million by the year 2050.[71] Of the anticipated increase in numbers of 'older people', it is generally expected that older women are more likely than older men to live alone.[72] It is also anticipated that two thirds of older people globally are resident in developing regions with close to eight out of ten older people living in developing regions by 2050.[73]

Similar trends of ageing can be seen around the world. In fact, it is anticipated that there will be an increase in the number of people aged sixty years and over between now and 2050 in all 201 countries and/or areas with more than 90,000 residents,[74] with the ageing population being the 'most advanced' in North America and Europe. It is estimated that by 2050 older persons will account for 35% of Europe's population, 28% in Northern America, 25% in Latin America and the Caribbean, 24% in Asia,

---

[60] United Nations, Ageing; Parliamentary Budget Office (2019), 3.
[61] United Nations, Ageing.
[62] Ibid.
[63] Ibid.
[64] Parliamentary Budget Office (2019), iv.
[65] Kornfeld-Matte (2016), 5: 17.
[66] United Nations Department of Social and Economic Affairs, Population Division (2017).
[67] United Nations, Ageing.
[68] World Health Organization (2018).
[69] United Nations, Ageing.
[70] United Nations Department of Social and Economic Affairs, Population Division (2017).
[71] Ibid.; United Nations, Ageing.
[72] United Nations Department of Social and Economic Affairs, Population Division (2017).
[73] Ibid.
[74] Ibid.

23% in Oceania and 9% in Africa.[75] In Australia, for example, the number of people aged eighty-five years and over is expected to increase from 400,000 in 2010 to 1.8 million by 2050 with twice as many women than men in the eighty-five year and over age bracket.[76] In fact, the Asian-Pacific region is said to be at the forefront of the ageing phenomenon with one in four people anticipated to be aged sixty years or over by 2050.[77]

In Europe, the number of older people is expected to increase while the number of working-age people declines which results in an increase to the old-age dependency ratio. Notably, the rates of people aged eighty-five years and over is expected to increase to 40 million by approximately 2050 (up from 14 million presently).[78] In the United Kingdom, there are approximately 12 million people aged sixty-five years and over with 1.6 million aged eighty-five years and over.[79] It is estimated that one in five people will be aged sixty-five years and over by 2030 with 3.2% of people in the old, old category. In fact, the eighty-five years and over age range is thought to be the fastest growing cohort with an expected 3.2 million by 2041.[80] Canada and the United States of America tell a similar story. Older people are expected to comprise 23% of Canadians (9.5 million) by 2030.[81] By the same year, one in every five residents in the United States of America will be aged sixty-five years and over.[82] The precise legal implications of these changes will of course vary from one jurisdiction to another, influenced by a range of factors and requiring responses tailored to each society and its legal system. However, variations of a number of common legal issues are emerging around the world, and the shared experience of ageing demographics allows for valuable lessons to be learned from other jurisdictions. A select range of key issues are addressed in this book which will be introduced in the following sections.

1.4 Select Legal Challenges

The ageing population offers a wealth of opportunity.[83] Opportunities exist for individuals to engage in further education, to spend more time with loved ones and to work longer if they so wish and are physically and cognitively able. Opportunities also exist to contribute to society in general, and for society to learn from the wisdom

---

[75]Ibid.

[76]Australian Bureau of Statistics (2010), 3201.0; Australian Government (2010), 56.

[77]United Nations Population Fund Asia and the Pacific (n.d.).

[78]World Health Organization Regional Office for Europe (2020), Demographic Trends, Data and Statistics on Ageing.

[79]AgeUK (2019).

[80]Ibid.

[81]Government of Canada (2014).

[82]United States Census Bureau (2018).

[83]Vienna International Plan of Action on Ageing (1982), 23–24 ('*VIPAA*').

and experiences of older people.[84] Indeed, older people often 'serve as the transmitters of information, knowledge, tradition and spiritual values: this important tradition should not be lost.'[85] However, several challenges emerge in ensuring the enjoyment of these opportunities, and in fostering the ability of older people to contribute to their local communities and society more broadly, however they so choose. First, people are living longer, but it is often without the ability to be able to make their own, legally recognised decisions. Therefore, questions of capacity are significant as individuals age, particularly for those in the 'old, old' age category. Issues also arise in relation to financial management and what is to happen in anticipation of and upon retirement, as well as what financial and other services are available through government-supported welfare systems. Connected to this is the issue of accommodation, as well as the issue of access to quality health and aged care. Again, people are living longer but whether it is with a high quality of life is a separate, and significant, question. It is also important to acknowledge what it means to age as a woman given the various gender-based issues associated with employment, income, pensions, and unpaid family and care work. Significant issues are also arising in relation to elder abuse, which is occurring at increasing rates and can take many forms. Across all of these areas the issue of access to justice is a consistent consideration. The following discussion is only intended to highlight some of these main themes that are currently emerging. Several select elder law-specific issues will then be discussed further throughout the remainder of the book within the overarching human rights framework developed in Chap. 3.

1.4.1 Capacity and Decision-Making

Capacity is an invaluable legal and social construct.[86] This is because it is demonstrative of individual autonomy within familial and wider societal, including legal, circumstances. It is also the legal construct by which an individual is assessed to be able to make legally recognised decisions. Notions of legal and clinical capacity are interconnected, with assessments being complex.[87] As people grow older, and particularly for people in the 'old, old' category, they are often, and frequently erroneously, viewed as being vulnerable and incapable of making their own decisions. Consequently, their dignity is affronted as they are marginalised and excluded from decision-making processes.[88] Appropriate laws and policies around capacity are therefore essential to addressing this problem.

---

[84]Ibid., 22.

[85]Ibid., 74.

[86]Carney (1997), 1.

[87]Purser (2017).

[88]VIPAA, 61.

Capacity and its effect on individual decision-making is garnering increasing attention.[89] As people age, and the incidents of mentally disabling conditions increase, the need for capacity assessments will grow.[90] Empirical evidence is limited in relation to the number of, and reasons for, capacity assessments being undertaken, with a particular shortage of recent empirical research. However, in the United States of America in the 1980s there was a 50% increase in the total number of referrals for capacity assessments for older people.[91] Although age is not automatically indicative of a lack of capacity, there are age-related conditions which mean that issues of capacity tend to escalate as individuals age, particularly for those in the 'old, old' category. This is because for the people in this category, capacity will start to wane—it is a question of the rate at which cognitive decline will occur, not if it will happen. Having said this, it is important to remember that in legal contexts there is a general (rebuttable) presumption of capacity. That is, every person over the age of eighteen is assumed to have capacity unless proven otherwise.[92] In ascertaining whether or not the presumption of capacity has been successfully rebutted, the courts – the final arbiters of capacity—may consider lay evidence from family members, friends and carers as well as 'expert' evidence from independent parties such as health, allied health and legal professionals.[93]

Capacity assessments are currently conducted on an ad hoc basis lacking guidance as to best practice.[94] There is often a (necessary) multidisciplinary approach to assessments.[95] However, legal, health and allied health professionals are not necessarily well-versed in working collaboratively to assess clinical concepts of capacity, for example the nature and stage of dementia, within the requisite legal framework. Capacity is time and decision specific, and different considerations and standards may apply depending upon the particular capacity concerned, for example capacity to make a will, marry, execute a contract, or drive.[96] An incorrect assessment of capacity has significant ramifications because it curtails an individual's autonomous ability to make legally recognised decisions.

As stated, autonomy is an inherently interconnected concept with capacity.[97] This is because in order to exercise one's autonomous decision-making ability, a person has to have the requisite legal capacity to make the decision in question at the specific time. Autonomy is also one of the fundamental principles underlying modern human rights jurisprudence. Therefore, given that capacity is an increasing issue for

[89] See, for example: Triebel et al. (2018), 219–235. Purser and Sullivan (2019); and Purser and Lonie (2019).

[90] Purser and Sullivan (2019).

[91] Sullivan (2004), 134.

[92] Re Caldwell [1999] QSC 182, [12] (Mackenzie J). Law Reform Committee, Parliament of Victoria (2010), 109–110.

[93] Cockerill et a1. (2005), 29.

[94] Purser (2017).

[95] Purser and Rosenfeld (2014), 483–5.

[96] Lonie and Purser (2017); Purser and Sullivan (2019).

[97] Purser and Sullivan (2019).

older people, and the ageist assumptions that currently exist within society, there is a real danger that unsatisfactory assessment processes may endanger a person's fundamental human rights.

Individual, familial and societal attitudes about what autonomous behaviour is, and should look like, also have the potential to impact upon a person's capacity and its assessment.[98] For instance, a person who lacks legal capacity can be hidden within a family structure. This can be a supportive environment for the person, or one which has the potential to foster unchecked abuse. External perceptions of a person and whether they are capable or not can also have an impact on how people see themselves.[99] If someone is constantly told they lack capacity, for example because they have been diagnosed with dementia, this can in effect become a self-fulfilling prophecy with the individual eventually coming to believe that they do indeed lack capacity.[100] This may be despite the fact that the diagnosis is actually early stage dementia and the individual can generally continue to make their own decisions. The concept of relational autonomy clearly has relevance here, in recognition that people often exercise autonomy in the context of relationships of care and support.[101] This is a complex area of scholarship and the interplay between relational autonomy and capacity, particularly in the context of supported and substitute decision-making, warrants further research that is beyond the scope of this work. The application of a human rights framework to capacity and decision-making is the focus of Chap. 6.

1.4.2 Elder Abuse

Definitional issues also attach to the concept of 'elder abuse'.[102] The uncertainty surrounding the terms 'older', 'elder' and 'elder abuse' noted above has meant that there is a limited evidence base from which to assess the prevalence of abuse leading to international calls for rigorous and national prevalence studies.[103] Elder abuse is defined by the WHO as 'a single, or repeated act, or lack of appropriate action, occurring within any relationship where there is an expectation of trust which causes harm or distress to an older person'.[104] The WHO definition identifies that a 'lack of appropriate action' can constitute elder abuse. Consequently, it must be recognised that elder abuse can occur in situations of neglect as well as deliberate harmful action. There are two types of neglect: active neglect and passive neglect. Active neglect is the deliberate withholding of basic care and passive neglect is the failure to provide

98 Darzins et al. (2000), 1.

99 Purser and Sullivan (2019).

100 Ibid.

101 CRPD, art 12.

102 Australian Law Reform Commission (2016), 17.

103 Ibid., Recommendation 3–5, 37,93.21–22.

104 World Health Organization (2002).

proper care due to carer stress, lack of knowledge or ability.[105] There is an element of unintentionality with passive neglect in that the carer may simply require additional support or knowledge. There are also a number of different types of abuse which can occur. In addition to neglect, the main categories include physical, psychological, sexual and financial abuse.[106]

It is proposed here that a comprehensive definition of elder abuse should incorporate any situation wherein the older person, or their human rights, are substantively and negatively affected. There should not however be a requirement for an element of 'trust'[107] as set out in the WHO definition because this appears to unnecessarily narrow the definition of abuse. While the problem of abuse within existing relationships is of obvious concern, the prevalence of ageism and the 'invisibility' of older persons in our community also leaves them vulnerable to abuse and exploitation from people with whom they do not have an existing, trust-based relationship, for example, in the form of online finance scams.

There is also discussion as to whether elder abuse should be conceptualised as a category of family violence.[108] While it is crucial to recognise the problem of interpersonal elder abuse, and the particular dimensions and subtleties this entails, it is also important to acknowledge that elder abuse occurs in a wider range of settings which must also be addressed. This is especially important when thinking about public understanding of elder abuse, particularly as education of the public is vital in helping to address the incidents of elder abuse that are believed to be occurring.[109] Whilst it might be practically convenient to tap into the public awareness of family and domestic violence, there is a concern that adopting this approach could overlook and further marginalise those older persons who are suffering abuse in other contexts, for example, where it is a singular incident or episodic, involving strangers, or linked more to institutionalised ageism. This last issue is particularly relevant in the context of aged care as has been identified by, for example, the Royal Commission into Aged Care Quality and Safety in Australia. Arguably, elder abuse may be distinguished from domestic violence because, whilst elder abuse often occurs as a result of greed, the motivator for domestic violence is often control.[110] Elder abuse may also not be as 'gendered' as family violence. Further, as noted above, elder abuse perpetrators do not come solely from within familial units.[111]

Elder financial abuse in particular can lead to a marginal existence for older victims as there is little or no time to replace depleted assets. Victims can therefore find themselves with limited options in terms of alternative sources of accommodation as well as health and aged care. Consequently, there is an increased demand on the welfare system, with a growing need for governments to support housing and care services

[105] Neglect is discussed in the Australian Law Reform Commission (2016), 17.

[106] Lindenberg et al. (2013).

[107] Australian Law Reform Commission (2016), 17.

[108] Ibid., 33.

[109] See, for example: Purser et al. (2018).

[110] Australian Law Reform Commission (2016), para 1.16.

[111] Ibid.

for those older people who lack the income and/or wealth security to afford them. Furthermore, with an increase in the pervasiveness of elder abuse and ageism comes an increased need for appropriately trained professionals to deal with these cases across a broad range of disciplines, not only in relation to the welfare system. For example, education and effective communication are also issues in the health, legal, law enforcement and social services sectors. Therefore, a comprehensive human rights-based approach, one which respects and protects all human rights, requires a sufficiently broad conceptualisation of elder abuse, distinct from domestic violence, and a comprehensive approach to education, identification and victim-support. Elder abuse will be discussed in Chap. 7.

1.4.3 Financial Security

Currently, particularly with available and ever-advancing healthcare and technology, many people will live well beyond 'retirement age'. Consequently, the notion of preparing for retirement should not be something that is delayed and dealt with only when imminent and unavoidable. Retirement should also not be a negative experience, but rather something that is a natural part of life and, ideally, an activity which is promoting future well-being.[112] Some of the issues which require consideration include: what individuals want to do with their working life; when individuals want that working life to cease and retirement to begin; what they want that retirement to look like; what happens if an older person is not in a position to provide for themselves once they do retire, that is, what pensions or other government assistance are and should be available; and what kind of financial and/or estate/future planning, if any, should be undertaken, including giving thought to the utility, effect and appropriateness of intergenerational transfers.

The importance of financial considerations as people age cannot be disputed—particularly as individuals reach a time where they may not want, or be able, to continue in paid employment. This is because, when considering the global population, only a relatively small minority of people will accumulate enough resources, including savings and property, to be able to provide financial security for themselves as they age.[113] The danger of inadequate financial security as people grow older is significant, potentially leading to homelessness and/or poverty. Poverty is recognised as one of the main drivers of human rights violations, for example by impacting access to healthcare and housing, while financial security empowers individuals to enjoy their human rights to the greatest possible extent.[114]

[112]VIPAA, 31(e).

[113]United Nations Department of Economic and Social Affairs Programme on Ageing (2016).

[114]See the work of the United Nations Special Rapporteur on extreme poverty and human rights, Prof Philip Alston, for example his annual report of 2017 on the subject of universal basic income Alston (2017).

Without adequate financial resources many older people will have to work until they are physically incapable of doing so or become reliant upon either government pensions and/or family support, assuming such resources are even available.[115] Further, inequalities can exist not only between older people but also between older people and other generations which can affect not only retirement but also housing, resources and general prosperity.[116] In addition to ageist attitudes, access to financial resources as individuals age can also be affected by factors of gender, socio-economic status and education. Location is another strong contributing factor to inequality as income security in older age is, unsurprisingly, stronger in higher-income than lower-income countries.[117] These inequalities can, in turn, reinforce obstacles to older people participating in society and accumulating financial security.[118] Therefore, providing financially for older people who may not be in a position to fund their own retirement is a significant issue. This is not only for the individual in question but also society more generally when considering the broader impact this can have on welfare systems and available resources. It is also an essential component of ensuring the enjoyment of an older person's human rights and is therefore a key pillar of a human rights-based approach to elder law.

Older people often need to rely upon either families and/or, more commonly, government pension schemes as they age. This is often the case in older age, particularly for the 'old, old' cohort, where people can no longer work and where health concerns, including the availability and affordability of healthcare, may be increasing.[119] Noting that social security is a universally recognised human right, two broad approaches to pension schemes exist internationally.[120] One is based upon a government payment to an older person that is often means tested in order to qualify. It does not require any contributions from the older person throughout their life ('non-contributory schemes'). The second is a contributory pension scheme wherein individuals contribute to the pension throughout their life and that level of contribution will then be used to determine their entitlements on retirement. Many countries now have such schemes including, for example, Australia. As mentioned, older people in higher-income countries may therefore have better income security than those people in low-income countries. However, in order to balance viability with the rising costs of implementing pension schemes, governments in higher-income countries are now raising the retirement age and/or prohibiting or disincentivising early retirement in favour of a prolonged working life.[121]

Although social security is a recognised human right, statistics show that between 2010 and 2012, only 51.5% of the world's eligible population received a government funded, age-based pension, with this figure expected to decrease to around 42% in the

---

[115] United Nations Department of Economic and Social Affairs Programme on Ageing (2016), 1.

[116] Ibid.

[117] Ibid., 3.

[118] Ibid., 1.

[119] United Nations Department of Economic and Social Affairs Programme on Ageing (2016), 1.

[120] UDHR, art 22 and ICESCR, art 9.

[121] International Labour Organization (2014); United Nations Secretary-General (2012).

future.[122] Therefore, access to economic support as people age is unequal and often unreliable, being determined largely by location, strength of the respective national economy and gender.[123] Europe and North America have the highest proportion of pension coverage globally with an estimated 90%.[124] Moderate coverage is seen in Latin America and the Caribbean (56%), and the Asian-Pacific region (47%).[125] Low coverage can be seen in North Africa (37%), the Middle East (30%) and in sub-Saharan Africa (17%).[126] In Australia, for example, it is anticipated that approximately 80% of people aged sixty-five years and over rely, to some degree, on the age pension.[127] Further, more than one in four Australians aged sixty-five years and over are thought to live in poverty with older people comprising an estimated 7% of the homeless population.[128] Significantly, the majority of these older Australians are experiencing homelessness for the first time after having a conventional housing and employment history, with almost one quarter of people finding themselves in this position because of family breakdown, estrangement and/or carer stress.[129]

In considering contributory pension schemes, obviously the policy behind their implementation varies markedly between countries but the general premise, that of an individual contributing throughout their lifetime in anticipation of retirement, is broadly similar. Again, manifest inequalities can exist in the accumulation of such funds owing to, again, a person's location and gender, and their country's economic and political status. The nature of a person's work is also significant. This includes not only the position held, but also the wage received, the nature of employment (full time or casual, cash-in-hand or salaried), and whether the work is office-based or manual labour, understanding that the latter can potentially take a greater toll on a person's body and ability to continue working.

Financial planning and/or estate/future planning can therefore also be a significant issue as people age and plan for retirement from paid employment. Financial planning generally involves advisers (financial planners) who will help clients set financial goals and achieve them. They often require a licence and need to be distinguished from lawyers.[130] Estate or future planning, on the other hand, are terms which are used interchangeably to describe the work that legal professionals do in preparing a plan for their client to make provision for wealth growth, protection, management and transmission. The planning is generally in anticipation of death or a loss of capacity but can also use *inter vivos* mechanisms during a person's life, in addition to asset transmission upon a person's death. Common estate planning tools therefore include wills and enduring documents, such as enduring powers of attorney and advance

[122] International Labour Organization (2014).

[123] United Nations Department of Economic and Social Affairs Programme on Ageing (2016), 1.

[124] International Labour Organization (2014).

[125] Ibid.

[126] Ibid.

[127] Australian Human Rights Commission (2015).

[128] Ibid.

[129] Petersen and Jones (2011); Homelessness Australia (2016).

[130] See, for example: Australian Securities and Investment Commission (2019).

health directives, as well as trusts and companies. Planning for retirement can also be a significant part of estate planning, often concentrating on financial security when paid work stops through accrued retirement benefits such as the compulsory superannuation scheme in Australia. In fact, it is estimated that a significant amount of wealth resides in such superannuation schemes.[131] Indeed, the socio-economic implications for countries lacking formal retirement benefit schemes are largely negative, potentially impacting upon the well-being of the older individual.[132]

Dealing with the transmission of assets between generations can give rise to more negative issues around 'early inheritance syndrome', that is younger generations feeling entitled to the assets of older generations, and elder financial abuse which will be introduced below.[133] A further connected issue is the growing frequency of assets for care arrangements—property or other financial arrangements struck between an older person and their carer(s)—and what happens when they fail.[134] Underpinning any financial and/or estate planning is the question of whether the individual has the legal capacity to be able to make such decisions.

Therefore, the role of financial security when ageing demonstrates not only how inherently interconnected the aspects of 'elder law' are, but also how fundamental financial security is to protecting human rights in the context of 'ageing well'. Homelessness and poverty are problems worldwide but particularly in developing countries where the necessary resources and supports are not available for older people.[135] Consequently, there will need to be a global approach to ageing, and one in which there is a shared approach to resources and technology, so as to not infringe upon the human rights of older people.[136] Financial security will be discussed further in Chap. 8.

1.4.4 Accommodation

The right to adequate housing was recognised as part of the 'the right to live somewhere in security, peace and dignity' in the UDHR and is part of the right to an adequate standard of living in the ICESCR.[137] This is particularly significant in light of the previous discussion on financial security and the impact it can have on older people, with a lack of financial security potentially resulting in either homelessness and/or poverty. The ability to reside in adequate accommodation and have a safe

[131] Australian Bureau of Statistics (2016).

[132] VIPAA, 48.

[133] Purser et al. (2020).

[134] Australian Law Reform Commission (2016), 21–2.

[135] VIPAA,14.

[136] Ibid., 15.

[137] UDHR, art 25; ICESCR, art 11; Office of the United Nations High Commissioner for Human Rights (2014).

physical environment is fundamental to the health and well-being of people generally, and especially older people, with housing having a significant impact on an individual's quality of life.[138]

Healthcare and housing are two increasingly interconnected concepts. There is a growing push for people to be able to remain at home—to age and to die in their own homes or 'ageing in place'.[139] Significant in this is not only reducing systemic pressures on hospitals and healthcare providers, but also an acknowledgement of the power of being in a safe and familiar environment. Consequently, adaptations to the home may be necessary—be they physical modifications in the form of, for example, handrails for balance and to prevent falls, or provision being made for in-home healthcare and/or domestic assistance.[140]

Assistive technologies also have a role to play in providing support for older people to remain in their homes. There are, however, justified questions asking what this role should be and to what extent technology should be utilised.[141] For example, work has been undertaken developing a messaging kettle wherein family members can know if an older person is well through the use of a kettle—when the older person puts the kettle on, the family members are notified.[142] Whilst this can offer security (for instance against falls) and comfort knowing that the older person is active, it raises potentially serious questions about possible infringements of the right to privacy if such technology is adopted more broadly.[143] The use of assistive technologies in housing is just one example of the tension between security and privacy.

Currently, retirement villages are another widespread housing option for older people.[144] However, there is increasing discussion around whether this is a conduit to 'ageing well' and enjoying human rights in older age. Further, the legalities around entering—and exiting—retirement villages can be incredibly complex. Certainly in Australia, for example, recent adverse publicity has highlighted difficulties for the retirement village sector as prospective residents are signing contracts without necessarily having a complete understanding of the terms of these agreements.[145] This is because of a number of contributing factors. Not only are there complex legislative requirements, there can also be difficulty in accessing expert legal advice which can result from prohibitive professional fees, the uneven distribution of legal professionals with relevant expertise, and/or a lack of understanding among lawyers of older people's needs. Prospective residents may also find it challenging to address the estate planning necessary for ageing; feel intimidated by legal professionals and the 'law'; lack the financial, legal, and/or technological literacy necessary to be able to effectively engage in the process to increase their understanding; and/or not

[138] VIPAA, 64.
[139] Wiles et al. (2012).
[140] VIPAA, 64.
[141] Bennett et al. (2017); Bennett (2019).
[142] Brereton et al. (2015).
[143] ICCPR, art 17.
[144] See, for example: Petersen et al. (2017); Hu et al. (2017).
[145] Choice (2017).

have the ability to both locate and then access appropriate legal advice because of, for example, prohibitive cost and/or geography. There is also potentially the ever-present issue of legal capacity. This includes not only whether the person has the capacity to enter into the relevant transaction, but also whether they have the capacity to give instructions.

People may find the process overwhelming or experience pressure from a spouse, family members, or salespeople to make the decision to move into a retirement village. There may also be societal pressure on older people to vacate their homes to help with housing—that is, the sense that they should make space for young families or first home-buyers and that, in not doing so, they may be accused of being selfish or greedy.[146] This last point again raises the issue of intergenerational harmony, or disharmony as may be the case. Issues in the aged accommodation sector can be further compounded by a lack of clear communication and restrictive business models prioritising profit. Thus, not only is accommodation clearly a significant issue facing people as they age, it also demonstrates the interconnectedness of a host of legal, social, health and financial issues in the ageing context. Accommodation will be discussed in Chap. 9.

1.4.5 Health and Aged Care

The ageing population indicates a biological and technological feat for humankind. However, although many older people do enjoy a good quality of life, there is an increasing risk of poor (or poorer) health as individuals grow older.[147] Common conditions associated with older age include loss of hearing and eye-sight, joint pain, back pain, osteoarthritis, heart disease, diabetes, age-related cognitive decline such as dementia, and depression or other mental illness.[148] These conditions can occur separately or in combination with each other. Older people can also experience 'geriatric syndromes' which can be a mixture of 'frailty, urinary incontinence, falls, delirium and pressure ulcers'.[149] In fact, age is one of the main determinants of health.[150] Health in this context includes not only physical well-being, but also the interdependence of mental, emotional, spiritual and social well-being, and is influenced by both natural and physical environmental factors.[151] Other relevant factors in ageing well include: genetics, which are believed to be responsible for approximately 25% of differences in health and function as people age; whether the

146 Bakewell and Foster (2015).

147 United Nations Department of Social and Economic Affairs, Population Division (2015b).

148 World Health Organization (2018).

149 Ibid.

150 United Nations Department of Economic and Social Affairs Programme on Ageing (2018), 1.

151 VIPAA, 52; United Nations Department of Social and Economic Affairs, Population Division (2015b); World Health Organization (2015).

individual engages in risky behaviours such as smoking and/or not getting any exercise; and individual characteristics, for instance level of income and education.[152] Healthcare thus includes not only medicine but also social and familial elements. Disparities in other areas, particularly socio-economic status, can serve to have a cumulative and negative effect when considering opportunities for older people to age 'well', particularly in relation to accessing healthcare.

International human rights law recognises that health is the product of a number of interconnecting factors, some of which can be improved while others can only be mitigated. This is reflected in the right to the highest attainable standard of health. This right is understood as a holistic concept requiring provision of affordable and appropriate healthcare services as well as attention to a range of determinants of health, including clean air and water, sanitation, adequate nutrition and health-related information.[153]

Healthcare and accommodation are two particularly interrelated areas. For instance, healthy ageing strategies should continue to question the approach of removing an older person from their home and placing them in care, instead fostering a system wherein the person is able to remain in their home—and their community—for as long as they choose and is feasible given their specific circumstances.[154] This is not, however, to deny that medical conditions, including both physical and cognitive decline, increase with age, especially in the 'old, old' age range, which can also increase risk factors resulting in adverse health events such as falls. There will be situations where older people will require care—either in-home or in residential-based facilities. In fact, given the ageing population and the current approach to ageing, there currently are growing numbers of people in need of in-home nursing and residential care.[155]

This then raises the issue of access to affordable and quality health and aged care. A significant number of older people globally have inadequate access to healthcare.[156] In fact, a 2010 survey found that 63% of participants had trouble accessing quality healthcare when it was needed.[157] As stated, a significant contributing factor to accessibility is affordability. Absent government funded universal healthcare, older people may have to choose between the basics of food and shelter, and healthcare.[158] If available, and accessible, issues then arise in relation to quality of care.[159] Location can have a significant impact on this. Availability and accessibility of quality healthcare

---

[152]World Health Organization (2015).

[153]ICESCR, art 12; Committee on Economic, Social and Cultural Rights (2000).

[154]VIPAA, 54. Note that various concepts have been devised as ways of reframing ageing, including healthy ageing, productive ageing, successful ageing or active ageing. Various critiques of these concepts exist and will be discussed in more detail in Chap. 4 in relation to their potential to encourage greater participation and inclusion of older people.

[155]Australian Government (2015), 21.

[156]United Nations Secretary-General (2012).

[157]HelpAge International (2011a, b).

[158]United Nations Department of Economic and Social Affairs Programme on Ageing (2018), 2.

[159]Ibid.

is a concern not only in developing countries, but also in higher-income countries. For example, in rural, regional and remote areas, especially where geographical distances can be significant (such as in Canada and Australia), the cost of even reaching healthcare can be prohibitive.[160] Additionally, there are significant concerns around the training of people who are administering the care with recognition of the fact that a better systemic approach is needed to health and aged care generally. This requires multi-stakeholder involvement including not only the care facilities but also families, policy-makers and governments.

Ageist attitudes and individual perceptions about the stigma that attaches to ageing can also be a preventative factor in accessing quality healthcare. Take, for example, dementia. In addition to the potential legal ramifications associated with being diagnosed with dementia, such as the loss of legal capacity and decision-making autonomy discussed above, the personal, familial and social effects can be significant.[161] A person's capacity generally may also be called into question by their family, friends and/or society.[162] Therefore, the potential ramifications of accessing healthcare may derail a person from doing so, thus further compounding the health issue at hand given the lack of appropriate treatment.

Further, preconceived and discriminatory attitudes about ageing and towards older people can sometimes result in rationing of the care being offered.[163] That is, treatment is withheld, curtailed or changed in some way because of the age of the patient in question and anticipated treatment outcomes. For example, of the 200 members of the British Geriatrics Society surveyed in 2009, 72% indicated their belief that older people were less likely to be referred for chemotherapy or surgery, and 66% felt that there was less likely to be an investigation into the symptoms of older people.[164] Significantly, depression may also be disregarded in older people given the all too common attitude that the associated symptoms are merely a part of the ageing process.[165] This is particularly significant in relation to capacity as depression can have an impact on whether the person has the requisite legal capacity, for example, to make a will.[166]

Therefore, as populations age worldwide, there is increased demand on the health and aged care systems.[167] It is important to remember that there is no 'typical' older person, with ageing influenced by the social and physical environments of the person throughout their life as well as other economic and demographic factors.[168]

---

[160] United Nations Secretary-General (2012); (HelpAge International (2011a, b); Simpson and McDonald (2017), 72.

[161] Alzheimer's Disease International (2012).

[162] Purser and Sullivan (2019).

[163] United Nations Department of Economic and Social Affairs Programme on Ageing (2018), 2.

[164] See BBC News (2009), cited in United Nations Secretary-General (2012); United Nations Department of Economic and Social Affairs Programme on Ageing (2018), 2.

[165] United Nations Department of Economic and Social Affairs Programme on Ageing (2018), 2.

[166] Lonie and Purser (2017).

[167] Australian Government (2015), 21.

[168] World Health Organization (2018).

Consequently, any health and aged care system must be equipped for a wide variety of experiences and requirements.[169] It is, however, fundamental to the protection of the human rights of older persons to achieve a balance not only between respecting the autonomy of the individual and protecting them from harm, but also in relation to the role of external agencies and families in providing support and care.[170] Issues arising in the health and aged care contexts will be discussed in Chap. 10.

1.5 Access to Justice

Access to justice is a significant issue, particularly when there are instances of abuse or neglect. Considerable challenges can arise for older persons and their families, carers and advocates in attempting to navigate the often expensive and intricate legal landscape, if it becomes necessary to do so. As noted by the UN, 'in the absence of access to justice, people are unable to have their voice heard, exercise their rights, challenge discrimination or hold decision-makers accountable.'[171]

As a term, 'access to justice' is somewhat nebulous in nature. Does it mean the ability of an individual and/or group to access formal (the courts) and/or less formal dispute resolution methods (such as tribunals and mediation)? Alternatively, does it focus on the amorphous question of 'justice' and whether it has been achieved, and what kind of justice is it concerned with—distributive, remedial or procedural, or something else? Is it a question of accessing legal professionals? There is also an interrelated issue of the 'quality' of justice that can be accessed. For example, a legal professional who practises exclusively in a specific substantive area may be able to secure preferred outcomes for their clients compared with a general legal practitioner unfamiliar with the area but who has been retained to act. Access to justice has traditionally been more narrowly defined as relating to an individual's ability to access formal resolution processes and legal representation.[172] Given the increasing focus on less formal dispute resolution processes, a more general conceptualisation of the term has emerged focusing on an individual's ability to exercise their recognised legal rights.

In human rights law, access to justice is recognised as an important right, connected both to the right to equality before the law and also the right to obtain a remedy where one's rights have been violated.[173] Governments are obliged to ensure not only that rights are protected and promoted, but that adequate mechanisms are in place for a person to pursue justice. This therefore highlights both the necessity and appropriateness of adopting a human rights framework for elder law. Not only does a human rights approach set the standard for laws which respect the autonomy,

[169] World Health Organization (2015).

[170] VIPAA, 58.

[171] United Nations (n.d.).

[172] Law Council of Australia (2018).

[173] United Nations (n.d.).

dignity, liberty and other rights of older persons, but it also demands that work be undertaken to overcome the various barriers to accessing justice which older persons can face, including financial, language, geographical and health concerns. Further, avenues must be put in place for older persons to seek legal support and an appropriate remedy when their rights have been breached. Access to justice is thus a significant issue cutting across all legal areas examined in this book and will be discussed in those specific contexts in later chapters.

1.6 Women and Ageing

The particular experiences of older women and the gender dimensions of ageing and 'elder law' also need to be recognised. Women increasingly constitute the majority within ageing populations worldwide, as life expectancies are typically longer for women than men.[174] Women will therefore require greater access to health and aged care services. They also need income for a longer period of time to ensure that they have sufficient financial security in older age. Yet, at the same time, there are frequently significant disparities in both the income and social security benefits received by women and men.[175] For example, Egypt and Jordan are two of the countries with the greatest pension gaps, with men being approximately seven to eight times more likely than women to be in receipt of a pension.[176] However, even in higher-income countries where there is generally a limited gender-gap in the number of people receiving pensions, the disparity in benefit level can be significant. For example, the pension benefits in the European Union tend to be approximately 40% greater for men than for women.[177]

Women may also not have the financial security necessary to ensure independence as they age given, for example, time out of the workforce to have children or other periods of unpaid care, and the implications this can have on accrued assets and retirement entitlements.[178] Leaving the workforce, albeit temporarily, can have a double effect. Not only can it make it harder to return, it can also significantly impact a woman's ability to save for retirement. For example, for the years 2009–2010, the average superannuation payout for a woman in Australia was just under half of that for a man earning roughly the equivalent amount of money.[179] Women with disabilities, migrant women and women from racial or ethnic minorities often experience greater

[174] VIPAA, 20(c).

[175] United Nations Department of Economic and Social Affairs Programme on Ageing (2016), 5.

[176] Ibid.

[177] United Nations Entity for Gender Equality and the Empowerment of Women (2015); United Nations Department of Economic and Social Affairs Programme on Ageing (2016), 5.

[178] United Nations Secretary-General (2014); United Nations Department of Economic and Social Affairs Programme on Ageing (2016), 5.

[179] Australian Human Rights Commission (2015).

financial insecurity, and thus are at greater risk of homelessness and poverty.[180] This therefore places older women in an especially vulnerable position.

Older women are also more likely to experience neglect, abuse and/or violence in older age, with this often going undetected. A 2013 report, *Neglect, Abuse and Violence against Older Women*,[181] presents the different risk factors and forms of abuse suffered by older women. Women are more likely than men to be victims of elder abuse, especially as women tend to live longer than their male partners which can lead to increased risks of vulnerability.[182] There is, however, a lack of comprehensive data on older women's experiences of neglect, violence and abuse. Such data is essential for the development of comprehensive evidence-based policies and for measuring the effectiveness of legislation or a national plan targeting elder abuse. The former Secretary-General in his 2014 Report to the United Nations General Assembly stated that most studies on violence against women survey only women *under* the age of fifty years, thereby omitting the views and experiences of older women altogether.[183] The Secretary-General suggested that this situation reflects the lack of an agreed definition of what constitutes violence against older women. The Report concluded by recommending that member states 'consider developing an explicit reference and policy framework for addressing neglect, violence and abuse against older women'.[184] Therefore, the particular issues facing older women is a recurrent theme discussed throughout the analysis presented in this book and one which requires both acknowledgement and a dedicated response.

1.7 Conclusion

The select issues discussed above all highlight the relevance and importance of human rights as both an analytical framework and a set of guiding principles for improving elder law and, ultimately, the lived experiences of older people. The fundamental connection between human rights, dignity, autonomy, liberty and equality make it an apt framework for approaching any of the individual issues discussed above, but also for understanding the various ways in which these issues compound, contradict or reinforce each other. The commitment which human rights makes to respecting the value and dignity of each individual person directly challenges many of the ageist assumptions which have typically driven elder abuse and other shortcomings within elder law. A human rights-based approach therefore promises to promote a society

---

[180] Australian Human Rights Commission (2019); United Nations Department of Economic and Social Affairs Programme on Ageing (2016), 5; Westwood (2012).

[181] United Nations Department of Economic and Social Affairs Division for Social Policy and Development (2013).

[182] Australian Law Reform Commission (2016), para 1.16.

[183] United Nations Secretary-General (2014), 11.

[184] Ibid.

in which the opportunities represented by ageing are recognised and celebrated, and in which discriminatory behaviour and ageism are eliminated.[185]

As discussed above, the globally ageing society presents questions about how to respond to associated social, economic and legal challenges, but also how (and whether) to define and categorise 'older' people. Recognising that ageing is a life-long process and not something that is triggered at a particular chronological age is essential to changing socially constructed ideas about older people. As noted by the UN in the *Vienna International Plan of Action of Ageing*, determining 'old' merely in terms of a number and whether someone is in employment or not is a 'sad paradox' emerging as a result of the socio-economic focus of some countries.[186] The fact that a person has reached a specific chronological age should not see them 'sidelined in their own society'.[187] Understanding this will assist in addressing preconceived notions that ageing is a negative thing and that once people become older, they are therefore vulnerable and in need of care. Rather, ageing provides opportunities to learn from the cumulative wisdom of older people, which is predicated upon acknowledging the vital role they have played, and continue to play, in society. Nevertheless, it is important to accept that with age, especially in the 'old, old' category, comes an increased chance of disability, cognitive impairment, health problems, social isolation and/or financial dependence, all of which operate as risk factors for abuse and negative human rights impacts.[188]

As seen in the above discussion, preparing for ageing is not just restricted to the physical and health aspects of growing older. Financial, mental, emotional, psychological, religious, spiritual, cultural and relational factors all have a vital role to play.[189] Consequently, it is suggested that it is imperative for people to begin to prepare for the process of ageing throughout their entire life course. This will necessarily involve not only the individuals but also their families, communities, society more generally, and governments. Older individuals in particular should be given a voice in the development and actioning of any policies and systems which affect them. Conversely, the governments and policy makers responsible for designing and implementing the policies should be aware of their responsibility to older people generally, but especially in relation to specific groups of older people, such as women, those from culturally and linguistically diverse backgrounds, those who lack capacity and people living in poverty, all of whom may experience heightened levels of vulnerability.

It is important to recognise that the response to ageing should be phrased in terms of what is appropriate for the specific country having regard to their socio-economic capacities and cultural differences.[190] Having said this, there are a number of essential considerations that must be taken into account irrespective of different cultures,

---

185 VIPAA, 25(h).

186 Ibid., 28.

187 Ibid.

188 World Health Organization (2017c).

189 VIPAA, 26.

190 Ibid.

religions and social or economic status. These flow from the universal nature of human rights, which insists that certain basic needs and rights must be afforded to all people regardless of nationality. These include the basic human rights to work in just and favourable conditions;[191] social protection and an adequate standard of living;[192] the highest attainable standards of physical and mental health;[193] equality before the law and freedom from discrimination;[194] participate in public affairs and elections;[195] privacy;[196] and the protection of minority rights.[197] Further, human rights are understood as interdependent and indivisible, and a violation of one right is likely to impact other rights or coincide with multiple violations.[198] This helps to encourage a multidisciplinary, coordinated approach which appreciates the interconnectedness of the various challenges associated with ageing populations. As noted above, issues like ageism, capacity and abuse cut across elder law and need to be addressed appropriately to ensure all human rights can be enjoyed.

This is not to say that looking to a human rights framework will address all the challenges associated with ageing—it obviously will not. One of the biggest drawbacks of adopting a human rights framework is enforcement. While most countries have signed and ratified the major international human rights conventions, domestic implementation of those conventions is more fragmented and enforcement options for affected individuals can be limited. Nevertheless, a human rights approach does provide a comprehensive and well-developed set of principles against which to analyse existing elder laws, articulate shortcomings and develop more appropriate approaches. This book advocates for improving the domestic protection of human rights so that older persons in all jurisdictions have adequate recourse to legal protections. However, even in the absence of such human rights legislation, human rights remain one of the most practical and positive ways to reframe elder law in order to better promote the dignity, autonomy and liberty of older people.

It is for these reasons that this book both adopts and advocates a human rights-based approach to elder law. In the following chapters we outline and apply a human rights framework to a series of legal areas and policy challenges. We begin by examining, in Chap. 2, existing human rights obligations which apply to older persons, before then articulating a novel, multidimensional human rights framework which draws on these laws and on various international instruments as well as multidisciplinary scholarship (Chap. 3). We then apply this critical human rights lens to a series of issues and areas wherein older people interact with the law and legal actors. These include barriers to economic and social participation (Chap. 4); ageism and age-discrimination (Chap. 5); legal capacity (Chap. 6); elder abuse (Chap. 7); financial

[191] ICESCR, arts 6,7.
[192] ICESCR, arts 9 and 11.
[193] ICESCR, art 12.
[194] ICCPR, arts 20 and 26.
[195] Ibid., art 25.
[196] Ibid., art 17.
[197] Ibid., art 27.
[198] Office of the United Nations High Commissioner for Human Rights (2014).

management, retirement and estate planning (Chap. 8); accommodation (Chap. 9); and health and aged care (Chap. 10). The book will conclude by drawing together recommendations for improving elder law in both specific fields and more generally, drawing on best practice from various jurisdictions and leading scholarship in the area.

This book identifies individual autonomy in decision-making and respect for dignity and liberty as core values of a human rights-based approach to elder law, arguing that these ought to underpin all laws affecting older people. No single work has previously adopted a human rights lens to these issues and addressed them altogether, making this a unique contribution to the emerging body of elder law literature. The analysis offered here adopts an innovative approach to furthering the existing dialogue around what it means to 'age well'. It will advocate for the utility of a human rights framework in protecting the rights of older people in the major areas of law which can impact them acknowledging that, as people age, they should be valued as legitimate rights holders rather than passive beneficiaries of younger generations' benevolence. It is this paradigm shift which this book hopes to advance, arguing that a human rights perspective can help to combat ageism and value the dignity and autonomy of each individual.

References

AGE Platform Europe (2018) Submission based on seminar with Europe-based experts on the human rights of older persons. Submitted to the 9th session of the open-ended working group on ageing. https://social.un.org/ageing-working-group/documents/ninth/Inputs%2520NGOs/AGE_Brussels_seminar_April2018_FINAL.pdf. Accessed 6 Feb s2020

AgeUK (May 2019) Later life in the United Kingdom 2019. https://www.ageuk.org.uk/globalassets/age-uk/documents/reports-and-publications/later_life_uk_factsheet.pdf. Accessed 6 Feb 2020

Alzheimer's Disease International (2012) World Alzheimer report 2012: overcoming the stigma of dementia. https://www.alz.co.uk/research/world-report-2012. Accessed 6 Feb 2020

Australian Digital Inclusion Index (2020). https://digitalinclusionindex.org.au/. Accessed 6 Feb 2020

Australian Bureau of Statistics (2010) Population by age and sex, Australian states and territories (cat. no. 3201.0)

Australian Bureau of Statistics (2016) Household income and wealth, Australia, 2013–14: superannuation in Australia (cat. no. 6523.0)

Australian Government (2015) Intergenerational report 2015: overview. https://static.treasury.gov.au/uploads/sites/1/2017/06/2015IGR_Overview.pdf. Accessed 6 Feb 2020

Australian Government (2010) Australia to 2050: future challenges. https://treasury.gov.au/sites/default/files/2019-03/IGR_2010_Overview.pdf. Accessed 6 Feb 2020

Australian Human Rights Commission (2015) Face the facts: older Australians. https://www.humanrights.gov.au/our-work/education/face-facts-older-australians. Accessed 6 Dec 2019

Australian Human Rights Commission (2016) Willing to work: national inquiry into employment discrimination against older Australians and Australians with disability. https://www.humanrights.gov.au/our-work/disability-rights/publications/willing-work-national-inquiry-employment-discrimination. Accessed 9 Dec 2019

Australian Institute of Health and Welfare (2017) Ageing. https://www.aihw.gov.au/ageing/. Accessed 19 Dec 2020

Australian Institute of Health and Welfare (2019a) Social isolation and loneliness. https://www.aihw.gov.au/reports/australias-welfare/social-isolation-and-loneliness. Accessed 19 Dec 2019

Australian Institute of Health and Welfare (2019b) Aged care for indigenous Australians https://www.aihw.gov.au/reports/australias-welfare/aged-care-for-indigenous-australians. Accessed 15 Jan 2020

Australian Law Reform Commission (2016) Elder abuse (Discussion Paper 83)

Australian Law Reform Commission (2017) Elder abuse: a national legal response. Report No 131

Australian Securities and Investment Commission (2019) ASIC professional standards for financial advisers—reforms. https://asic.gov.au/for-finance-professionals/afs-licensees/professional-standards-for-financial-advisers/. Accessed 6 Feb 2020

Avers D et al (2011) Use of the term "elderly". J Geriatr Phys Ther (34)(4):153–154

Bakewell J, Foster D (2015) Should older people downsize to help solve the housing crisis? The guardian. https://www.theguardian.com/commentisfree/2015/sep/28/should-older-people-downsize-housing-crisis-homes. Accessed 6 Feb 2020

Barbosa Neves B, Sanders A, Kokanović R (2019) 'It's the worst bloody feeling in the world': experiences of loneliness and social isolation among older people living in care homes. J Aging Stud 49:74-84

Beard JR, Officer AM, Cassells AK (2016) The world report on ageing and health. Gerontologist 56(S2):S163–S166

Bennett B et al (2017) Assistive technologies for people with Dementia: ethical considerations. Bull World Health Organ 95(11):749–755

Bennett B (2019) Technology, ageing and human rights: challenges for an ageing world. Int J Law Psychiatry 66:1–7. Article 101449

Brereton M et al (2015) The messaging Kettle: prototyping connection over a distance between adult children and older parents. In: Proceedings of the 33rd annual ACM conference on human factors in computing systems.Seoul, South Korea. 713–716

Butler RN (1989) Dispelling ageism: the cross-cutting intervention. Ann Am Acad Polit Soc Sci 503:138–147

Carney T (1997) Competence. Int J Law Psychiatry 20(1):1–4

Choice (2017) The hidden costs of retirement village contracts. https://www.choice.com.au/money/property/buying/articles/retirement-village-contracts. Accessed 6 Feb 2020

Cockerill J, Collier B, Maxwell K (2005) Legal requirements and current practices. In: Collier B, Coyne C, Sullivan K (eds) Mental capacity, powers of attorney and advance health directives. Federation Press, Annandale.

Dārziņš P, Molloy DW, Strang D (eds) (2000) Who can decide? The six step capacity assessment process. Memory Australia Press, Adelaide

Doron I, Apter I (2010) The debate around the need for an international convention on the rights of older persons. The Gerontologist 50(5):586–593

Fredvang M, Biggs S (2012) The rights of older persons: protection and gaps under human rights law. Centre for Public Policy and Brotherhood of Saint Laurence, Social Policy Working Paper no 16

Government of Canada (2014) Action for seniors report. https://www.canada.ca/en/employment-social-development/programs/seniors-action-report.html#tc2a. Accessed 6 Feb 2020

Greengross S (2019) Human rights and ageing: are we doing it right? keynote address to Australian association of gerontology. Brisbane, Australia (15 Nov 2019)

HelpAge International (2011) Insights on ageing: a survey report

HelpAge International (2011) The right to health in old age: unavailable, inaccessible and unacceptable. https://www.helpage.org/silo/files/the-right-to-health-in-old-age-unavailable-inaccessible-and-unacceptable.doc. Accessed 6 Feb 2020

Hill A (2019) Over a third of britons admit ageist behaviour in new study. The guardian. https://www.theguardian.com/society/2019/aug/19/over-a-third-of-britons-admit-ageist-behaviour-in-new-study. Accessed 6 Feb 2020

Homelessness Australia (2016) Homelessness and older people. https://www.homelessnessaustralia.org.au/sites/homelessnessaus/files/2017-07/Homelessness_and_Older_People.pdf. Accessed 6 Feb 2020
Hu X et al (2017) Retirement villages in Australia: a literature review. Pac Rim Property Res J 23(1):101–122
International Labour Organization (2014) World social protection report 2014/15: building economic recovery, inclusive development and social justice. https://www.ilo.org/global/research/global-reports/world-social-security-report/2014/WCMS_245201/lang--en/index.htm. Accessed 6 Feb 2020
International Expert-Conference on the Human Rights of Older Persons (2018) Vienna declaration on the human rights of older persons. https://ageing.at/site/pub_html/sites/default/files/ICHRoP%2520Conference%2520Declaration%2520Final_0.pdf. Accessed 19 Dec 2019.
Kornfeld-Matte R (2016) Report of the independent expert on the enjoyment of all human rights by older persons. UN Doc A/HRC/33/44
Lagacé M et al (1989) The silent impact of ageist communication in long term care facilities: elders' perspectives on quality of life and coping strategies. J Aging Stud Spec Section: Innovative Approaches Int Comparisons 26(3):335–342
Law Council of Australia (2018) The justice project final report. https://www.lawcouncil.asn.au/files/web-pdf/Justice%2520Project/Final%2520Report/Justice%2520Project%2520_%2520Final%2520Report%2520in%2520full.pdf. Accessed 19 Dec 2019
Law Reform Committee, Parliament of Victoria (2010) Inquiry into powers of attorney final report of the victorian law reform committee
Lewis B (2018a) Environmental human rights and climate change: current status and future prospects. Springer Nature, Singapore
Lewis B et al (2018b) Submission by the Australian centre for health law research to the Australian human rights commission human rights and technology issues paper. https://eprints.qut.edu.au/122938/. Accessed 7 Feb 2020
Lindenberg J et al (2013) Elder Abuse an international perspective: exploring the context of Elder Abuse. Int Psychogeriatr 25(8):1213–1215
Lonie J, Purser K (2017) Assessing testamentary capacity from the medical perspective. Aust Bar Rev 44(3):297–317
O'Connor D (2010) Personhood and dementia: toward a relational framework and assessing decisional capacity. J Mental Health Training, Educ Pract 5(3):22–30
Office of the United Nations High Commissioner for Human Rights (2014) The right to adequate housing (Fact Sheet No. 21/Rev. 1). https://www.ohchr.org/Documents/Publications/FS21_rev_1_Housing_en.pdf. Accessed 6 Feb 2020
Parliamentary Budget Office (2019) Australia's ageing population: understanding the fiscal impacts over the next decade. Report no. 02/2019
Petersen M, Jones A (2011) Homelessness and older Australians: developing a research evidence base. Parity 24(9):31
Petersen M, Tilse C, Cockburn T (2017) Living in a retirement village: choice, contracts, and constraints. J Housing Elderly 31(3):229–242
Purser K, Rosenfeld T (2014) Evaluation of legal capacity by doctors and lawyers: the need for collaborative assessment. Med J Aust 201(8):483–485
Purser K (2017) Capacity assessment and the law: problems and solutions. Springer International, Switzerland
Purser K et al (2018) Alleged financial abuse of those under an enduring power of attorney: an exploratory study. Br J Soc Work 48(4):887–905
Purser K, Sullivan K (2019) Capacity assessment and estate planning—the therapeutic importance of the individual. Int J Law Psychiatry 64:88–98
Purser K, Lonie J (2019) Mapping dementia and cognitive decline in testamentary capacity. Int J Law Psychiatry 66:101450

Purser K, Cockburn T, Ulrick E (2020) Examining access to formal justice mechanisms for vulnerable older people in the context of enduring powers of attorney. Elder Law Rev 12(2)
Relationships Australia (2018) Is Australia experiencing an epidemic of loneliness? Working paper. https://www.relationships.org.au/what-we-do/research/an-epidemic-of-loneliness-2001-2017. Accessed 6 Feb 2020
Simpson C, McDonald F (2017) Rethinking rural health ethics. Springer, Berlin
Stanford Center on Longevity (n.d.) The sightlines project. https://longevity.stanford.edu/the-sightlines-project/. Accessed 6 Feb 2020
Sullivan K (2004) Neuropsychological assessment of mental capacity. Neuropsychol Rev 14(3):131–142
Sunlife (2019) Ageist Britain? https://www.sunlife.co.uk/siteassets/documents/ageist-report-2019.pdf. Accessed 19 Dec 2019
Sutin AR et al (2018) Loneliness and risk of dementia. J Gerontol. Series B:1–9
Triebel KL, Gerstenecker A, Marson DC (2018) Financial and medical decision-making capacity in mild cognitive impairment and Dementia. In: Smith GE, Farias ST (eds) APA handbook of dementia. American Psychological Association, Washington, DC, pp 219–235
United Nations (n.d.) Access to justice. https://www.un.org/ruleoflaw/thematic-areas/access-to-justice-and-rule-of-law-institutions/access-to-justice/. Accessed 6 Feb 2020
United Nations Department of Economic and Social Affairs. Division for Social Policy and Development and Programme on Ageing (2013) Neglect, abuse and violence against older women. https://www.un.org/esa/socdev/documents/ageing/neglect-abuse-violence-older-women.pdf. Accessed 2 July 2020
United Nations Department of Economic and Social Affairs, Population Division (2015a) World population ageing: highlights. https://www.un.org/en/development/desa/population/publications/pdf/ageing/WPA2015_Highlights.pdf. Accessed 6 Feb 2020
United Nations Department of Economic and Social Affairs, Population Division (2015b) World population ageing. UN Doc ST/ESA/SER.A/390
United Nations Department of Economic and Social Affairs, Population Division (2017) World population ageing: highlights. https://www.un.org/en/development/desa/population/publications/pdf/ageing/WPA2017_Highlights.pdf. Accessed 6 Feb 2020
United Nations Department of Economic and Social Affairs Programme on Ageing (2016) Economic inequalities in old age. https://www.un.org/development/desa/ageing/wp-content/uploads/sites/24/2016/08/Briefing-Paper_Economic-Inequalities_Final.pdf. Accessed 6 Feb 2020
United Nations Department of Economic and Social Affairs Programme on Ageing (2018) Health inequalities in old age. https://www.un.org/development/desa/ageing/wp-content/uploads/sites/24/2018/04/Health-Inequalities-in-Old-Age.pdf. Accessed 6 Feb 2020
United Nations Entity for Gender Equality and the Empowerment of Women (2015) Protecting women's income security in old age: toward gender-responsive pension systems, Policy Brief, No. 3. https://www.unwomen.org/en/digital-library/publications/2015/12/women-income-security-in-old-age. Accessed 6 Feb 2020
United Nations Population Fund Asia and the Pacific (n.d.) Ageing. https://asiapacific.unfpa.org/en/node/15208. Accessed 6 Feb 2020
United Nations Secretary-General (2012) Report on follow-up to the second world assembly on ageing. UN Doc A/67/188 (26 July 2012)
United Nations Secretary-General (2013) Report on follow-up to the second world assembly on ageing. UN Doc A/68/167 (19 July 2013)
United Nations Secretary-General (2014) Report on follow-up to the second world assembly on ageing. UN Doc A/69/180 (24 July 2014)
United States Census Bureau (2018) Older people projected to outnumber children for the first time in U.S. history. https://www.census.gov/newsroom/press-releases/2018/cb18-41-population-projections.html. Accessed 6 Feb 2020
Westwood S (2012) I may be older, but i ain't no 'elder': a critique of 'elder law'. Temp Pol Civ Rights Law Rev 21(2):485–510

Wiles JL et al (2012) The meaning of "ageing in place" to older people. The Gerontologist 52:357–366
Williams J (2003) When i'm sixty-four: lawyers, law and 'old age'. Cambrian Law Rev 34:103–119
Winick BJ (1996) The side effects of incompetency labeling and the implications for mental health law. In: Wexler DB and Winick BJ (eds) Law in a therapeutic key: developments in therapeutic jurisprudence. Carolina Academic Press, Durham
World Health Organization Regional Office for Europe (2020) Demographic trends, data and statistics on ageing. https://www.euro.who.int/en/health-topics/Life-stages/healthy-ageing/data-and-statistics/demographic-trends,-statistics-and-data-on-ageing. Accessed 6 Feb 2020
World Health Organization (2002) The Toronto declaration on the global prevention of elder abuse. https://www.who.int/ageing/publications/toronto_declaration/en/. Accessed 6 Feb 2020
World Health Organization (2004) Ageing and health. Technical report vol.5: a glossary of terms for community health care and services for older persons. https://www.who.int/kobe_centre/publications/Ageing_Tech_Report_5/en/. Accessed 19 Dec 2019
World Health Organization (2015) World report on ageing and health. https://www.who.int/ageing/events/world-report-2015-launch/en/. Accessed 19 Dec 2019
World Health Organization (2017a) Elder abuse fact sheet. https://www.who.int/mediacentre/factsheets/fs357/en/. Accessed 6 Dec 2019
World Health Organization (2017b) Ageing and life-course: commit to action. https://www.who.int/ageing/commit-action/en/. Accessed 2 Dec 2019
World Health Organization (2018) Ageing and health. https://www.who.int/news-room/fact-sheets/detail/ageing-and-health

International Instruments

Committee on Economic, Social and Cultural Rights (1995) *General Comment No. 6: The Economic, Social and Cultural Rights of Older Persons* (8 December 1995)
Committee on Economic, Social and Cultural Rights (2000) *General Comment No. 14: The right to the highest attainable standard of health (Art 12)* (11 August 2000)
Convention on the Rights of Persons with Disabilities (2006) Opened for signature 13 December 2006, entered into force 3 May 2008. 2515 UNTS 3
International Covenant on Civil and Political Rights (1966) Opened for signature 16 December 1966, entered into force 23 March 1976. 999 UNTS 171
International Covenant on Economic, Social and Cultural Rights (1966) Opened for signature 16 December 1966, entered into force 3 January 1976. 993 UNTS 3
United Nations General Assembly (2013) *Towards a Comprehensive and Integral International Legal Instrument to Promote and Protect the Rights and Dignity of Older Persons.* Resolution 67/139. UN Doc A/RES/67/139 (13 February 2013)
United Nations Principles for Older Persons (1991) Adopted by United Nations General Assembly Resolution 46/91. UN Doc A/RES/46/91 (16 December 1991)
Universal Declaration of Human Rights (1948) Adopted by United Nations General Assembly Resolution 217A (III). UN Doc A/810 (10 December 1948)
Vienna International Plan of Action on Ageing (1982) Adopted by World Assembly on Ageing, Vienna, 26 July – 6 August 1982. Endorsed by UN General Assembly Resolution 37/51. UN Doc A/RES/37/51 (3 December 1982)

Cases

Re Caldwell [1999] QSC 182

Chapter 2
The Existing Framework for Protecting the Human Rights of Older Persons

A human rights-based approach to elder law draws heavily on the rights that are guaranteed to all people under international and regional human rights law. While there is no dedicated treaty protecting the rights of older people in particular, numerous other human rights are useful in setting standards and articulating obligations in the various fields of elder law. This chapter outlines some of the most pertinent of these rights and introduces some of the implications that will be analysed further in later chapters, drawing as well on a number of soft-law instruments on this topic. It also addresses the *Convention on the Rights of Persons with Disabilities* (CRPD) noting that, while some older persons will experience disability and will therefore fall within the scope of the CRPD, it is important not to equate older age with impairment or assume that the CRPD is sufficient. With this in mind, this chapter outlines some of the work done to date in advocating for a dedicated convention on the rights of older persons.

2.1 Introduction

As the previous chapter identified, promoting and protecting the rights of older persons is a challenge which spans a wide range of socio-economic and legal issues. In particular, ageism and elder abuse are significant human rights issues which undermine the dignity and autonomy of older persons and threaten their physical, emotional, social and financial well-being. As the number of older persons in our societies increases, so too does the challenge of ensuring that legal systems support the full enjoyment of human rights and protect older persons from human rights violations. A key objective of this book is to identify ways in which human rights laws and principles can be better integrated into the framework of elder law with a view to meeting this challenge. A starting point for this analysis is to identify relevant human rights protections which are already recognised within human rights law in order to consider how these rights should shape elder law at the domestic level. This

© Springer Nature Singapore Pte Ltd. 2020

B. Lewis et al., *The Human Rights of Older Persons*,
https://doi.org/10.1007/978-981-15-6735-3_2

examination also affords the opportunity to pinpoint areas where greater protections are required within international human rights law, and this chapter argues in particular that a dedicated international convention on the rights of older persons (CROP) is a crucial element required for a comprehensive rights-based framework.

First however, the chapter begins by considering the major international and regional human rights treaties. It identifies a number of specific rights that are particularly relevant to the experiences and needs of older persons, while noting that older persons are entitled to enjoy the full range of human rights without discrimination. It then examines a number of international soft-law instruments which, while not having the force of binding law, go some way towards articulating the specific obligations which governments owe towards older people under more generic human rights law. It will then provide an overview of recent efforts to pursue a dedicated CROP, addressing the key criticisms of the current framework and arguments in favour of stronger, more targeted protections that address the particular needs and vulnerabilities of older persons. This critical analysis of the current framework and potential future directions helps to clarify the scope of existing obligations and identify their likely evolution. It also informs the human rights-based approach to elder law which is developed in Chap. 3 and applied throughout the remainder of this book.

2.2 International Law Protecting the Human Rights of Older Persons

International human rights law consists of a number of multilateral treaties, mostly established and monitored by the United Nations (UN) human rights bodies, and particularly by the Office of the High Commissioner for Human Rights. In addition to the UN treaties, a number of regional inter-governmental organisations have also adopted human rights treaties. These include the Council of Europe's *European Convention for the Protection of Human Rights and Fundamental Freedoms*,[1] the Organisation of American States' *Declaration on the Rights and Duties of Man*,[2] and *American Convention on Human Rights*,[3] and the African Union's *Charter on Human and Peoples' Rights*.[4] Two of these regional bodies have enacted instruments specifically focused on the rights of older persons: the *Inter-American Convention on Protecting the Rights of Older Persons*,[5] and the *Protocol to the African Charter on the Rights of Older Persons in Africa*.[6] To date, seven states have ratified the

[1] *European Convention for the Protection of Human Rights and Fundamental Freedoms* (1950) ('*ECHR*').

[2] *American Declaration on the Rights and Duties of Man* (1948).

[3] *American Convention on Human Rights* (1969).

[4] *African Charter of Human and Peoples' Rights* (1981).

[5] *Inter-American Convention on Protecting the Rights of Older Persons* (2015).

[6] *Protocol to the African Charter on Human and Peoples' Rights on the Rights of Older Persons* (2016).

American Convention, which entered into force in 2017, while only two states have ratified the African Protocol. The latter instrument requires fifteen ratifications in order for it to enter into force.[7]

Among the body of human rights law are treaties of broad application, such as the *International Covenant on Civil and Political Rights* (ICCPR)[8] and the *International Covenant on Economic, Social and Cultural Rights* (ICESCR),[9] as well as treaties which protect the rights of particular classes of vulnerable people, like the *Convention on the Rights of the Child*[10] and the *Convention on the Elimination of All Forms of Discrimination Against Women* (CEDAW).[11] The *Convention on the Rights of Persons with Disabilities* (CRPD)[12] is the most recently adopted UN human rights treaty and, as will be discussed in more detail below, has some particular relevance for the rights of older persons (as well as providing a useful model for the adoption of a dedicated convention for older persons). However, for the reasons discussed, the CRPD ought not to be viewed as an adequate substitute for a dedicated CROP, and international human rights law currently contains no binding and widely adopted agreement that deals specifically with the human rights of older persons.

Despite the lack of a dedicated convention, existing human rights law does guarantee a number of important rights. The discussion below details the key human rights found in international human rights law and the corresponding obligations they create with respect to older persons. They represent important standards applying to a wide range of relevant issues, including housing, health, mobility, social and cultural participation, financial independence and decision-making. As discussed in the previous chapter and in more detail in Chap. 7, many of these rights are violated when elder abuse occurs and are frequently undermined by ageist attitudes and behaviours.

Under international human rights law, states bear three levels of obligations: to respect, protect and fulfil. The duty to respect requires that states refrain from any action which would directly or indirectly interfere with the enjoyment of human rights. The duty to protect entails a duty to regulate the actions of private actors, including corporate entities, to ensure that they do not impinge upon human rights. The third level, the duty to fulfil, is the most demanding level of obligation, and requires that states take the necessary steps to achieve full realisation of the right. These steps would include legislative, administrative, judicial, financial and other measures required to facilitate and provide for fulfilment of the right.

As will be seen below, the specific requirements for each of these levels of duty will vary according to the subject matter of each right. For economic, social and cultural rights protected within the ICESCR, states' obligations will also vary according to their particular circumstances. Article 2(1) states that:

---

[7] Ibid., art 26.

[8] *International Covenant on Civil and Political Rights* (1966) ('*ICCPR*').

[9] *International Covenant on Economic, Social and Cultural Rights* (1966) ('*ICESCR*').

[10] *Convention on the Rights of the Child* (1989).

[11] *Convention on the Elimination of All Forms of Discrimination Against Women* (1979) ('*CEDAW*').

[12] *Convention on the Rights of Persons with Disabilities* (2006) ('*CRPD*').

> Each State Party to the present Covenant undertakes to take steps, individually and through international assistance and co-operation, especially economic and technical, to the maximum of its available resources, with a view to achieving progressively the full realization of the rights recognized in the present Covenant by all appropriate means, including particularly the adoption of legislative measures.

Article 2 implies a recognition that many of the rights found in the ICESCR, including the right to an adequate standard of living and the right to the highest attainable standard of health (discussed below), demand the investment of considerable time and resources in order to achieve their full realisation. The standard of obligation required of states is therefore one of 'progressive realisation', such that states will be held to have discharged their obligations under the Covenant provided that they are actively making efforts to progress the fulfilment of the rights. What will be considered 'adequate' in this regard will also be assessed relative to the state's specific circumstances, and particularly its available resources.

The Committee on Economic, Social and Cultural Rights (CESCR) has clarified, however, that article 2.1 does include two obligations which require immediate implementation. The first is that states must actually 'take steps' towards progressive realisation by 'all appropriate means'. As the Committee has said, 'such steps should be deliberate, concrete and targeted as clearly as possible towards meeting the obligations recognized in the Covenant.'[13]

The other immediate obligation is of particular relevance to older persons, as it requires that whatever measures a state chooses to implement, it must do so without discrimination.[14] This requires that states address both formal and substantive discrimination.[15] Eliminating formal discrimination alone will not be sufficient: states must 'immediately adopt the necessary measures to prevent, diminish and eliminate the conditions and attitudes which cause or perpetuate substantive or de facto discrimination.'[16] For older persons this requires that states understand the nature and impact of ageism in their societies, identify the particular needs of older persons, and ensure that laws and policies are developed both to address ageism and to achieve the full realisation of economic, social and cultural rights. The obligation to address discrimination is not only relevant to economic, social and cultural rights but also applies to all rights, and is a fundamental principle underpinning human rights law generally. It is also articulated in the right to equality before the law, discussed below.

---

[13]Committee on Economic, Social and Cultural Rights (1990), 1–3.

[14]Ibid., 2.

[15]Committee on Economic, Social and Cultural Rights (2009), 8.

[16]Ibid.

2.2.1 *Right to Equality Before the Law*

International human rights law guarantees that all individuals are to be treated equally before the law.[17] Article 26 of the ICCPR states:

> All persons are equal before the law and are entitled without any discrimination to the equal protection of the law. In this respect, the law shall prohibit any discrimination and guarantee to all persons equal and effective protection against discrimination on any ground such as race, colour, sex, language, religion, political or other opinion, national or social origin, property, birth or other status.

The associated obligation to remove and protect against discrimination extends both to formal or explicit discrimination, as well as substantive discrimination or other inequalities which operate to restrict the rights of marginalised or vulnerable groups.[18]

The link between equality and non-discrimination goes to the heart of human rights law. International human rights treaties all emphasise that, because people possess human rights by virtue of their being human, it follows that there can be no grounds to discriminate against any person in the fulfilment of their human rights. That is, human rights belong to all persons equally and all persons are entitled to the full range of human rights.

For older people, the right to equality before the law has a number of interrelated dimensions. While the specific grounds of discrimination listed in article 26 and other provisions do not include age, they do refer to 'other status' and the list is not intended to be exhaustive.[19] The Human Rights Committee (HRC), which has responsibility for supervision and enforcement of the ICCPR, has noted that, while the listed grounds of discrimination are not comprehensive, the notion of 'discrimination' ought to be interpreted to include 'any distinction, exclusion, restriction or preference … which has the purpose or effect of nullifying or impairing the recognition, enjoyment or exercise by all persons, on an equal footing, of all rights and freedoms'.[20] As will be explained in more detail in Chap. 5, the phenomenon of ageism is antithetical to equality and non-discrimination. The interdependent nature of human rights means that ageism is not only a human rights violation in itself, but also leads to a range of other more specific rights breaches. It is clear then that discrimination on the basis of age which operates to restrict or undermine the enjoyment of human rights would be incompatible with human rights law.

The right to equality before the law also demands that governments address other forms of discrimination which intersect with ageism, including discrimination on the basis of disability, gender or race. The need to ensure equality between men and women is of particular relevance to older persons as discrimination between men and

---

[17]ICCPR, art 2.1; ICESCR, arts 2.2, 3. Also American Convention, art 24; African Charter, art 3.

[18]In relation to non-discrimination and economic, social and cultural rights, see Committee on Economic, Social and Cultural Rights (2009), 20.

[19]Committee on Economic, Social and Cultural Rights (1995); Committee on Economic, Social and Cultural Rights (2009), 15; Human Rights Committee (1989), 7.

[20]Human Rights Committee (1989), 7.

women throughout the life course can lead to increased disadvantage in later life. Multiple factors combine to produce these differential impacts for women. Most notably, these include the disproportionate time spent by women caring for family or performing other unpaid work, during which period women generally earn lower or no income and make limited or no contributions to superannuation or pension funds.[21] The resulting reduction in resources available at the time of retirement from paid employment is further compounded by the fact that women on average live longer than men.[22] The impact of financial insecurity on the human rights of older women will be explored in more detail in Chap. 8.

2.2.2 Right to an Adequate Standard of Living, Incorporating Rights to Adequate Food, Water and Housing

International law guarantees the right to an adequate standard of living in article 11 of the ICESCR, which reads:

> The States Parties to the present Covenant recognize the right of everyone to an adequate standard of living for himself and his family, including adequate food, clothing and housing, and to the continuous improvement of living conditions. The States Parties will take appropriate steps to ensure the realization of this right, recognizing to this effect the essential importance of international co-operation based on free consent.

Interpretation of this provision by the CESCR has confirmed that it guarantees to all people rights to adequate food, water and housing, and that an assessment of what is 'adequate' must incorporate considerations of affordability, accessibility, and appropriateness.[23]

As with the right to equality before the law, the right to an adequate standard of living must be afforded to all people without discrimination. While age is not mentioned as an explicit ground of discrimination in the Covenant, the Committee has included it when explaining the requirement that states avoid discrimination in relation to food, water and housing.[24]

It is included in the equivalent provision in the *Universal Declaration of Human Rights*, which recognises that:

> Everyone has the right to a standard of living adequate for the health and well-being of himself and his family, including food, clothing, housing and medical care and necessary

---

[21] United Nations Department of Economic and Social Affairs Programme on Ageing (2016) 1, 5; Cox (2015), 44; CEDAW, art 2. The CEDAW addresses the importance of equality before the law in a number of fields, including employment and social security (art 11).

[22] United Nations Department of Economic and Social Affairs Programme on Ageing (2016), 5; Australian Law Reform Commission (2017), 2.5.

[23] Committee on Economic, Social and Cultural Rights (1991).

[24] Ibid., para 6; Committee on Economic, Social and Cultural Rights (1999), 18; Committee on Economic, Social and Cultural Rights (2003), 13.

> social services, and the right to security in the event of unemployment, sickness, disability, widowhood, old age or other lack of livelihood in circumstances beyond his control.[25]

In relation to the right to housing, the Committee has explained that the right requires not just that all persons have access to shelter, but that they are able 'to live somewhere in security, peace and dignity.'[26]

In her recent report to the UN Secretary-General, the UN Special Rapporteur on the Right to Adequate Housing explained that housing policy which is consistent with human rights:

> must ensure that no one is left behind. In other words, it must aim to change societies in which significant numbers of people are deprived of the right to adequate housing, into societies in which everyone has access to adequate housing and in which housing is a means to ensure dignity, security and inclusion in sustainable communities.[27]

This creates an obligation for governments to ensure that all older persons have access to accommodation which is affordable and appropriate for their needs, and which balances their needs for security and support against their rights to dignity and autonomy. The particulars of a human rights-based approach to accommodation for older people will be addressed in more detail in Chap. 9.

International law also recognises that older people may have special difficulties in accessing appropriate food and water, and that in such circumstances governments are under an obligation to ensure that these needs are provided for.[28] The CESCR considers that the core content of the right to food involves 'availability of food in a quantity and quality sufficient to satisfy the dietary needs of individuals, free from adverse substances, and acceptable within a given culture.'[29]

Satisfaction of dietary needs is of particular relevance to older persons, whose nutritional requirements may change as they age or as they experience health issues.[30] The CESCR's General Comment No. 12 recognises that:

> *Dietary needs* implies that the diet as a whole contains a mix of nutrients for physical and mental growth, development and maintenance, and physical activity that are in compliance with human physiological needs at all stages throughout the life cycle (emphasis in original).[31]

This has clear implications for the provision of health and aged care and will be discussed in more detail in Chap. 10.

The reference to family in article 11 is also noteworthy. The provision recognises the right of each individual 'to an adequate standard of living for himself and his

---

[25] *Universal Declaration of Human Rights* (1948), art 25(1). See also: *European Social Charter (Revised)* (1996), art 31 ('*ESC*'); *Additional Protocol to the European Social Charter* (1988), art 4 ('*Protocol to the ESC*').

[26] Committee on Economic, Social and Cultural Rights (1991), para 7.

[27] Farha (2018), 4.

[28] Committee on Economic, Social and Cultural Rights (2003), para 16(h).

[29] Committee on Economic, Social and Cultural Rights (1999), para 8.

[30] World Health Organization (n.d.).

[31] Committee on Economic, Social and Cultural Rights (1999), para 9.

family'. This language clearly dates the ICESCR to its drafting in 1966, and it has since been accepted that the implied presumption of a patriarchal family structure has been replaced with a wider understanding of 'family'.[32] Applied to older persons it should also encompass recognition that many households include several generations of family-members, and highlights the importance of inter-generational relations. It is not uncommon, for example, for older persons to live with their children and grand-children, to receive care from them and/or to provide care to younger children.[33] In certain cultural contexts the roles of different family members may be more significant. Family living arrangements can also at times create particular vulnerabilities for older persons, especially where financial arrangements (sometimes called 'assets for care arrangements') create reliance or potential for exploitation, a point which will be explored in more detail in Chap. 8.[34] Thus, the right to an adequate standard of living needs to be understood on a household basis such that all members of a household, young and old, are guaranteed the necessary elements of housing, food and water that is affordable, accessible and appropriate.

2.2.3 *Right to the Highest Attainable Standard of Health*

The human right to the highest attainable standard of health (commonly shortened to 'the right to health', though as noted below the distinction is noteworthy) is a key component of a human rights-based approach to elder law and ageing. It is found primarily in article 12 of the ICESCR, which guarantees the enjoyment of the highest attainable standard of both physical and mental health.[35] The provision enumerates a number of specific steps to be taken by states, including working towards improving all aspects of environmental and industrial hygiene, and taking the necessary steps towards the prevention, treatment and control of epidemic, endemic, occupational and other diseases. It also requires that states work towards creating conditions that would ensure the provision of medical assistance to all persons in the event of sickness.

Significantly, the right found in article 12 of the ICESCR is not a right to *be healthy*. The concept of the 'highest attainable standard of health' takes into account the various factors that determine a person's condition of health and recognises that no government can guarantee good health, even when providing the best possible healthcare services. It can be thought of in a similar sense to the notion of 'healthy ageing', which aims not to guarantee perfect health, but to ensure conditions in which older persons can enjoy the best possible health as they age and avoid the limiting consequences of poor health as far as possible. Thus, article 12 is understood as guaranteeing the right to enjoy a wide range of health services and underlying

[32]Ibid., 1; Committee on Economic, Social and Cultural Rights (1991), para 6.

[33]Pew Research Center's Social & Demographic Trends Project (2010); Olsberg (2004).

[34]Hall (2002); British Columbia Law Institute (2002).

[35]See also African Charter, arts 16 and 24; ESC, art 11; Protocol to the ESC, art 4(1)(b).

conditions necessary to ensure each individual can enjoy the highest standard of health available to them.[36]

In its General Comment 14, the CESCR has elaborated on the terms of article 12, clarifying that it includes an obligation to work on improving the underlying determinants of health. The Committee explains that the right to health is defined as:

> an inclusive right extending not only to timely and appropriate health care but also to the underlying determinants of health, such as access to safe and potable water and adequate sanitation, an adequate supply of safe food, nutrition and housing, healthy occupational and environmental conditions, and access to health-related education and information, including on sexual and reproductive health.[37]

As with the right to an adequate standard of living, the right to health must be provided in a way which is non-discriminatory. Healthcare services, medicines, equipment and public health initiatives must be accessible, affordable and acceptable, that is 'respectful of the culture of individuals, minorities, peoples and communities, sensitive to gender and life-cycle requirements, as well as being designed to respect confidentiality and improve the health status of those concerned.'[38]

Removing any discrimination in the provision of healthcare services is one of the immediate obligations imposed on governments under article 12. While the more substantive steps towards providing the right to health are to be progressively realised, the Committee interprets this as imposing 'a specific and continuing obligation to move as expeditiously and effectively as possible towards the full realization of article 12.'[39] The steps taken by states must be 'deliberate, concrete and targeted towards the full realization of the right to health.'[40]

Specific action that a state may be required to take under the right to health includes:

i. Ensuring that all people, particularly those from marginalised or disadvantaged groups, have equal access to preventative, curative and palliative health services which are culturally appropriate, available and affordable. This includes ensuring that any privatisation of the health sector does not diminish accessibility or quality of facilities, goods or services;
ii. Providing healthcare, particularly emergency medicine and immunisation against major infectious diseases, sufficient number of hospitals and other health facilities, counselling and mental health services;
iii. Ensuring equal access to the underlying determinants of health, including safe water and food, basic sanitation and adequate housing and living conditions;
iv. Supporting medical research and education and promoting public information about health issues; and

[36]Committee on Economic, Social and Cultural Rights (2000), paras 8–9.

[37]Ibid., para 11.

[38]Ibid., para 12(3).

[39]Ibid., para 30.

[40]Ibid.

v. Ensuring that health practitioners and health professionals receive appropriate education and training, including in relation to ethical codes of conduct and cultural competency.[41]

The right to health has numerous implications for older persons, and specific dimensions in relation to healthcare decision-making, aged care, palliative care and end-of-life will be explored in detail in Chap. 10. In general terms, the obligation set out in article 12 requires states to take steps to ensure that older people have physical access to the health services they require within a reasonable distance, including in rural areas. States must ensure that healthcare services, facilities, medicines and equipment are available in sufficient number to meet the needs of all older people and that they are affordable for all, even the most disadvantaged. Health services must also be provided in a manner which is respectful of the individual, taking into account factors of culture, age, gender and sexuality. The right provides a useful framework for achieving the goal of 'healthy ageing' and is a key pillar of a human rights-based approach to elder law.

2.2.4 Freedom of Movement

The right to freedom of movement is protected within the ICCPR in article 12, and includes the freedom to choose one's place of residence.[42] The HRC has explained that liberty of movement is essential to the free development of the person and supports the enjoyment of a range of other human rights.[43] Article 12 permits restrictions on freedom of movement only in limited circumstances. These are where the restrictions are imposed by law and where they are necessary to protect national security, public order, public health or morals, or the rights and freedoms of others. Such restrictions must also be consistent with other human rights found in the ICCPR.[44] In interpreting this aspect of the right, the HRC has explained in its General Comment 27 that these limitations must not nullify liberty of movement – they must be imposed on a basis which presumes liberty of movement and supports it to the greatest possible extent, such that restrictions will only be permissible in 'exceptional circumstances' where they are truly necessary to fulfil one of the listed grounds.[45] Further, those restrictions must be proportionate to achieving that objective.[46]

The right to freedom of movement has implications for elder law in a number of ways. The right can be significant if a time comes when an older person needs to move into aged care or more supported living arrangements. As is noted in General Comment 27, the right to choose one's place of residence implies freedom to choose

[41] Ibid., paras 34–7.

[42] See also African Charter, art 12; American Convention, art 22; ECHR, art 2.

[43] Human Rights Committee (1999), 1.

[44] ICCPR, art 12(3).

[45] Human Rights Committee (1999), para 11.

[46] Ibid., para 14.

without being subject to the decision of another person, including a relative. For older persons, this means that decisions relating to accommodation or residential aged care need to be made freely and independently, and ought not be made on their behalf by another member of the family unless the individual older person lacks the requisite capacity to make that decision for themselves. Governments need to ensure that legal frameworks are in place to support independent or supported decision-making, where possible, with respect to these important choices.

Freedom of movement also applies to transportation and mobility. Older persons must be able to enjoy the freedom to move around as much as possible, and governments are obliged to support mobility to the greatest extent reasonably possible, including through the provision of appropriate public and personalised transportation. The importance of transportation and accessibility of public spaces will be discussed in more detail in Chap. 4, where it will be linked to older persons' participation in society more generally.

As noted above, article 12(3) of the ICCPR does provide limited circumstances in which restrictions on freedom of movement will be permissible. For older persons, the question arises as to whether restrictions on movement might be legitimately imposed for the individual's own safety. For example, would it be lawful for a person with advanced dementia to be either physically or chemically restrained in order to protect them from potential harm? The issue of how to balance individual liberty and personal safety or security is a significant question for a human rights-based approach to elder law, and particular manifestations of this issue will be identified in later chapters. Under a strict interpretation of article 12(3), protection of the individual's own safety is not listed as one of the grounds for a lawful restriction of movement. Further, as the HRC has noted, the requirements of necessity and proportionality mean restrictions on freedom of movement must be the least intrusive option available to achieve the desired outcome—if there is a less harmful measure available then that must be pursued instead. This means that even where there might be legitimate concerns about an older person's potential to harm themselves or others, restrictions on their freedom of movement could only ever be lawful as a last resort. In most cases it is likely that some less intrusive, more supportive measure could be devised to address the risk of harm without restricting liberty.

2.2.5 *Right to Liberty and Security of the Person and Freedom from Cruel, Inhuman or Degrading Treatment*

The questions raised above about what measures might be acceptable in protecting the safety and security of older persons, particularly those in aged care or suffering from cognitive impairments such as advanced dementia, also raise issues relating to the right to liberty and security of the person and freedom from cruel, inhuman or degrading treatment. These rights are guaranteed in articles 7 and 9 of the ICCPR.[47]

The distinction between liberty and security of the person is clarified by the HRC in General Comment 35, which explains that liberty of the person relates to 'freedom from confinement of the body' (rather than a general freedom of action) while security of the person concerns freedom from physical or mental injury, and bodily and mental integrity.[48] Security of the person is therefore closely related to the right to be free from cruel, inhuman or degrading treatment, although that right is more broadly defined to prohibit not just physical or mental harm, but also treatment which undermines human dignity.[49]

For older persons, the rights to security of the person and to be free from cruel, inhuman or degrading treatment establish an important standard for the treatment and care received in aged or residential care facilities. These standards would apply both to publicly and privately-run facilities, where the government would be obliged to ensure that proper regulations are in place to prevent mistreatment. These rights correspond to an obligation on states to adopt the necessary laws, policies and other measures to prevent and punish cruel, inhuman or degrading treatment of older persons living with family or in other private arrangements.[50] This has clear relevance to the problem of elder abuse which, as noted in Chap. 1, has become more prevalent (or at least more commonly reported) in recent decades. At its most serious, elder abuse can involve physical violence, psychological harm or humiliation, which is in clear contravention of the rights to security of the person and freedom from cruel, inhuman or degrading treatment.

Article 7 of the ICCPR also guarantees that 'no one shall be subjected without his free consent to medical or scientific experimentation.' The HRC has identified that this requires special protections for persons who are not capable of giving valid consent.[51] For older persons who require high levels of care and who may lack independent decision-making capacity, special processes are required to ensure that they are not exposed to the risk of medical or scientific experimentation to which they cannot consent. The issues associated with legal capacity will be explored in detail in Chap. 6.

---

[47]See also African Charter, art 5; American Convention, art 5; ECHR, art 3.

[48]Human Rights Committee (2014), para 3.

[49]Human Rights Committee (1992), para 2.

[50]Ibid., para 8. These rights have particular implications for older persons in prisons or other corrective facilities. They are not discussed in detail in this work, but for more detail see for example: Stevens et al. (2018); Zinger (2019).

[51]Human Rights Committee (1992), para 7.

2.2.6 Right to Privacy

The right to privacy is guaranteed in international law in article 17 of the ICCPR. It is also protected under article 8 of the European Convention and article 11 of the American Convention. Being able to freely choose what personal information is shared, and with whom, is an important exercise of autonomy, and protection of privacy is essential to respecting individuals' dignity. Safeguarding individual privacy can become more challenging as people age, as more people and organisations become involved in our lives through, for example, delivering various forms of support and assistance. The nature of the information required to be shared can become more personal, resulting in personal and intimate spaces and activities becoming increasingly difficult to keep private. The right to privacy therefore provides an essential standard in ensuring that laws and policies affecting older persons maintain high levels of respect for dignity and autonomy. The right to privacy should help lawmakers, institutions and individual agents determine which information is necessary and strike an appropriate balance to protect privacy as far as possible.

There are a number of areas where the right to privacy has implications for elder law. For those individuals who are living in residential facilities, the right to privacy must be safeguarded in their day-to-day experiences. This is essential to respect the dignity of each individual resident. The right to privacy requires appropriate design of facilities and policies and adequate training of staff to maintain adequate levels of privacy.

Older persons are also entitled to privacy regarding their decision-making, including the ability to make decisions without family members' knowledge or intervention. For example, concerns around privacy rights and elder abuse have been raised in relation to proposals to establish public registers of enduring powers of attorney or other instruments. In such circumstances, it has been pointed out that in some cases older persons may feel under pressure to appoint a particular family member if they are unable to make such appointments confidentially, or may face abuse or reprisals if their decision is not accepted.[52]

In the modern era, the right to privacy is closely linked to 'informational self-determination'.[53] This is the right of each individual to decide for themselves what information they disclose and in what circumstances. It also includes the right to be forgotten by digital systems, to have control over one's digital 'legacy' after death, and to rectify records in the case of error. This has obvious implications for older persons' health information. As health records are increasingly digitised and centralised, the right to privacy requires appropriate safeguards in place to ensure that such information is not misused for other purposes.[54]

Privacy concerns also arise in relation to technological innovations in aged care. New technologies, particularly assistive technologies, have great potential to improve the way that care and support are provided for older persons, and have the potential

[52] Purser et al. (2017), 9.

[53] Kornfeld-Matte (2017), 11.

[54] See, for example: Karp and Knaus (2018).

to increase independence, social interaction and freedom of movement. However, where these technologies involve surveillance of older persons or gathering and storage of data, significant privacy issues exist in relation to how that information is used and shared.[55] As noted above, the right to privacy encompasses a right to informational self-determination, and regulations must ensure transparency around the way that information is gathered and shared, and strike an appropriate balance between privacy and other legitimate interests.[56]

2.2.7 *Right to Work and Education*

Age-discrimination in the workforce is one of the major barriers to older persons' full enjoyment of their human rights. Being able to remain in employment is important for older persons' financial security, social inclusion and independence. Laws and policies are therefore required which assist older persons to find and remain in employment. A key component of this is to provide access to appropriate education and training, particularly as new technologies disrupt traditional workforces and many professions transition to new modes of work. The human rights to work and to education are key to achieving these objectives, and in turn help promote the enjoyment of a wider range of rights.[57]

The right to work is guaranteed in the ICESCR, which provides in article 6 the right of every person to have an opportunity to gain their living by work which they freely choose or accept.[58] Article 7 expands the right to work to include the enjoyment of just and favourable conditions of work, specifically:

i. fair wages and equal remuneration for work of equal value;
ii. remuneration which is sufficient to provide a decent living for workers and their families;
iii. safe and healthy working conditions;
iv. equal opportunity for promotion and other recognition; and
v. reasonable working hours with adequate allocations for rest and leisure.

To ensure that older workers are able to remain in employment for as long as possible, a number of issues need to be addressed by states. These include addressing the attitudes of employers and promoting the benefits of employing older workers, many of whom will have a broad skill set and wealth of experience.[59] Similar work may be required to ensure employers provide equal opportunities to their older employees to pursue training, development and promotion opportunities. Financial

---

[55] Lewis et al. (2018); Bennett (2019).

[56] Kornfeld-Matte (2017).

[57] Kornfeld-Matte (2016), 11.

[58] See also African Charter, art 15; Protocol to the ESC, arts 1–4.

[59] The barriers presented by ageism in terms of the right to work will be examined in more detail in Chap. 5.

incentives or subsidies may be appropriate to encourage businesses to employ older workers, or to assist workplaces create an enabling environment and remove barriers for older persons, especially those with reduced mobility or other needs.[60]

Availability of education and training is a core component of promoting the right to work for older persons. The Independent Expert on the Enjoyment of All Human Rights by Older Persons has identified lifelong learning as a precondition for longer participation in the workforce, and an important part of ensuring social inclusion for older persons.[61] The right to education is guaranteed in article 13 of the ICESCR.[62] While the focus of this provision is on guaranteeing primary and secondary education for children, at a minimum, international law requires that education be accessible to all persons without discrimination. To guarantee that older persons are fully able to enjoy their rights to education and work, education programs should be designed to take into account their learning needs and interests. At the same time, such programs should not be designed in a way which limits opportunities for interaction with younger generations, as this would lead to social exclusion. The implications of the rights to work and education will be explored in more detail in Chaps. 4 and 5.

2.2.8 Right to Social Security

The right to social security, guaranteed in article 9 of the ICESCR,[63] has been recognised as a crucial right for ensuring that older persons can maintain an adequate standard of living and enjoy the full range of other human rights. As discussed in Chap. 1, older persons worldwide are over-represented amongst those living in extreme poverty, and this can have serious detrimental effects on their social inclusion and well-being.[64] Social security ideally guarantees that older persons can enjoy financial security and independence, and is an essential safeguard given the prevalence of age-discrimination and the limited employment and education opportunities which older persons can face.[65] It therefore underpins the full enjoyment of all human rights, and ought to be a corner-stone of governments' approaches to implementing a rights-based approach to elder law.[66]

While states are able to implement varying forms of social security, at a minimum they must guarantee non-contributory benefits to those who are unable to access

[60] Kornfeld-Matte (2016), 12; Committee on Economic, Social and Cultural Rights (1995), para 6.

[61] Kornfeld-Matt (2016), 14.

[62] See also African Charter, art 17; ECHR, art 2; ESC, art 10.

[63] The right to social security is also guaranteed in the African Charter, art 18, ESC, arts 12–15; the Protocol to the ESC, art 4; the American Declaration, art, XVI; and the *Additional Protocol to the American Convention on Human Rights in the Area of Economic, Social and Cultural Rights* (1988) (*San Salvador Protocol*), arts 9 and 17. It is also provided for in International Labour Organization *Convention 102* (1952) and *Convention 128* (1967).

[64] Sepúlveda Carmona (2010).

[65] Committee on Economic, Social and Cultural Rights (2008).

[66] Kornfeld-Matte (2016), 10.

an insurance-based or other form of contributory scheme.[67] Mechanisms such as a universal basic income[68] or 'social protection floor'[69] could be effective in guaranteeing basic income security and essential services for all older persons. The importance of social security for the fulfilment of all human rights will be discussed in more detail in Chap. 8.

2.2.9 Rights to Social and Cultural Participation

The rights discussed above all play a part in ensuring that older persons are able to participate fully in the social and cultural life of their community. Rights to work, education, social security, freedom of movement and equality before the law all operate to increase older persons' independence and inclusion, and can therefore be thought of as culminating in achieving social and cultural participation. As will be discussed in more detail in the next chapter, a human rights-based approach to elder law is one which positions social inclusion as a key guiding principle. In this way, older persons are positioned as active rights-holders whose contributions to the community are valued, and social and cultural participation is understood to go beyond the physical ability to participate in the workforce.[70] Furthermore, a human rights-based approach maintains that older persons must participate in the formulation of public policy which affects them, and this in turn helps to strengthen social integration and protect against isolation and exclusion.[71]

In addition to this mutually supportive relationship between participation and human rights, there are a number of specific rights guaranteed within international human rights law which focus on enhancing participation in social and cultural life. Article 15 of the ICESCR requires states to protect and promote the right to take part in cultural life. Article 25 of the ICCPR recognises individuals' rights to vote and participate in public affairs. Both the European and African human rights frameworks recognise the rights to participate in the cultural life of the community, along with article 27 of the ICCPR which recognise the rights of people who belong to minority groups to be able to continue practising the customs, traditions and languages of those groups.[72] The enjoyment of these rights requires that states identify and address barriers which might prevent older persons from being able to participate meaningfully.[73] More detailed discussion of measures to ensure greater

[67] Office of the High Commissioner for Human Rights (2012), 16; Committee on Economic, Social and Cultural Rights (1995).

[68] Kornfeld-Matte (2016).

[69] International Labour Organization (2012).

[70] Grover (2011), 13.

[71] *United Nations Principles for Older Persons* (1991), Principle 7; Kornfeld-Matte (2016), 13.

[72] Protocol to the ESC, art 4; see also African Charter, art 17.

[73] See United Nations Principles, principle 16.

social and cultural inclusion for older persons will be provided in later chapters, particularly Chap. 4, which will focus on economic, social and cultural rights.

2.2.10 Convention on the Rights of Persons with Disabilities

In addition to the rights guaranteed in the human rights treaties discussed above, older persons may also be entitled to specific protections under the CRPD.[74] The CRPD expands on many of the rights contained in the ICCPR and ICESCR, articulating what those rights require for persons with disabilities. Where an older person has an impairment which meets the definition of 'disability' found in article 1 of the CRPD they will be protected by the provisions contained in the Convention.[75] However, as will be discussed, the factors which cause disadvantage and vulnerability for older persons may differ from those which affect persons with disabilities, and the CRPD ought not be viewed as necessarily providing adequate protection for the human rights of older persons.

The approach taken in the CRPD is to articulate a number of fundamental principles which ground a human rights-based approach to disability. These include respect for dignity and autonomy, substantive equality and non-discrimination, inclusion and participation, and accessibility.[76] Article 4 sets out a number of general obligations which states must fulfil to ensure and promote the full realisation of human rights for persons with disabilities. These include adopting appropriate legislation to support and promote the enjoyment of rights and abolishing all discriminatory laws and policies. States are also obliged to support research and development for goods, services, technologies and facilities which enable the full enjoyment of rights by persons with disabilities, and to ensure that professionals and others working with persons with disabilities receive appropriate training in order to implement the provisions of the Convention.

An influential aspect of the CRPD is its emphasis on participation. States are obliged to take effective and appropriate measures to ensure all persons with disabilities can enjoy full inclusion and participation in the social, cultural and political life of their community.[77] Article 4 also obliges states to ensure the meaningful participation of persons with disabilities in decision-making and policy formulation which affects them. This emphasis on participation is reflective of the clarion call heard during the negotiations for the CRPD: 'nothing about us without us.'[78]

---

[74] CRPD.

[75] Article 1 defines a person with a disability to include: 'those who have long-term physical, mental, intellectual or sensory impairments which in interaction with various barriers may hinder their full and effective participation in society on an equal basis with others'.

[76] CRPD, art 3.

[77] Ibid., arts 3, 9, 19, 24, 26, 29, 30.

[78] Harpur (2017).

In addition to these general principles, the CRPD addresses the many specific rights found in other international human rights law and makes clear how these are to be implemented with respect to persons with disabilities. This coverage includes rights relating to physical security, such as the rights to liberty and security of the person; freedom from torture and cruel, inhuman or degrading treatment or punishment; freedom from exploitation, violence and abuse; integrity of the person and freedom of movement and personal mobility.[79] It also emphasises legal rights, including the right to equality before the law and equal access to justice,[80] and rights to privacy, education, and work.[81] Considerable detail is provided in relation to the right to health, with article 25 outlining a number of specific steps that states must take to ensure that persons with disabilities are able to access the same quality of care which is available to the community in general, while at the same time ensuring that their specific needs are provided for.

A major contribution of the CRPD is that it seeks to shift attitudes and approaches towards disability away from a medical model, where disability is viewed as flowing from a person's inherent limitations, to a social model, where it is understood that disability is the product of a person's interactions with various barriers present in and created by society.[82] The focus of the CRPD is therefore on removing or reducing those barriers to promote full inclusion and participation in society. This shift in perspective is equally warranted with respect to older persons.

In thinking about how the CRPD applies to older persons, a number of points need to be made. As people age, their abilities inevitably reduce, and many older persons, particularly those in the 'old, old' category explained in Chap. 1, will therefore be covered by the provisions of the CRPD. This has the effect of providing additional human rights protections for some of the most vulnerable older persons, for example those whose level of impairment necessitates their living in residential care.[83]

However, it is important to recognise that the barriers which interfere with the rights of older persons are not the same as those affecting persons with disabilities. Much of the disadvantage and exclusion which older persons experience is caused by ageism, which, as discussed in Chap. 5, operates to exclude them from full participation in society and devalues their contributions. As Harpur explains, age discrimination is often based on the *appearance* of age, and not the *fact* of age.[84] While it is closely linked to assumptions about ability, age discrimination is not the same as disability discrimination. There are other factors influencing age discrimination, including society's fear of ageing and cultural norms which reinforce notions that youth equates to beauty and strength. Further, age discrimination is present in and enacted through different social structures, for example mandatory retirement ages and rules around superannuation. Ageism in the workforce, for instance, can be

---

[79]CRPD, arts 14–20.

[80]Ibid., arts 12–3.

[81]Ibid., arts 22, 24, 27.

[82]Lord and Stein (2008), 460; Harpur (2015).

[83]Harpur (2015), 1044.

[84]Ibid., 1053.

based on assumptions about older persons' attitudes, political views, or the capability to integrate well with younger workers, rather than their physical ability to perform the work.

The very assumption that older persons are impaired or disabled is itself a form of ageism.[85] Even positive stereotypes and assumptions about older persons, for example that they require assistance and deserve respect, can be damaging to their human rights. As Mitchell has argued, 'linking old age with disability feeds hostile and benevolent prejudice, stereotypes both groups, and typifies intersectional discrimination.'[86] These distinctions between age and disability must be understood in order to ensure that the rights of both groups can be adequately protected and promoted.

While the CRPD is clearly not adequate to protect the rights of older persons, there are some points to note about the process of adopting the CRPD which are instructive for the project of enacting a human rights-based approach to elder law. The process of convincing states that there was a need for a dedicated CRPD was arduous, and many states argued that existing laws were enough.[87] Now that the CRPD is in place, it may be even more difficult to convince states of the necessity of a CROP, in part because of common ageist assumptions that older persons are impaired.[88] However, as will be discussed below, existing laws have not been able to adequately protect and promote the human rights of older persons, and the CRPD in particular does not cover the full range of issues which can affect older persons' enjoyment of their rights.

There are, however, lessons we can take from the experience of adopting the CRPD which ought to inform and assist efforts to enact a dedicated instrument for older persons. These include the important role for civil society groups in advocating for law reform, and especially the need for meaningful participation by older persons themselves. In terms of the content and structure of a new instrument, the CRPD also shows the value of adopting a principled-based approach which focuses on inclusion and participation.

2.3 Soft-Law Instruments and Special Procedures

In addition to the international and regional human rights instruments discussed above, the international human rights regime has for some time been committed to a program of work which advances the rights of older persons.[89] This work has produced a number of soft-law instruments and other initiatives which, while not legally binding, help to articulate how states ought to conceive of and implement their

---

[85] Ibid., 1052.

[86] Mitchell (2017), 2.

[87] Ibid.; Harpur (2015), 1031; Harpur (2017).

[88] Harpur (2015), 1032; Mitchell (2017), 2.

[89] Open-ended Working Group on Ageing (2013); Kornfeld-Matte (2016).

human rights obligations in relation to older persons and establish internationally agreed upon policies. They also signify the momentum which is growing towards the adoption of a dedicated CROP (see discussion below).

2.3.1 Vienna International Plan of Action on Ageing

The imperative to address the specific needs of older persons was first affirmed by the international community in 1982, when the UN convened the World Assembly on Ageing in Vienna. The outcome of the meeting was the *Vienna International Plan of Action on Ageing*. The Vienna Plan of Action focuses on the implications of development for the ageing population, and it lists its primary aims as strengthening the capacity of states to 'deal effectively with the ageing of their populations and with the special concerns and needs of their elderly.'[90] It goes on to articulate a number of specific objectives, principles and actions in support of these aims, including encouraging international cooperation and exchange of expertise, preventing discrimination against older persons, enhancing the health and well-being of older persons and ensuring that their needs are met. It reflects a welfare model in response to the phenomenon of an ageing population, emphasising the need for governments to adopt policies and practices which address the needs of individual older persons and society in general. As such it positions older persons as beneficiaries of state action, rather than autonomous individuals possessing human rights and entitled to minimum, enforceable guarantees of those rights from their government.

2.3.2 United Nations Principles for Older Persons

Following on from the Vienna Plan of Action, the United Nations General Assembly adopted the *United Nations Principles for Older Persons* in 1991 ('the UN Principles').[91] One aim of the UN Principles was to articulate guidelines to help states implement the Vienna Plan of Action, and these are organised around five key areas:

i. independence;
ii. participation;
iii. care;
iv. self-fulfilment; and
v. dignity.

[90] *Vienna International Plan of Action on Ageing* (1982), Foreword, para 2 ('*VIPAA*').
[91] United Nations Principles.

The language adopted in the UN Principles is aspirational rather than obligatory, and the instrument 'encourages governments to incorporate the … principles into their national programmes whenever possible.'[92]

While the UN Principles align more closely with the range of human rights discussed above, they do not stray far from the welfare model adopted in the Vienna Plan of Action and the focus remains on what governments 'should' provide for older persons, rather than the rights which older persons possess, and which governments are obliged to fulfil.

2.3.3 *Madrid International Plan of Action on Ageing*

In 2002 the UN convened the Second World Assembly on Ageing in Madrid.[93] The states which attended the meeting adopted two key instruments: a Political Declaration and the Madrid International Plan of Action on Ageing ('the Madrid Plan of Action').[94] The Madrid Plan of Action addresses issues of ageing primarily from a development perspective and is organised around three priority areas:

i. older persons and development;
ii. advancing health and well-being into old age; and
iii. ensuring enabling and supportive environments.

Within each of these priority areas the Madrid Plan of Action specifies a number of objectives and actions which states should undertake.

In the way in which it conceptualises the challenges of the ageing population, the Madrid Plan of Action is consistent with human rights, but it falls short of adopting a human rights-based approach.[95] It recognises the contributions that older persons can make to development, including through being part of the work-force, and the need for them to participate in decision-making. It also identifies a number of cohorts within the older population who have particular needs, including migrants, women, persons with disabilities, people living in rural areas and people living in poverty. However, the Madrid Plan of Action is best described as an aspirational policy document, and it has not specifically been designed to enhance protection of the rights of older persons.[96] Further, as will be discussed below, implementation of the Madrid Plan of Action since its adoption has been slow and many gaps exist between aspiration and reality.[97]

---

[92] Ibid., preamble.

[93] *Madrid International Plan of Action on Ageing* (2002).

[94] Ibid.

[95] Mitchell (2017), 14; Kornfeld-Matte (2016), 14–7.

[96] Kornfeld-Matte (2016), 15; Mitchell (2017), 14.

[97] Kornfeld-Matte (2016), 18.

2.3.4 *Open-Ended Working Group on Ageing*

One of the key contributions of the Madrid Plan of Action is that it has catalysed follow-up action within the UN human rights regime, particularly the General Assembly and Human Rights Council. In 2010 the General Assembly created the Open-ended Working Group on Ageing,

> for the purpose of strengthening the protection of the human rights of older persons by considering the existing international framework of the human rights of older persons and identifying possible gaps and how best to address them, including by considering, as appropriate, the feasibility of further instruments and measures.[98]

The Working Group is open to all states to join, and it regularly receives submissions from both state and non-state actors relating to the current state of protections for the human rights of older persons.

An important development occurred in 2013 with the adoption of General Assembly Resolution 67/139, entitled 'Towards a comprehensive and integral international legal instrument to promote and protect the rights and dignity of older persons.'[99] By this resolution the General Assembly expanded the mandate of the Working Group to 'consider proposals for an international legal instrument to promote and protect the rights and dignity of older persons.'[100] The resolution requested the Working Group to present a proposal containing 'inter alia, the main elements that should be included in an international legal instrument to promote and protect the rights and dignity of older persons, which are not currently addressed sufficiently by existing mechanisms and therefore require further international protection.'[101]

The Working Group's annual sessions have provided a valuable forum for non-government organisations and civil society actors to contribute to the debate about the need for, and contours of, a dedicated instrument to protect and promote the human rights of older persons. Despite the clear mandate to bring a proposal to the General Assembly for an international legal instrument, progress towards that objective has been slow. In part this is because of a lack of consensus among states as to the need for, or desirability of, a new, legally binding instrument.[102]

In an effort to progress this work further, the Chair of the Working Group identified a number of thematic areas for more detailed consideration and called for submissions from experts to help identify appropriate normative content which could be included in a legal instrument.[103] Sessions of the Working Group focus on particular thematic areas. Over recent years, these focus areas have included equality and

[98]United Nations General Assembly (2011), para 28.

[99]United Nations General Assembly (2013).

[100]Ibid., 1.

[101]Ibid., 2.

[102]Open-ended Working Group on Ageing (2017); Mitchell (2017), 16.

[103]Mitchell (2017), 16–18.

non-discrimination, and violence, neglect and abuse (eighth session);[104] autonomy and independence, and long-term and palliative care (ninth session);[105] education, training life-long learning and capacity building, as well as social protection and social security (tenth session);[106] and access to the labour market and access to justice (eleventh session).[107] Numerous submissions are received and discussed in relation to these thematic areas, including proposals for the key elements of a new legal instrument and how normative standards could be defined.[108] While the Working Group has been able to bring together states and civil society, and has made some progress in identifying gaps in legal protections and potential areas for legal reform, the objective of adopting a consensus-based proposal for a new legal instrument remains some way off.

2.3.5 United Nations Independent Expert

In parallel with the Open-ended Working Group established by the General Assembly, the Human Rights Council has also been pursuing its own work in relation to protecting the rights of older persons. In 2013 it created a special mandate for an Independent Expert on the Enjoyment of All Human Rights by Older Persons,[109] and in 2014 Rosa Kornfeld-Matte was appointed to the role.[110] The mandate of the Independent Expert is to assess the implementation of existing international instruments with respect to older persons, including specifically the Madrid Plan of Action. The two key objectives in this respect are to identify implementation best practice and gaps in implementation of existing law. The Independent Expert is also mandated to consult with states and stakeholders, to raise awareness of the human rights of older persons, and to work together with states to develop measures which advance the protection and promotion of those rights.

A particular focus of the Independent Expert's work is to explore the intersectionality of human rights issues related to age and other factors. The mandate requires the Independent Expert to 'integrate a gender and disability perspective into his/her work, and to pay particular attention to older women, persons with disabilities, persons of African descent, individuals belonging to indigenous peoples, persons belonging to national or ethnic, religious and linguistic minorities, rural persons, persons living on the streets, and refugees, among other groups.'[111] This work is to be carried out in

[104] Open-ended Working Group on Ageing (2017).

[105] Open-ended Working Group on Ageing (2018).

[106] Open-ended Working Group on Ageing (2019a).

[107] Open-ended Working Group on Ageing (2019b).

[108] AGE Platform Europe (2018); AGE Platform Europe et al. (2017a, b, 2018).

[109] Human Rights Council (2013).

[110] In 2016 the mandate of the Independent Expert was extended by a further 3 years, see Human Rights Council (2016).

[111] Human Rights Council (2013), para 5(e).

close coordination with the Open-ended Working Group, while avoiding unnecessary duplication.[112]

In fulfilment of her mandate, Kornfeld-Matte has to date produced a number of annual reports, including a series of thematic reports focused on particular issues relating to the protection and promotion of the human rights of older persons. For instance, they have included reports on autonomy,[113] social exclusion,[114] and the implications of new technologies for the human rights of older persons.[115] These reports have identified a number of significant challenges in ensuring that all older persons are able to enjoy the full range of human rights, and that these are likely to expand as the ageing population increases.

In her first comprehensive report under her mandate, the Independent Expert regarded the actions contained in the Madrid Plan of Action, while not framed in rights-language, as nonetheless consistent with human rights. On this basis, if states were to implement the Action Plan fully, they would go a long way towards achieving full protection and fulfilment of human rights for older persons. However, after surveying many states and conducting a number of state visits, the Independent Expert concluded that progress towards achieving these objectives has been uneven, and that much remains to be done before the Action Plan could be said to have fulfilled its aspirations.[116] Further, the analysis of existing laws found that current protections are inadequate, and that work should continue on developing a dedicated legal instrument on the rights of older persons.[117]

2.4 The Case for a Convention on the Rights of Older Persons

The work of the Independent Expert and the Open-ended Working Group confirms the view of many scholars in the field that a dedicated treaty or other legal instrument is required to adequately ensure the human rights of older persons.[118] The analysis presented above highlights the many human rights which are relevant to the field of elder law, but there are a number of issues where specific, legally enforceable standards and guidelines are required. The discussion in later chapters will expound on these in more detail, but a short outline is provided here.

Some have argued that the breadth of existing laws, coupled with the specific rights found in the CRPD for persons experiencing disability, are adequate to protect

[112]Ibid., para 5(g).

[113]Kornfeld-Matte (2015).

[114]Kornfeld-Matte (2018).

[115]Kornfeld-Matte (2017).

[116]Kornfeld-Matte (2016), paras 17–8.

[117]Ibid., para 21.

[118]Fredvang and Biggs (2012), 21; Mégret (2011); Doron and Apter (2010); International Expert-Conference on the Human Rights of Older Persons (2018).

older persons, and that singling them out as a cohort deserving special protection itself constitutes ageism.[119] However, recognising that older persons can experience human rights interferences in particular ways or that they may have particular needs and vulnerabilities does not amount to singling them out as a distinct cohort with different rights. On the contrary, it recognises that older persons are entitled to the very same human rights that all other individuals possess, but that the law as it currently stands does not adequately acknowledge the particular threats to those rights, which become more common as we age.

To begin with, current international human rights law fails to recognise older persons as a particularly disadvantaged or vulnerable group. While there are dedicated instruments to promote and protect the rights of women, children, migrant workers, and persons with disabilities, the particular vulnerabilities, needs and interests of older persons are currently only recognised in very limited contexts. To use Mégret's language, there are specificities in the way that the rights of older persons are impacted in their lived experiences which are not captured by existing human rights law.[120]

Further, current laws fail to recognise the pervasive and damaging impacts of ageism. Non-discrimination provisions in almost all current human rights treaties do not list age as a ground of prohibited discrimination. Human rights treaty bodies have clarified that age is captured by wording such as 'other status' in non-discrimination clauses.[121] This expanded interpretation has been part of arguments that specific protections are not needed. However, as Mitchell has pointed out, the historical failure of states to include age as a specific ground of discrimination in previous drafting exercises points to the relative invisibility of older persons within human rights protections, and should in fact serve to reinforce calls for specific recognition.[122]

As noted above, the CRPD is inadequate to protect the rights of older persons fully, and was not intended to do so. Not only does this indicate gaps in the law which need to be filled, the CRPD also demonstrates a new approach to human rights protections which is highly appropriate in the context of older persons. For instance, the emphasis on participation in the CRPD ought to be replicated for older persons to ensure that they are given opportunities for meaningful input into laws and policies which affect them, and to facilitate their full inclusion and participation in society. This will help to achieve the necessary paradigm shift which positions older persons as autonomous rights-holders, rather than beneficiaries of care.

A dedicated instrument to protect the rights of older persons is also required to ensure that appropriate monitoring and enforcement mechanisms are in place. This will help to ensure that states are accountable for their human rights obligations and

---

[119] Greengross (2019).

[120] Mégret (2011), 68ff.

[121] Committee on Economic, Social and Cultural Rights (1995); Committee on Economic, Social and Cultural Rights (2009), 15; Human Rights Committee (1989), 7.

[122] Office of the High Commissioner for Human Rights (2012); United Nations Department of Economic and Social Affairs Division for Social Policy and Development and Programme on Ageing (2009); Mitchell B (2017), 7–8; AGE Platform Europe et al. (2017a), 5.

provide a process for individuals to seek redress in the event that their rights are violated. A monitoring and enforcement process, for example through the creation of a dedicated treaty committee, would also help to reinforce the human rights-based framework which ought to underpin all elder law.

This book argues that a dedicated instrument to protect and promote the human rights of older persons is required at the international level, and that this would support and encourage states in developing a framework of elder law which is consistent with human rights. Even without a dedicated instrument, however, it is possible to pursue a human rights-based approach to elder law which draws on the fundamental principles of human rights and the specific rights guaranteed in existing law. Moreover, adopting a human rights-based approach using existing law could serve as an important interim step towards the enactment of a dedicated covenant. Once states commit to a human rights-based approach and integrate it into their laws and policies, the move to a dedicated treaty ought to be seen as a more natural, less onerous undertaking.

2.5 Conclusion

From the above discussion it is clear that, despite the absence of a dedicated treaty on the rights of older persons, international and regional human rights law offers a rich tapestry of rights which are relevant to the broad field of elder law. Soft-law instruments like the Madrid Plan of Action and the work of the Independent Expert have been extremely useful in fleshing out these rights and translating them into specific actions and objectives for governments and policy-makers to pursue. As noted, however, there is still a very strong case for a dedicated CROP, which would enable a greater recognition of the specific experiences of older persons and facilitate their enhanced participation in policy-making in support of their rights.

Even without such a dedicated instrument, the existing framework of human rights provides a deep and nuanced set of norms for improving the various fields of elder law. The following chapter will explain the human rights-based approach which has been developed as the basis for analysis in the remaining chapters of the book. It draws on the specific rights which have been discussed here, and integrates them with a number of core values and principles which together can be used both to identify threats to human rights as well as to develop appropriate laws and policies in response.

References

AGE Platform Europe (2018) Submission based on seminar with Europe-based experts on the human rights of older persons. Submitted to the 9th session of the open-ended working group on ageing. https://social.un.org/ageing-working-group/documents/ninth/Inputs%2520NGOs/AGE_Brussels_seminar_April2018_FINAL.pdf. Accessed 6 Feb 2020

AGE Platform Europe et al (2017a) Equality and non-discrimination. https://social.un.org/ageing-working-group/documents/eighth/Inputs%2520NGOs/Joint_Paper_Equality.pdf. Accessed 9 Dec 2019
AGE Platform Europe et al (2017b) Freedom from violence, abuse and neglect. https://social.un.org/ageing-working-group/documents/eighth/Inputs%2520NGOs/Joint_Paper_Violence.pdf. Accessed 9 Dec 2019
AGE Platform Europe et al (2018) Autonomy, independence, long-term care and palliative care. https://social.un.org/ageing-working-group/documents/ninth/Joint_Discussion_paper.pdf. Accessed 9 Dec 2019
Australian Law Reform Commission (2017) Elder abuse: a national legal response. Report No 131
Bennett B (2019) Technology, ageing and human rights: challenges for an ageing world. Int J Law Psychiatry 66(101449):1–7
British Columbia Law Institute (2002) Private care agreements between older adults and friends or family members. https://www.bcli.org/sites/default/files/Private_Care_Agreements_Between_Older_Adults_and_Friends_or_Family_Members.pdf. Accessed 9 Dec 2019
Cox CB (2015) Social policy for an aging society: a human rights perspective. Springer, New York
Doron I, Apter I (2010) The debate around the need for an international convention on the rights of older persons. Gerontologist 50(5):586–593
Farha L (2018) Report of the special rapporteur on adequate housing on the implementation of effective rights-based housing strategies. UN Doc A/HRC/37/53
Fredvang M, Biggs S (2012) The rights of older persons: protection and gaps under human rights law. Centre for public policy and brotherhood of saint laurence, social policy working paper no 16
Greengross S (2019) Human rights and ageing: are we doing it right? Keynote address to Australian association of gerontology, Brisbane, Australia, 15 Nov 2019
Grover G (2011) Thematic study on the realization of the right to health of older persons by the special rapporteur on the right of everyone to the enjoyment of the highest attainable standard of physical and mental health. UN Doc A/HRC/18/37
Hall M (2002) Care agreements: property in exchange for the promise of care for life. Australian Law Reform Comm Reform J 81:29–33
Harpur P (2015) Old age is not just impairment: the crpd and the need for a convention on older persons. Univ Pennsylvania J IntLaw 37:1027–1047
Harpur P (2017) Nothing about us without us: the un convention on the rights of persons with disabilities. In: Leithner A, Libby K (eds) Oxford research encyclopedia of politics. Oxford University Press, Oxford, pp 1–20
International Expert-Conference on the Human Rights of Older Persons (2018) Vienna declaration on the human rights of older persons. https://ageing.at/site/pub_html/sites/default/files/ICHRoP%2520Conference%2520Declaration%2520Final_0.pdf. Accessed 19 Dec 2019
International Labour Organization (2012) Social protection floors. Recommendation No. 202
Karp P, Knaus C (2018) GPs and social service providers demand my health record protections. The Guardian (27 July 2018) https://www.theguardian.com/australia-news/2018/jul/27/gps-and-social-service-providers-demand-my-health-record-protections. Accessed 9 Dec 2019
Kornfeld-Matte R (2015) Report of the independent expert on the enjoyment of all human rights by older persons. UN Doc A/HRC30/43
Kornfeld-Matte R (2016) Report of the independent expert on the enjoyment of all human rights by older persons. UN Doc A/HRC/3344
Kornfeld-Matte R (2017) Report of the independent expert on the enjoyment of all human rights by older persons. UN Doc A/HRC/36/48
Kornfeld-Matte R (2018) Report of the independent expert on the enjoyment of all human rights by older persons. UN Doc A/HRC/39/50
Lewis B et al (2018) Submission by the Australian centre for health law research to the australian human rights commission human rights and technology issues paper. https://eprints.qut.edu.au/122938/. Accessed 7 Feb 2020

Lord J, Stein M (2008) The domestic incorporation of human rights law and the united nations convention on the rights of persons with disabilities. Washington Law Rev 83:449–479
Mégret F (2011) The human rights of older persons: a growing challenge. Human Rights Law Rev 11(1):33–66
Mitchell B (2017) Towards a convention on the rights of older persons. Paper presented at the ADA Australia national conference, gold coast, Australia, 23–24 March 2017
Office of the High Commissioner for Human Rights (2012) Normative standards in international human rights law in relation to older persons: analytical outcomes paper: https://social.un.org/ageing-working-group/documents/OHCHRAnalyticalOutcomePaperonOldePersonsAugust2012.doc. Accessed 9 Dec 2019
Olsberg D (2004) Ageing-in-place? Intergenerational and intra-familial housing transfers and shifts in later life, Australian Housing and Urban Research Institute, UNSW-UWS Research Centre
Open-ended Working Group on Ageing (2013) Compilation of existing international legal instruments, documents and programmes that directly or indirectly address the situation of older persons, including those of conferences, summits, meetings or international or regional seminars convened by the united nations and intergovernmental and non-governmental organizations. https://social.un.org/ageing-working-group/documents/fourth/compilation.pdf
Open-ended Working Group on Ageing (2017). Report of the open-ended working group on ageing on its eighth working session. UN Doc A/AC.278/2017/2
Open-ended Working Group on Ageing (2018) Report of the open-ended working group on ageing on its ninth working session. UN Doc A/AC.278/2018/2
Open-ended Working Group on Ageing (2019a) Report of the open-ended working group on ageing on its tenth working session. UN Doc A/AC.278/2019/2
Open-ended Working Group on Ageing (2019b) Overview of 11th session. https://social.un.org/ageing-working-group/eleventhsession.shtml. Accessed 15 Jan 2020
Pew Research Center's Social & Demographic Trends Project (2010) The return of the multi-generational family household. https://www.pewsocialtrends.org/2010/03/18/the-return-of-the-multi-generational-family-household/. Accessed 15 Jan 2020
Purser K et al (2017) Submission responding to australian law reform commission discussion paper 83: elder abuse. Submission No 298
Sepúlveda Carmona M (2010) Report of the independent expert on the question of human rights and extreme poverty on social protection and old age poverty. UN Doc A/HRC/14/31
Stevens BA et al (2018) Systematic review of aged care interventions for older prisoners. Australasian J Ageing 37(1):34–42
United Nations Department of Economic and Social Affairs. Division for Social Policy and Development and Programme on Ageing (2009) Report of the expert group meeting "rights of older persons". https://www.un.org/esa/socdev/ageing/documents/egm/bonn09/reportofegm.pdf. Accessed 9 Dec 2019
United Nations Department of Economic and Social Affairs Programme on Ageing (2016) Economic inequalities in old age. https://www.un.org/development/desa/ageing/wp-content/uploads/sites/24/2016/08/Briefing-Paper_Economic-Inequalities_Final.pdf. Accessed 6 Feb 2020
United Nations Secretary-General (2009) Report on follow-up to the second world assembly on ageing. UN Doc A/64/127 (6 July 2009)
World Health Organization (n.d.) Nutrition for older persons. https://www.who.int/nutrition/topics/ageing/en/. Accessed 7 Feb 2020
Zinger I (2019) Confronting the problems faced by the elderly in prison. Paper delivered to the Canadian elder law conference, Vancouver, Canada, 15 Nov 2019

International Instruments

Additional Protocol to the American Convention on Human Rights in the Area of Economic, Social and Cultural Rights (1988) Opened for signature 17 November 1988, entered into force 16 November 1999. OAS Treaty Series No 69

Additional Protocol to the European Social Charter (1988) Opened for signature 5 May 1988, entered into force 5 September 1992. ETS 128

African Charter of Human and Peoples' Rights (1981) Opened for signature 27 June 1981, entered into force 21 October 1986. 21 ILM 58

American Declaration on the Rights and Duties of Man (1948) Adopted 2 May 1948. OAS Res XXX

American Convention on Human Rights (1969) Opened for signature 22 November 1969, entered into force 18 July 1978. 1144 UNTS 123

Committee on Economic, Social and Cultural Rights (1990) *General Comment No. 3: The nature of States parties' obligations (Art 2 par 1)* (14 December 1990)

Committee on Economic, Social and Cultural Rights (1991) *General Comment No. 4: The right to adequate housing (Art 11 par 1)* (13 December 1991)

Committee on Economic, Social and Cultural Rights (1995) *General Comment No. 6: Economic, Social and Cultural Rights of Older Persons* (8 December 1995)

Committee on Economic, Social and Cultural Rights (1999) *General Comment No. 12: The right to adequate food (Art 11)* UN Doc E/C.12/1999/5 (12 May 1999)

Committee on Economic, Social and Cultural Rights (2000) *General Comment No. 14: The right to the highest attainable standard of health (Art 12)* UN Doc E/C.12/2000/4 (11 Aug 2000)

Committee on Economic, Social and Cultural Rights (2003) *General Comment No. 15: The Right to Water (Arts 11 and 12)* UN Doc E/C.12/2002/11 (20 January 2003)

Committee on Economic, Social and Cultural Rights (2008). *General Comment No. 19: The Right to Social Security (Art. 9)* UN Doc E/C.12/GC/19 (4 February 2008)

Committee on Economic, Social and Cultural Rights (2009) *General Comment No. 20: Non-discrimination in Economic, Social and Cultural Rights (Art 2 par 2)* UN Doc E/C.12/GC/20 (2 July 2009)

Convention on the Elimination of All Forms of Discrimination Against Women (1979) Opened for signature 18 December 1979, entered into force 3 September 1981. 1249 UNTS 13

Convention on the Rights of Persons with Disabilities (2006) Opened for signature 13 December 2006, entered into force 3 May 2008. 2515 UNTS 3

Convention on the Rights of the Child (1989) Opened for signature 20 November 1989, entered into force 2 September 1990. 1577 UNTS 3

European Convention for the Protection of Human Rights and Fundamental Freedoms (1950) Opened for signature 4 November 1950, entered into force 3 September 1953. 213 UNTS 221

European Social Charter (Revised) (1996) Opened for signature 3 May 1996, entered into force 1 July 1999. ETS 163

Human Rights Committee (1989) *General Comment No. 18: Non-Discrimination* (10 November 1989)

Human Rights Committee (1992) *General Comment No. 20: Prohibition of Torture, or Other Cruel, Inhuman or Degrading Treatment or Punishment (Art 7)* (10 March 1992)

Human Rights Committee (1999) *General Comment No. 27: Freedom of Movement (Art 12)* UN Doc CCPR/C/21/Rev.1/Add.9 (2 November 1999)

Human Rights Committee (2014) *General Comment No. 35: Liberty and Security of Person (Art 9)* UN Doc CCPR/C/GC/35 (23 October 2014)

Human Rights Council (2013) *The Human Rights of Older Persons*. Resolution 24/20. UN Doc A/HRC/24/L.37/Rev.1 (25 September 2013)

Human Rights Council (2016) *The Human Rights of Older Persons*. Resolution 35/5. UN Doc A/HRC/33/L.9 (26 September 2016)

Inter-American Convention on Protecting the Rights of Older Persons (2015) Opened for signature 15 June 2015, entered into force 11 January 2017. 55 ILM 985

International Covenant on Civil and Political Rights (1966) Opened for signature 16 December 1966, entered into force 23 March 1976. 999 UNTS 171

International Covenant on Economic, Social and Cultural Rights (1966) Opened for signature 16 December 1966, entered into force 3 January 1976. 993 UNTS 3

International Labour Organization (1952) *Convention 102 (Social Security (Minimum Standards))*

International Labour Organization (1967) *Convention 128 (Invalidity, Old-Age and Survivors' Benefits)*

Madrid International Plan of Action on Ageing (2002) Adopted by Second World Conference on Ageing, Madrid, 8–12 April 2002

Protocol to the African Charter on Human and Peoples' Rights on the Rights of Older Persons (2016) Opened for signature 30 January 2016 (not yet in force)

Universal Declaration of Human Rights (1948) Adopted by United Nations General Assembly Resolution 217A (III). UN Doc A/810 (10 December 1948)

United Nations General Assembly (2013) *Towards a Comprehensive and Integral International Legal Instrument to Promote and Protect The Rights and Dignity of Older Persons.* Resolution 67/139. UN Doc A/RES/67/139 (13 February 2013)

United Nations Principles for Older Persons (1991) Adopted by United Nations General Assembly Resolution 46/91. UN Doc A/RES/46/91 (16 December 1991)

Vienna International Plan of Action on Ageing (1982) Adopted by World Assembly on Ageing, Vienna, 26 July–6 August 1982. Endorsed by UN General Assembly Resolution 37/51. UN Doc A/RES/37/51 (3 December 1982)

Chapter 3
A Human Rights-Based Approach to Elder Law

A framework for pursuing a human rights-based approach consists of many interrelated concepts ranging from the fundamental philosophical underpinnings through to practical principles which aid in implementation. This chapter presents the framework which has been developed for the human rights-based analysis presented in this book. It begins by outlining the core values of human rights, demonstrating their relevance to elder law, before addressing the specific rights and duties which can be found in human rights law. It then articulates a number of framework principles which assist in the development and operationalisation of a human rights-based approach to particular legal challenges. This framework is employed in later chapters.

3.1 Introduction

As the previous chapter identified, international human rights law guarantees a number of specific rights which are particularly relevant to the experiences and needs of older persons, and these have been developed further in a series of soft-law instruments relating to the rights of older persons. These rights flow from the fundamental principles of human rights, which include respect for dignity, equality, liberty and autonomy. A human rights-based approach to elder law is one which is guided by and champions these principles, and which ensures promotion and protection of specific human rights.

As the Independent Expert on the Enjoyment of All Human Rights by Older Persons has explained, a human rights-based approach is 'a conceptual framework normatively based on international human rights standards and operationally directed towards promoting and protecting the rights of older persons.'[1] In this chapter we outline the contours of a human rights-based approach to elder law. We explain how the fundamental principles of human rights, and the specific rights which they

[1] Kornfeld-Matte (2016).

© Springer Nature Singapore Pte Ltd. 2020

B. Lewis et al., *The Human Rights of Older Persons*,
https://doi.org/10.1007/978-981-15-6735-3_3

give rise to, can be used as both an analytical framework for identifying issues with existing law, and a set of principles for law reform. This work will inform the analysis presented in subsequent chapters, and practical applications have been identified as recommended reforms where appropriate.

A human rights-based approach to elder law offers the potential to reconceptualise our responsibilities towards older generations and reposition older persons as active rights-holders, not merely passive beneficiaries of care who are unable to look after themselves and are therefore in constant need of protection. This represents a major shift in the way that health and aged care for older persons, and elder law more broadly, have traditionally been approached within our society, and marks a move away from medical or charity-based models which have influenced law and policy in the past. A human rights-based approach has the potential not only to ensure minimum standards within the various substantive disciplines which comprise elder law, but, if approached systematically and supported by an appropriate legal architecture, could reshape community attitudes towards our ageing population and reframe our legal responsibilities, policies and processes. This would positively influence multidisciplinary approaches to ageing and lead to genuine, practical improvement.

In part, the value of a human rights-based approach flows from the moral force of human rights and the common understanding of rights as inalienable entitlements. There is therefore a normative weight which comes from describing issues relating to elder law in rights language. This moral and normative value is reinforced by the attendant legal obligations that rest with governments under international and (in some jurisdictions) domestic law.

Of course, the impact of these legal obligations is contingent on effective enforcement, and this has been a criticism levelled at human rights frameworks. This is one reason that this book advocates for the adoption of an international treaty dedicated to the rights of older persons, equipped with effective supervisory and complaints processes. It is also necessary to ensure strong domestic human rights architecture to guarantee rights are upheld. However, even in the absence of strong enforcement mechanisms, there are numerous benefits to be gained from a human rights-based approach to elder law, and it remains the most appropriate conceptual framework for addressing the significant threats to dignity, autonomy and liberty which have historically plagued this area of law.

The following sections outline the key components of a human rights-based approach which can be applied to elder law. These elements are strengthened by robust legal guarantees and procedures, but can be used to guide best practice even in the absence of fully developed human rights laws. This chapter will also explain the way such an approach has been employed as a framework for analysis in this book to examine the key legal issues which arise as people age (identified in Chap. 1) and to develop the recommendations for legal reform (Chap. 11).

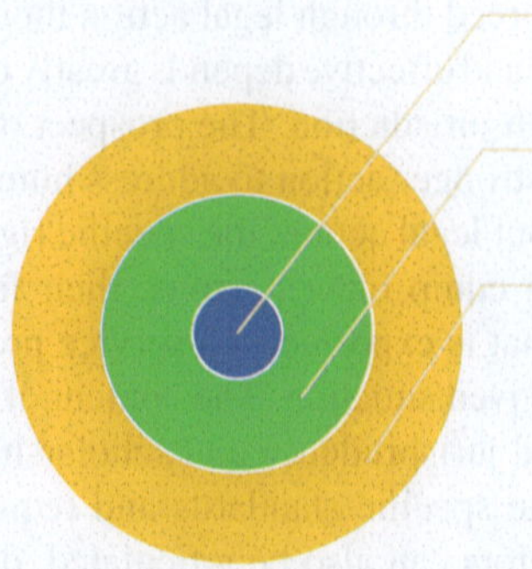

Fig. 3.1 Overview of human rights framework

3.2 The Elements of a Human Rights-Based Approach to Elder Law

The human rights-based approach developed in this chapter and employed throughout this book moves beyond the specific rights discussed in the previous chapter to incorporate other principles and values drawn from human rights theory and scholarship on the rights of older persons. The result is a novel multi-tiered framework that brings together fundamental concepts and principles, which both underpin and operationalise specific rights as well as corresponding duties (see Fig. 3.1).

This human rights-based approach is built on a foundation of core values which should guide all law, policy, decisions and actions, which relate to older persons' lives. The values include respect for individual dignity, autonomy, equality and liberty—the fundamental principles informing human rights law.[2] These core values underpin and rationalise the specific human rights found in international and domestic law, which are understood to be universal, inalienable, indivisible and interdependent. This understanding can be incorporated into a human rights-based approach to help us achieve a more systematic, coordinated network of rights-based laws which operate in a mutually supportive fashion. A human rights-based approach is most capable of addressing the main challenges facing people as they age which, as identified in Chap. 1, include issues in relation to loss of autonomy (dealt with in relation to capacity and decision-making in Chap. 6), dignity (relevant to discrimination, discussed in Chap. 5, and health and aged care, dealt with in Chap. 10) and security (which relates to elder abuse (Chap. 7), financial security (Chap. 8), and accommodation (Chap. 9)).

The core values which form the foundation of a human rights-based approach are translated into the specific rights outlined in Chap. 2. These rights provide guaranteed minimum standards of treatment and targets for what every government must do to fulfil those rights. Importantly, the rights within international human rights law are

---

[2]Preambles to the *Universal Declaration of Human Rights* (1948), the *International Covenant on Civil and Political Rights* (1966) ('*ICCPR*') and the *International Covenant on Economic, Social and Cultural Rights* (1966) ('*ICESCR*').

able to be enforced through legal action though, as noted above, the extent to which this is possible and effective depends greatly on the domestic human rights protections in place in each jurisdiction. The prospect of legal enforcement ought to encourage states to take stronger action to address human rights issues within elder law.

Even without legal action, the specific rights provided under international human rights law can equip older persons, their families and advocates with a language to describe what is expected of a service provider, government agency or decision-maker in any given situation. The content of rights can be interpreted with reference to international jurisprudence and scholarship, allowing the core values to be translated into quite specific standards and required actions. The corresponding duties of different actors can also be articulated, drawing on human rights law's tripartite framework of obligations discussed in the previous chapter. These consist of the duties to respect, protect and fulfil human rights, which provide a roadmap for full enjoyment of all human rights by all individuals.

Scholarship and soft-law relating to the human rights of older persons have generated a number of framework principles which align with the core values of human rights law and the specific rights it provides. These have a more pragmatic character and are specifically directed at the rights and needs of older persons, such that their implementation can help to ensure compliance with specific rights and achievement of core values across the field of elder law. They include principles of participation, non-discrimination, respect for preferences and access to justice, and when these are implemented in practice, they produce an approach that is both compatible with and facilitative of human rights.

Together, therefore, these values, rights, duties and principles create a comprehensive human rights-based framework which safeguards specific human rights in individual circumstances and decision-making contexts. As indicated, the framework is relevant across the full spectrum of elder law. It also applies across the whole life course, and is not limited to issues which typically arise later in life, such as retirement and estate planning, accommodation, health and aged care (both in-home and institutional as well as short and long-term care), and end of life matters, although these are obviously the focus of this book. Under a human rights-based approach, a failure to fulfil legal obligations or deliver services, or policies and services characterised by sub-standard care, neglect or abuse, would be viewed as violations of human rights which would trigger the availability of legal processes.

Importantly, a human rights-based approach can provide a set of principles to guide decision-making when the objectives of different human rights might be seen to conflict. For example, decisions relating to the provision of care might produce challenging questions of how to balance autonomy, self-sufficiency and privacy, on the one hand, with care, protection and security, on the other. A comprehensive human rights framework which encompasses both specific rights and underlying values and principles can help provide guidance on how to navigate these complex issues.

Ultimately, this framework builds a systematic and coherent human rights-based approach across all areas of elder law which views older persons as fully-fledged rights holders and positions them at the centre of all policy and legal reform. The following sections provide more detail on each of these components of the human rights-based approach as it applies to elder law.

3.2.1 *Core Values*

The philosophical and theoretical foundations of human rights include several values which underpin a human rights-based approach. These are reflected in the core international human rights instruments, particularly the *Universal Declaration of Human Rights* (UDHR) and the *International Covenants on Civil and Political Rights* and *Economic, Social and Cultural Rights* (ICCPR and ICESCR).

3.2.1.1 Dignity

As a starting point, these instruments recognise that human rights flow from the dignity of each human person.[3] Human rights are the things which are essential for each individual to live a life of dignity. The inextricable link between rights and dignity is core to natural rights theory, which is widely considered to be the philosophical underpinning of modern human rights law.[4] Not only is the connection with dignity the rationale for the rights which have been recognised in law, but respecting the dignity of each individual is instrumental in ensuring that all those rights are fulfilled. The paradigm shift which is sought within the various fields of elder law is strongly connected with dignity: many of the worst abuses of older persons are characterised by egregious disrespect for their dignity. At the same time, many of the recommendations that can be made across different legal areas are in essence practical manifestations of enhanced respect for dignity. Respect for dignity is therefore a core value of any human rights-based approach, and an especially crucial objective of a human rights-based approach to elder law.

3.2.1.2 Autonomy

Closely intertwined with respect for human dignity is the imperative to respect individual autonomy. This principle is fundamental to an alternative explanation of rights, one sometimes referred to as 'will theory', which emphasises that to have rights is to have the ability to determine what others may or may not do in relation to our lives.[5]

---

[3]The Preamble of the UDHR states: 'Whereas the inherent dignity and of the equal and inalienable rights of all members of the human family is the foundation of freedom, justice and peace in the world.' The inherent dignity of the person is also recognised in preambles of the ICCPR and ICESCR, which both state: 'Recognizing that these rights derive from the inherent dignity of the human person'.

[4]Donnelly (1982), 403; Alston (1988), 26–27.

[5]Wenar (2015).

It is reflected in an understanding of individuals as 'small-scale sovereigns', to use the language of HLA Hart,[6] and draws on Immanuel Kant's philosophical imperative that a person must not be used as a means to an end, but is an end in themselves.[7] The individual's autonomy is the foundation from which all other rights and moral rules can be derived. It is also a common assumption within various legal systems, which presume that any person over the age of majority possesses legal capacity to make decisions for themselves.

While will theory can be presented as an alternative to natural rights as an explanation of the origin of human rights, both approaches are reflected in international human rights law, and it is recognised that respect for independence and autonomy is a necessary condition for the enjoyment of a dignified life. Respect for autonomy has an especially powerful resonance with issues relating to elder law generally, and capacity assessment in particular, since many of these legal and policy issues concern an individual's ability to make legally recognised decisions for themselves, or have the potential to limit their autonomy, for example through the appointment of a third-party decision-maker. Where autonomy is diminished, for example through limited or lost capacity without adequate support or assistance, older persons may experience a heightened risk of vulnerability to abuse or mistreatment. Consequently, a human rights-based approach to elder law is one which promotes individual autonomy in decision-making to the greatest extent possible.

An important principle which draws together both dignity and autonomy in relation to elder law is the 'dignity of risk'.[8] This recognises that some individuals may choose to forego greater security, for example in a high-care residential environment, in favour of the greater independence which comes with remaining in the home. Respecting the dignity of risk is an important component of a human rights-based approach to elder law, but there are challenges in delivering this commitment within a legal setting, some of which will be explored in more detail in relation to specific areas of elder law in later chapters.

3.2.1.3 Liberty

A related core value also fundamental to human rights is respect for liberty. Protection of individual liberty is another necessary condition for the enjoyment of a dignified life, and has obvious links to autonomy, since a person who lacks autonomy to choose for themselves is at risk of restraints on their liberty. Many of the rights found in human rights law are framed in terms of freedoms or liberties, and these are commonly understood as 'negative' rights—rights which prohibit interference by the state.[9] Accordingly, elder law ought to be implemented in a way which minimises

---

[6] Hart (1982), 183.

[7] Kant (2002).

[8] Nay (2002); Morgan (2004); Ibrahim and Davis (2013).

[9] For example, the rights to liberty of the person, freedom from cruel, inhuman or degrading treatment, and freedom of movement, all guaranteed under the ICCPR.

restrictions on individual liberty. The importance of liberty in the context of aged care, for instance, extends to the physical, medical, social, cultural and economic dimensions of older persons' lives, and these will be explored in later chapters.

3.2.1.4 Equality

In addition to respect for dignity, autonomy and liberty, human rights are also underpinned by the core value of equality. Equality is a corollary of the fact that human rights flow from inherent human dignity—by virtue of this fact all humans are equally entitled to human rights. These rights must therefore be assured to all people without discrimination as to age, race, religion, nationality, gender, sex, political opinion or any other ground. As part of a human rights-based approach to elder law, equality entails that laws and policies must be substantively (and not just formally) supportive of rights for all persons, and in particular that ageism must be effectively combatted as part of a comprehensive anti-discrimination approach. As explained in Chap. 5, ageist attitudes are socially constructed and cut across a range of societal dimensions, with potential to impinge upon older persons' human rights in similarly varied ways. Equality must therefore be embedded in any human rights-based approach to ensure that all older persons receive equal protection of their human rights.

3.2.2 Specific Rights and Duties

The specific human rights which are included in international law were addressed in Chap. 2, which identified and expanded upon a number of rights which have particular significance for elder law. While all human rights are indivisible and interdependent, it is recognised that older persons can experience risks to particular human rights and, as such, a number of key rights are identified as being of particular importance to the field of elder law. These include rights to health, housing, an adequate standard of living, liberty and security of the person, freedom of movement and the right to privacy. These rights give content to the core values addressed above and provide a detailed structure to a human rights-based approach. What is emphasised here is that a human rights-based approach also relies heavily on the corresponding duties which attach to particular rights, and which are enshrined in international and domestic human rights law. A human rights-based approach also incorporates a number of more general characteristics of human rights which inform our understanding.

It is generally understood that governments bear three levels of duty with respect to human rights. These are the duties to respect, protect and fulfil. The duty to respect involves a duty to refrain from actions which would interfere with the enjoyment of human rights, and is viewed as the basic standard of human rights obligation. The duty to protect is the obligation that states bear to ensure, through legislative or other means, that other actors (individuals, corporations, or other non-state actors) do not interfere with human rights. The duty to fulfil is the third, most onerous level of obligation, and entails states' obligations to take positive measures to ensure that all persons are able to enjoy their human rights to the fullest extent possible. The specific rights guaranteed under human rights law form a core structure for a human rights-based approach, providing duties and standards which can be applied as minimum guarantees in individual circumstances.

These rights and duties are interpreted and applied within an understanding of the general characteristics of human rights as being universal, inalienable, indivisible and interdependent. The notion that human rights are enjoyed by all people universally is a cornerstone of the UN human rights framework. It is articulated in the UDHR, ICCPR and ICESCR, which also emphasise the fact that all persons are entitled to human rights without distinction or discrimination of any kind.[10] This reinforces the need for a human rights-based approach to elder law to address age-based discrimination, but also to ensure that older people can fully enjoy their human rights without discrimination on any other basis.

A corollary of universality is inalienability—because human rights flow from our inherent human dignity, they cannot be given or taken away any more easily than we can cease to be human. All people remain entitled to the full complement of human rights throughout the whole of their lives. This has particular relevance for elder law, as it insists that we must continue to support the full enjoyment of all human rights by older persons, even where that requires particular support or assistance as they age.

Human rights are also understood to be indivisible and interdependent. This means that all people are entitled to the full range of human rights, and that different rights are mutually supportive of each other. While there may be reasons to prioritise action on certain rights based on a person's individual circumstances or due to available resources or capacities, there is no intrinsic hierarchy among rights, and all rights are understood as being equally important and interdependent. Applied to elder law, certain rights will have a particular relevance—for instance rights to housing and healthcare are especially pertinent to questions about aged care—but a comprehensive rights-based approach means that we must also work equally hard to safeguard all economic, social, cultural, civil and political rights for all older persons. Doing so will help to ensure not only that older people enjoy a fully realised, rich and rewarding life free from abuse and neglect, but that they also have the opportunity to continue to contribute to society, which in turn benefits from their knowledge, experience and skills.

3.2.3 Framework Principles

In addition to the core human rights treaties which give content to human rights and duties under international law, a number of soft-law instruments have been developed to support states in implementing their human rights obligations owed to

[10] ICCPR, art 2 and ICESCR, art 2.

older persons. These instruments include the *Vienna International Plan of Action and Ageing*,[11] the *United Nations Principles for Older Persons*[12] and the *Madrid International Plan of Action on Ageing* (Madrid Plan of Action), discussed in more detail in Chap. 2.[13] As explained in that chapter, these documents are not legally binding, but they are generally consistent with a human rights-based approach and contain guidance on how such an approach should be implemented.[14] We can use this body of soft-law and other scholarship to inform our understanding of law and theory, and from this identify a number of framework principles which support the effective operationalisation of a human rights-based approach to elder law.

3.2.3.1 Participation

Ensuring participation of older persons should be a key principle behind all law and policy which affects their lives. There are three dimensions to participation. The first is that older persons should be able to participate in decisions which affect them directly. The second is that older persons should be able to participate in their communities in a meaningful and rewarding way. Finally, the third aspect to participation is that older persons should be able to participate in discussions around policies which affect them more generally.

The importance of participation was highlighted throughout the negotiation of the *Convention on the Rights of Persons with Disability* (CRPD), with persons with disabilities repeatedly stressing 'nothing about us without us.'[15] Participation underpins the framework of rights established in the CRPD, which emphasises that states must take action to remove barriers which prevent persons with disability from full and effective participation as equal members of society.[16] The CRPD further accentuates several specific dimensions of participation relating to decision-making and policy formulation, independent living, education, housing, political and public life, and cultural life, recreation and sport.[17]

As discussed in Chap. 2, the CRPD provides an appropriate model for the development of human rights law for older persons, both in terms of the conceptualisation of participation as a core substantive principle, and the commitment to participation in the process of negotiating and adopting the treaty. Participation should similarly be a core principle in the reform of all areas of elder law as it foregrounds the importance of autonomy, as well as ensuring that the law adequately reflects the lived experiences of older persons. It is recognised as one of the five principles articulated in the UN Principles, and the Madrid Plan of Action emphasises participation in numerous

[11] *Vienna International Plan of Action on Ageing* (1982).

[12] *United Nations Principles for Older Persons* (1991) ('*UN Principles*').

[13] *Madrid International Plan of Action on Ageing* (2002).

[14] United Nations Secretary-General (2009).

[15] Harpur (2017).

[16] *Convention on the Rights of Persons with Disabilities* (2006), arts 1, 3 ('*CRPD*').

[17] Ibid., arts 4, 9, 24, 26, 29, 30.

ways, including in article 5, which stresses that as people age they must be able to continue to participate fully in the economic, social, cultural and political lives of their society. To ensure these various forms of participation for older persons, we must appreciate the social and cultural environments in which older persons are situated, and seek to identify the barriers which prevent them from fully enjoying their human rights. Such barriers can be found in specific areas of employment, education, healthcare, social services, transportation and technology, as well as in community attitudes more generally.[18] A primary focus of any human rights-based reforms to elder law and policy should therefore be removing or minimising these barriers.[19]

Including older persons in discussions around law and policy reforms is essential. Not only is such participation to be valued in its own right, but it importantly ensures that laws and policies are ‘age-sensitive’, that they address the real concerns of older persons and learn from their experiences, and that they develop strategies which promote further social inclusion.[20] As discussed in more detail in Chap. 4, participation is essential to ensure that older persons are able to enjoy the full range of human rights, and it also helps to reinforce an appreciation of older persons as fully valued members of the community who are entitled to respect for dignity and autonomy.[21]

3.2.3.2 Respect for Will and Preferences

A second important framework principle is respect for will and preferences. This has obvious links to the core values of autonomy, liberty and dignity discussed above, and in this way underpins the enjoyment of other specific rights which give effect to these values. It also draws on the language of the CRPD which identifies respect for will and preferences as a key part of ensuring equality before the law for persons with disabilities.[22] It is also reflected in the UN Principles, which include independence and self-fulfilment.[23]

The extent to which elder law and policy can embed respect for older persons’ will and preferences depends in no small part on their ability to participate in decision-making: respecting their preferences is contingent on knowing what those preferences are. It is also crucial to ensuring that older persons are meaningfully able to enjoy their full range of human rights, as many of these rights depend on an individual being able to choose for themselves how they wish to live (for example rights to housing, freedom of movement, healthcare, and privacy). This therefore highlights

[18]European Network of National Human Rights Institutions (2017), 23; Office of the High Commissioner for Human Rights (2012).

[19]AGE Platform Europe (2018).

[20]Office of the High Commissioner for Human Rights (2012), 33.

[21]Ibid., 32; Kanter (2008–2009).

[22]CRPD, art 12.4.

[23]UN Principles, paras 1–6 and 15–6.

the interdependence of the various framework principles, values and rights that form the basis for a human rights-based approach to elder law.

In practice, the principle of respect of will and preferences will require careful consideration in terms of how it is reflected within specific areas of elder law. A key issue relates to how an individual's preference can be expressed and implemented when they may have impaired or lost decision-making capacity.[24] In this case it then becomes a question of the supported or substitute decision-maker knowing and respecting the will and preferences of the older person. The importance of supported decision-making will be explained in more detail in Chap. 6, which focuses on legal capacity and decision-making.

3.2.3.3 Non-discrimination

The principle of non-discrimination is closely linked to the core value of equality and the notion of universality discussed above. In relation to elder law it requires that governments address ageism—discrimination on the basis of age—which can operate both at individual and structural levels, and which operates as a barrier to the full enjoyment of human rights. As discussed in more detail in Chap. 5, ageism is often present in areas relating to education, employment, accommodation or social participation.

Ageism can be caused by assumptions about the capabilities, needs and preferences of older persons. Even where held with good intentions as an act of respect or sympathy towards older persons, these ageist attitudes, as Harpur explains, are still grounded in stereotypical notions of older persons' capabilities and vulnerabilities, and can operate to deny them agency and full enjoyment of their human rights.[25] Age-based discrimination can also be seen in the imposition of age-based criteria in areas of employment, education or social security, which can operate arbitrarily to limit older persons' opportunities and independence.[26]

It is also important to recognise the potential for intersectionality between ageism and other forms of discrimination.[27] Older persons may face discrimination on a range of grounds, including race, gender, sex, nationality, or religion, and this may compound and complicate the way that ageism affects their lives. Designing and implementing law and policy which is truly guided by the principle of non-discrimination requires that we appreciate and address the complex factors which contribute to inequality, and that we develop laws and policies which are culturally appropriate and celebrate diversity among the older population. A human rights-based approach to elder law would promote a better understanding of the connections between discrimination and the enjoyment of human rights, and help to confront the damaging effects of ageism.

---

[24] AGE Platform Europe (2018).

[25] Harpur (2015), 1055; Baum (2018).

[26] AGE Platform Europe (2018), 5.

[27] Ibid., 7.

3.2.3.4 Access to Justice

The fourth framework principle which forms part of a human rights-based approach is access to justice. This principle recognises that, despite the best attempts to protect and promote the enjoyment of human rights, violations may still occur. When this happens, it is essential that legal processes provide affordable and accessible avenues for victims to seek redress, including access to an appropriate remedy. This principle is found in international human rights law, which obliges governments to ensure that a person whose rights have been affected has access to an appropriate remedy as judged by a competent authority.[28]

This points to the need for an effective human rights-based legal framework at the domestic level which can enforce human rights generally, but which acknowledges that special measures are also required to address the particular situation of older persons. For instance, appropriate pathways must be available for older persons to report elder abuse, neglect or other mistreatment, including for people who may have difficulty reporting such incidents because, for example, they may be reliant upon the perpetrator. Institutions or agencies involved in the provision of services to older persons may be required to report potential human rights breaches or risks when they occur. Support services should be provided for victims and people at risk, including appropriate legal services, and these must be timely, effective and appropriate to the individual's circumstances.[29]

As noted above, ageism can operate as an obstacle to the enjoyment of human rights, and it can similarly impact the ability of older persons to seek justice when their rights have been abused. Special care must be taken that legal processes do not reinforce ageist attitudes, but instead respect the agency and dignity of all older persons, and support them in seeking justice for human rights interferences. Other barriers to justice may exist similar to those noted above as barriers to participation in society more broadly, and can be connected to financial, linguistic, cultural, gender, mobility or other factors. Operationalising the principle of access to justice requires proper information about what these barriers are and how they operate, as well as the strategies needed to be implemented to overcome them.

3.3 Conclusion

The discussion above provides an overview of the human rights-based approach that is employed in this book. It consists of core values of human rights, specific rights and corresponding duties, and framework principles which support implementation within the field of elder law. Figure 3.2 illustrates the way that the core values flow out into more specific rights and duties, and then further into pragmatic principles.

---

[28] ICCPR, art 2.

[29] AGE Platform Europe (2018), 8.

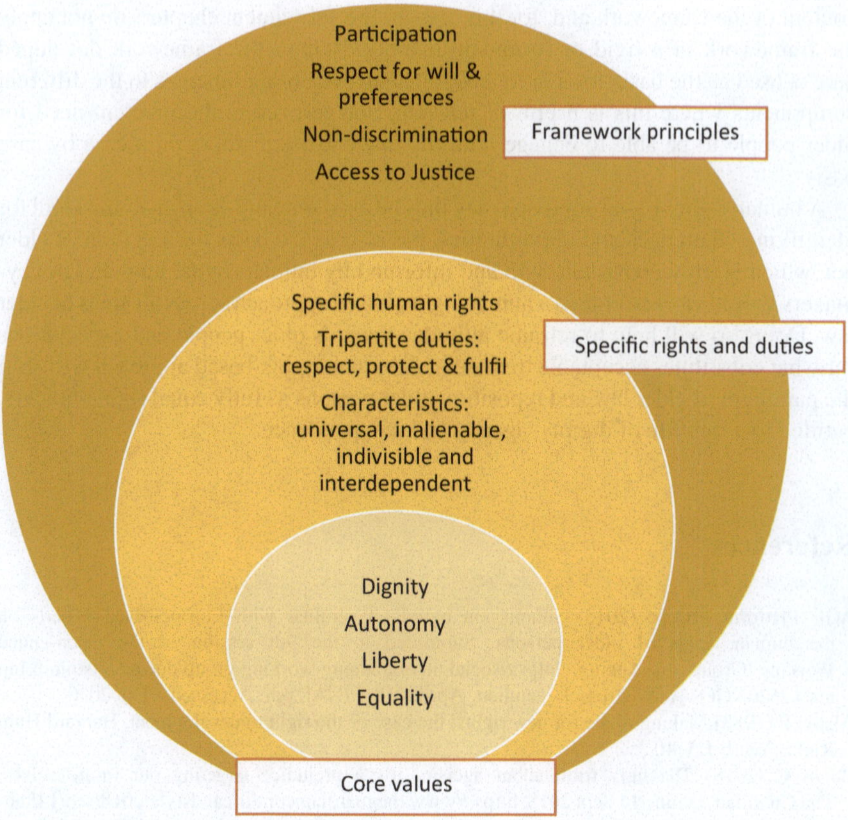

Fig. 3.2 Content of human rights framework

We have used this framework to conduct the analysis of specific issues that appears in later chapters. The interconnected and nuanced components of the human rights-based approach make it an appropriate framework for addressing the various human rights issues experienced by older persons, and the socially-constructed origins of many of those challenges. The approach can be applied to identify specific challenges and shortcomings which need to be addressed, as well as to identify appropriate recommendations for reform.

The chapters that follow address both substantive legal areas (financial security in Chap. 8, accommodation in Chap. 9, health and aged care in Chap. 10) as well as broader legal issues that cut across the spectrum of elder law (namely participation in Chap. 4, ageism in Chap. 5, legal capacity in Chap. 6 and elder abuse in Chap. 7). In each chapter the human rights framework is used to identify key issues and challenges which threaten to interfere with or limit the rights of older persons. It also provides a basis for developing strategies to address these issues, identifying best-practice examples or suggestions for reform which are inspired by human rights and associated values and principles. Not all issues or challenges are able to be connected to the full content of the framework and, for this reason, the subsequent chapters do not apply the framework in a rigid or formulaic manner. Rather, the framework developed here is used as the basis for a more holistic analysis, noting linkages to the different components where this is useful or relevant, and respecting the inherent need for older people to be able to engage with the law and legal actors on a case by case basis.

A human rights-based approach may thus be used not only as an analytical tool for identifying challenges and opportunities, but also as the basis for a system of elder law which is truly consistent with and informed by human rights. Law-and policy-makers should work to embed a human rights-based approach across all areas of elder law. Doing so will help to reframe attitudes towards older people and expectations for what constitutes acceptable treatment. A human rights-based approach will shift the paradigm of elder law and reposition older persons as fully equal rights-holders, entitled to a rich life of dignity, agency and independence.

References

AGE Platform Europe (2018) Submission based on seminar with Europe-based experts on the human rights of older persons. Submitted to the 9th session of the Open-ended Working Group on Ageing. https://social.un.org/ageing-working-group/documents/ninth/Inputs%2520NGOs/AGE_Brussels_seminar_April2018_FINAL.pdf. Accessed 6 Feb 2020

Alston P (1988) Making space for new rights: the case of the right to development. Harvard Hum Rights Yearb 1:3–40

Baum C (2018) The ugly truth about ageism: it's a prejudice targeting our future selves. The Guardian. Online 14 Sept 2018. https://www.theguardian.com/lifeandstyle/2018/sep/14/the-ugly-truth-about-ageism-its-a-prejudice-targeting-our-future-selves. Accessed 19 Dec 2019

Donnelly J (1982) Human rights as natural rights. Hum Rights Q 4(3):391–405

European Network of National Human Rights Institutions (2017) Human rights of older persons and long-term care project: the application of international human rights standards to older persons in long-term care. https://www.nnhri.org/IMG/pdf/ennhri_application_of_human_rights_to_ltc_feb_2017.pdf. Accessed 23 Dec 2019

Harpur P (2015) Old age is not just impairment: the CRPD and the need for a convention on older persons. Univ P J Int Law 37:1027–1047

Harpur P (2017) Nothing about us without us: the un convention on the rights of persons with disabilities. In: Leithner A, Libby K (eds) Oxford research encyclopedia of politics. Oxford University Press, Oxford, pp 1–20

Hart HLA (1982) Essays on bentham: studies in jurisprudential and political theory. Clarendon Press, Oxford

Ibrahim JE, Davis M (2013) Impediments to applying the 'dignity of risk' principle in residential aged care services. Australas J Ageing 32(3):188–193

Kant I (2002) (trans Wood A W) Groundwork for the metaphysics of morals. Yale University Press, New Haven. Translation of: Grundlegung zur Metaphysik der Sitten (1785)

Kanter AS (2008–2009) The United Nations convention on the rights of persons with disabilities and its implications for the rights of elderly people under international law. Georgia State Univ Law Rev 25(3):527–574

Kornfeld-Matte R (2016) Report of the independent expert on the enjoyment of all human rights by older persons. UN Doc A/HRC/3344

Morgan S (2004) Positive risk-taking: an idea whose time has come. Health Care Risk R 10(10):18–19
Nay R (2002) The dignity of risk. Aust Nurs J 9(9):33
Office of the High Commissioner for Human Rights (2012) Normative standards in international human rights law in relation to older persons: analytical outcomes paper. https://social.un.org/ageing-working-group/documents/OHCHRAnalyticalOutcomePaperonOldePersonsAugust2012.doc. Accessed 9 December 2019
United Nations Secretary-General (2009) Report on follow-up to the second world assembly on ageing. UN Doc A/64/127 (6 July 2009)
Wenar L (2015) Rights. In: Stanford encyclopaedia of philosophy. https://plato.stanford.edu/entries/rights/. Accessed 7 Feb 2020

International Instruments

Convention on the Rights of Persons with Disabilities (2006) Opened for signature 13 December 2006, entered into force 3 May 2008. 2515 UNTS 3
International Covenant on Civil and Political Rights (1966) Opened for signature 16 December 1966, entered into force 23 March 1976. 999 UNTS 171
International Covenant on Economic, Social and Cultural Rights (1966) Opened for signature 16 December 1966, entered into force 3 January 1976. 993 UNTS 3
Madrid International Plan of Action on Ageing (2002) Adopted by Second World Conference on Ageing, Madrid, 8–12 April 2002
United Nations Principles for Older Persons (1991) Adopted by United Nations General Assembly Resolution 46/91. UN Doc A/RES/46/91 (16 December 1991)
Universal Declaration of Human Rights (1948) Adopted by United Nations General Assembly Resolution 217A (III). UN Doc A/810 (10 December 1948)
Vienna International Plan of Action on Ageing (1982) Adopted by World Assembly on Ageing, Vienna, 26 July—6 August 1982. Endorsed by UN General Assembly Resolution 37/51. UN Doc A/RES/37/51 (3 December 1982)

Chapter 4
Economic, Social and Cultural Participation

Economic, social and cultural rights are an essential component of a human rights-based approach to elder law. They help to ensure that older persons are able to enjoy full lives of their choosing and also to combat the ageist attitudes which operate as barriers to full inclusion in society. This chapter explores the economic, social and cultural rights of older persons with a particular focus on the principle of participation. It identifies participation as both a means and an end of a human rights-based approach to elder law because, not only is maximising participation and inclusion a key aim of elder law, this objective can only be achieved when older people are given meaningful opportunities to participate in the design, implementation and enforcement of those laws. When older people cannot experience social, economic and cultural inclusion this can lead to a range of negative human rights impacts. This chapter examines what participation looks like for older persons in the context of economic, social and cultural rights and identifies strategies for maximising inclusion. One of the major barriers to greater participation is ageism, which will be identified here but analysed in more detail in Chap. 5.

4.1 Introduction

One of the major benefits of a human rights-based approach to elder law is that it broadens our thinking about the interests and needs of older persons beyond traditional notions of care and security that are imposed on older persons to encompass all aspects of life. Such an approach can therefore contribute to a legal framework which is more supportive of a comprehensive range of human rights and helps to ensure that older persons can lead full lives *of their own choosing*. Key to this is recognition of, and respect for, the range of economic, social and cultural rights that are guaranteed under human rights law.

The body of economic, social and cultural rights is a categorisation employed within international human rights law and is commonly used in contrast to civil and

© Springer Nature Singapore Pte Ltd. 2020

B. Lewis et al., *The Human Rights of Older Persons*,
https://doi.org/10.1007/978-981-15-6735-3_4

political rights. For various reasons, mostly relating to the geo-political influences at play at the end of the Second World War, international human rights law was split into two major treaties: the *International Covenant on Civil and Political Rights* (ICCPR), and the *International Covenant on Economic, Social and Cultural Rights* (ICESCR). The latter includes rights like an adequate standard of living,[1] the highest attainable standard of health,[2] employment and fair conditions of work,[3] social security[4] and participation in the cultural life of the community.[5] It is understood that both sets of rights are mutually supportive, and that the democratic rights and liberties recognised in the ICCPR are crucial to ensuring that everyone is able to enjoy the rights found in the ICESCR without discrimination or oppression.[6]

Perhaps even more than civil and political rights, economic, social and cultural rights go to the heart of why a human rights-based approach is so valuable in elder law. They encompass many of the areas where rights-informed policy is most needed: health, housing, social security, employment, education and an adequate standard of living. They are also oriented towards addressing ageism and elder abuse, because they counter social exclusion and promote participation in the economic, political, social and cultural life of the community, thereby helping to combat negative age-based stereotypes which can lead to marginalisation, isolation, neglect and mistreatment.

As explained in Chap. 3, participation is a framework principle of our human rights-based approach to elder law and has a key role to play in fulfilling economic, social and cultural rights. Participation is recognised as a specific right of older persons within the Inter-American human rights regime. Article 8 of the *Inter-American Convention on Protecting the Rights of Older Persons* (Inter-American Convention) states that 'older persons have the right to active, productive, full, and effective participation in the family, community and society with a view to their integration. State parties shall adopt measures to enable older persons to participate actively and productively in their community and to develop their capacities and potentialities.'[7] The ICESCR lacks such a detailed participatory right, a gap which would arguably be well-filled by the introduction of a dedicated international convention on the rights of older persons (CROP). However, even without a specific right, the ICESCR still guarantees to all people the right to participate in the cultural life of their communities, in addition to a range of other economic, social and cultural rights.[8]

Participation of older persons is both a means and an end of a human rights-based approach: the aim of elder law ought to be maximising older people's ability

[1] *International Covenant on Economic, Social and Cultural Rights* (1966), art 11 ('*ICESCR*').

[2] Ibid., art 12.

[3] Ibid., arts 6, 7, 13.

[4] Ibid., art 9.

[5] Ibid., art 15.

[6] Ibid., Preamble; Committee on Economic, Social and Cultural Rights (1990).

[7] *Inter-American Convention on Protecting the Rights of Older Persons* (2015) ('*IACROP*').

[8] *ICESCR*, art 15.

to participate in their communities for as long as possible and in ways of their choosing, but this is best achieved by ensuring that they are included in the design of those laws.[9] Economic, social and cultural rights play a crucial role here not only because they explicitly include rights to take part in culture but also because they are oriented towards removing barriers to greater participation relating to work, education, housing and health, including through ensuring equitable access to social security and other necessary supports. At the same time, a human rights-based approach demands that the value of older persons to society is not measured according to narrow views of economic contribution or participation in the paid workforce. Participation must be viewed in the fullest sense as something which advances and signifies the full enjoyment of human rights, and not as a measure which would exclude older persons or restrict their choices.[10]

This chapter therefore considers the relationship between elder law and economic, social and cultural rights through the prism of full and meaningful participation. It will be organised around a number of topics which draw together the key economic, social and cultural rights, namely education, employment, and social and cultural participation. The argument is advanced in this chapter that these should be priority areas within a human rights-based framework for elder law. Select rights which are relevant here will be discussed in more detail in later chapters (particularly rights relating to employment, social security, accommodation and healthcare), but this chapter will elaborate on a number of key issues relating to the promotion and protection of economic, social and cultural rights more generally, including social inclusion, age-friendly urban design and cultural participation. The chapter also identifies that ageism (discussed fully in Chap. 5) operates as a significant barrier to older people's participation in many areas. The discussion is also premised on the understanding that poverty operates as a significant barrier to social, cultural and economic participation, and the important relationship between financial security and human rights will be explored fully in Chap. 8. To begin though, this chapter considers what 'participation' means for older persons, drawing on discourse around the common narratives of ageing and showing that some of the alternative approaches are more conducive to a human rights-based approach than others.

4.2 What Does Participation Mean for Older Persons?

Ensuring participation in all areas of life is key to addressing ageism and related human rights issues. Rosa Kornfeld-Matte, the United Nations Independent Expert on the human rights of older persons, explains that truly promoting participation of older persons requires a range of measures to address exclusion across all areas of society, and that these are fundamental in achieving the paradigm shift that is

---

[9] *United Nations Principles for Older Persons* (1991), Principle 7; ('*UN Principles*'); Kornfeld-Matte (2016).

[10] Grover (2011), para 13.

required in elder law from an economic and development model to one grounded in human rights.[11] She says: 'It is arguable that without such guarantees the human rights framework could be in danger of perpetuating a view of older people as passive and in need of protection, not as rights-holders, with a right to inclusion in the life of the community, to freedom, dignity and autonomy.'[12]

What meaningful or adequate participation looks like for older people and how it should be pursued have been contested questions in the fields of public policy, gerontology and human rights law. Some of the early work of international bodies on the rights of older persons offers lessons for how participation should, and should not, be defined. In 1982, the *Vienna Plan of Action on Ageing* ('Vienna Plan of Action') conceptualised participation as a key means of ensuring that older persons could satisfy their need for personal fulfilment. The Vienna Plan of Action stated that:

> The principal ways in which older people find personal satisfaction are through continued participation in the family and kinship system, voluntary services to the community, continuing growth through formal and informal learning, self-expression in arts and crafts, participation in community organizations and organizations of older people, religious activities, recreation and travel, part-time work, and participation in the political process as informed citizens.[13]

The recognition of participation as important to self-fulfilment is significant, and policies which support inclusion and engagement are essential to any framework which respects the dignity and autonomy of older persons. Further, the participation principle in the Vienna Plan of Action does list a broad range of activities that older persons might choose to engage in, as noted in the above extract. However, it also presents a rather stereotypical view of how an older person will spend their time in retirement (that is, post-paid employment), with a focus on leisure activities and family and community service. The reference to part-time work, for example, implies an assumption that older persons will not want to be in full-time employment, while the rather passive idea of 'participating in the political process as informed citizens' seems to exclude more active (or activist) roles within the political system or policy development and advocacy groups.

The *United Nations Principles for Older Persons* ('UN Principles'), adopted in 1992, make similar assumptions about the means and modes of older persons' participation. Under the heading of 'Participation,' the UN Principles provide that: 'older persons should remain integrated in society, participate actively in the formulation and implementation of policies that directly affect their well-being and share their knowledge and skills with younger generations.'[14] This rather broad understanding of participation is given more specificity by the next principle, which states that 'older persons should be able to seek and develop opportunities for service to the community and to serve as volunteers in positions appropriate to their interests and

---

[11] Kornfeld-Matte (2016), para 126.

[12] Ibid., para 33.

[13] *Vienna International Plan of Action on Ageing* (1982), para. 31(j) ('*VIPAA*').

[14] UN Principles, Principle 7.

capabilities.'[15] This suggests a narrower understanding of older persons' appropriate participation in society, indicating an assumption that they will opt to contribute via voluntary work rather than continue in paid employment (or even to contribute to neither paid nor unpaid work).

In 2002, the *Madrid International Plan of Action on Ageing* ('Madrid Plan of Action') moved away from this assumption. The Political Declaration adopted along with the Madrid Plan of Action emphasised that 'older persons should have the opportunity to work for as long as they wish and are able to, in satisfying and productive work.'[16] However, the Political Declaration also states that: 'the expectations of older persons and the economic needs of society demand that older persons be able to participate in the economic, political, social and cultural life of their societies.'[17] This indicates that the reasons for encouraging older persons' participation go beyond the rights of individuals themselves, and include economic considerations related to their productivity (or lack thereof). The Madrid Plan of Action does acknowledge that older persons' non-economic contributions (for example through care for family, household maintenance or volunteering) must be valued, and that 'participation in social, economic, cultural, sporting, recreational and volunteer activities also contributes to the growth and maintenance of personal well-being.'[18] It is of course important to correct ageist assumptions about the ability of older persons to contribute to society as all the varied roles that can be undertaken by older people need to be valued. However, we need to be wary of placing too great an emphasis on what older persons do for society (rather than what society should do for them) as this can distort the way in which policies relating to participation are approached.

Older persons' engagement with paid and unpaid work has become a key point of debate within the scholarship on ageing. As will be discussed in the next chapter in relation to ageism, over the past several decades different narratives of ageing have been influential in the way that laws and policies have been framed. In the 1960s, Butler identified that a medicalised narrative of ageing was producing negative, ageist stereotypes, within both the medical profession and the wider community.[19] The dominant view at that time was that older persons were inevitably in a state of physical and cognitive decline which reasonably limited their expectations for participation in society.[20] Kesby explains that this has persisted as a narrative of 'ageing as pathology', in which older persons are diseased or malfunctioning, feeding into ideas of older persons being unproductive, dependent and ultimately a burden on the welfare system.[21] Through this kind of thinking, older persons are easily segregated into nursing homes or other medicalised institutions, with obvious negative consequences for their ongoing and meaningful participation in society. They are

[15]Ibid., Principle 8.

[16]*Madrid International Plan of Action on Ageing* (2002), art 12 ('*Madrid Plan of Action*').

[17]Ibid.

[18]Ibid., para 20.

[19]Butler (1975); Butler (1989).

[20]Townsend (2007); Weicht (2013); Friedan (1985).

[21]Kesby (2017), 376.

further cast as passive recipients of care and welfare, undermining their position as autonomous holders of human rights.

Other narratives have emerged which challenge the medicalisation of old age and which offer the potential to promote greater participation of older persons both in society and in the design of laws and policies which affect them. However, these narratives deserve critique too, particularly because the tacit values which underpin them have implications for the kind of participation which is valued and advanced.

One such narrative is 'successful ageing', first articulated by John Rowe and Robert Kahn.[22] Their concept consists of three intersecting pillars:

i. reducing the probability of disease and disease-related disability;
ii. maximising cognitive and physical functional capacity; and
iii. active engagement with life, especially in the form of interpersonal relations and productive activity.[23]

While the prominence of health measures hints at the influence of medicalised narratives of ageing, the tripartite model demonstrates that participation (particularly social and cultural participation) is not just an entitlement for every older individual but also both a marker and driver of 'healthy ageing', a term used by the World Health Organization (WHO—as discussed below).[24] In terms of human rights, this aligns with the understanding of different rights being interdependent and mutually supportive.

The third component encompasses 'sustained engagement with productive and social activities', and therefore is most relevant to social rights and participation in the community as understood within the ICESCR. The notion that successful ageing involves 'productive activity' warrants further scrutiny because it implicitly links 'productivity' with economic contribution of some kind. However, Rowe and Kahn explain that the concept actually embraces both paid and unpaid societal roles. As they say: 'an activity is productive if it creates societal value, whether or not it is reimbursed. Thus, a person who cares for a disabled family member or works as a volunteer in a local church or hospital is being productive, although unpaid.'[25]

Kesby has criticised the linking of 'success' with 'productivity' or social contribution. For one thing, she argues that it over-emphasises individual (rather than societal) responsibility for successful ageing, in a way that celebrates the 'third age' of active, independent older people who have the time and resources to engage in range of activities. This fails to account for social inequalities or make space for those older persons who may lack the independent means or ability to enjoy such participation.[26]

Moreover, it obscures another problem which arguably plagues the various fields of elder law—that is, the dependence of our communities on charitable, volunteer labour, much of which is performed by older persons themselves. 'Successful ageing'

---

[22]Rowe and Kahn (1997).

[23]Ibid.

[24]World Health Organization (2015), 28–9, 31–2, 100–3.

[25]Rowe and Kahn (1997), 434.

[26]Kesby (2017), 381.

accords value to the unpaid work that older people perform within their communities. This is, of course, appropriate where the alternative is not recognising that work at all. However, it risks legitimising that work as a source of unpaid labour upon which our societies come to depend, when the preferred approach would be to fund that work properly so that people performing it can be remunerated for their efforts. The increasing reliance on the charitable sector to provide care for older people is a consequence of this view, but can lead to lower standards of treatment and a lack of accountability.

Thus, scholars have critiqued the concept of 'successful ageing' for the way in which it equates success with a contribution to society, rather than prioritising participation for the benefit of the older individual themselves. Biggs, for instance, has questioned whether the notions of success which appear in scholarship and public policy are oriented towards the needs and interests of older persons or towards 'the demands of other groups within advanced capitalism and high modernity.'[27]

The linking of successful ageing with productive activity is emphasised even further in an alternative narrative of ageing known as 'productive ageing'. An early advocate of 'productive ageing' was Betty Friedan, who sought to overcome the 'mystique of age' by pointing to the productive potential of older persons. Friedan said: 'There's a breath of fresh air when we suddenly stop dealing with the aged as crocks and deal with them as people who might be productive.'[28]

The narratives of 'productive ageing' and 'successful ageing' can contribute to countering medicalised narratives of ageing and the associated 'crisis' of ageing demographics which construct older people as an economic burden, dependent on a finite and an increasingly stretched welfare system. However, there are problems with these narratives when considered from a human rights perspective. Productive ageing in particular risks characterising older persons as an 'unutilized resource for meeting community needs'[29] and lays the foundation for an evaluation of older persons' worth based on what they can contribute to society or by some measure of their 'output'. This conflicts with the essential premise of a human rights-based approach (as outlined in the previous chapter) that people have an inherent value worthy of respect, regardless of what they contribute to society or to their own family or other networks.

Another alternative framing is 'active ageing', which overcomes some of the problems with successful or productive ageing and places greater emphasis on quality of life and social participation for individual older persons. As will be seen, however, it too is susceptible to critique within a human rights-based approach. Active ageing is defined by the WHO as 'the process of optimizing opportunities for health, participation and security in order to enhance quality of life as people age.'[30] In contrast to

[27] Biggs (2001), 312, in Kesby (2017), 381.

[28] Friedan (1985); Kesby (2017).

[29] Minkler and Holstein (2008).

[30] World Health Organization (2002), 12; United Nations Economic Commission for Europe. Active Ageing Index.

productive or successful ageing, being 'active' in older age is understood as encompassing 'continuing participation in social, economic, cultural, spiritual and civic affairs, not just the ability to be physically active or to participate in the labour force.'[31] Active ageing is linked to the concept of 'healthy ageing', which is also a key part of the WHO's strategies on ageing and focuses on individuals' functional abilities and inherent capacity (as discussed in Chaps. 6 and 10).[32] The link between active ageing and the right to health is specifically recognised in the Inter-American Convention, which also recognises the right to participation of older persons more broadly, as noted above.[33] Importantly, however, participation as part of active ageing is not dependent on good health but can and should be supported for all older persons including those living with illness or disability.[34]

The WHO has emphasised the connections between active ageing and human rights, explaining that the approach:

> is based on the recognition of the human rights of older people and the United Nations Principles of independence, participation, dignity, care and self-fulfilment. It shifts strategic planning away from a 'needs-based' approach (which assumes that older people are passive targets) to a 'rights-based' approach that recognizes the rights of people to equality of opportunity and treatment in all aspects of life as they grow older. It supports their responsibility to exercise their participation in the political process and other aspects of community life.[35]

Active ageing in its broad sense is therefore closest to a human rights-based framing of ageing. However, there is a risk that in its interpretation and application it too places undue emphasis on economic productivity. Even the WHO's approach hints at the economic objectives underpinning the policy: 'the WHO argues that countries can afford to get old if governments, international organizations and civil society enact "active ageing" policies and programmes that enhance the health, participation and security of older citizens.'[36] In practice many such policies take a narrow focus aimed at increasing participation in the workforce.[37] For example, in his foreword to Australia's *Intergenerational Report* in 2015, then Treasurer, The Hon. Joe Hockey MP, stated that: 'to drive higher levels of prosperity through economic growth, we must increase productivity and participation. If we are to achieve these goals we need to encourage those currently not in the workforce, especially older Australians and women, to enter, re-enter and stay in work, where they choose to do so.'[38] Similar policies also exist in other jurisdictions, especially within Europe.[39] The emphasis

---

[31] World Health Organization (2002), 12.

[32] World Health Organization (2015), 28–9, 32.

[33] *IACROP*, arts 8, 19(b).

[34] Kesby (2017), 379.

[35] World Health Organization (2002), 13; Kesby (2017).

[36] World Health Organization (2002), 6.

[37] Kesby (2017), 379.

[38] Australian Government (2015), iii.

[39] Parliamentary Assembly of the Council of Europe (2011); Doron and Georgantzi (eds) (2018); Foster (2018).

on retaining older persons in the workforce can also be inferred within many age-discrimination laws, which focus on addressing ageism in the workforce but less so in other spheres. These are discussed in more detail in Chap. 5.

Katz argues that policies which advance the 'production and celebration of an active body in old age' are at risk of becoming a disciplinary strategy justified on economic grounds, because we determine that there is an economic imperative to 'activate' 'dependency prone' bodies on the basis that 'societies can afford older persons so long as they are active and productive.'[40] Active ageing approaches can therefore be subject to the same criticism as 'successful' and 'productive' ageing—that they too easily prioritise older persons' participation in the form of economic contribution as a means to offset the costs of an ageing population and allow societies to be able to 'afford' their older citizens. Even where older persons' contributions are unpaid, this can become a relied-upon means to fill gaps in the social security and social protection systems, so that even voluntary activities come to take on economic significance.[41] As Biggs explains: 'active aging defined as work and volunteering … balances the books nicely.'[42] For these reasons, narratives of successful, productive or active ageing have been referred to as the 'capitalist takeover of ageing.'[43]

These critiques demonstrate that prominent narratives of ageing frequently do not accord with a human rights-based approach and participation of older persons has taken on a narrow understanding of economic participation, rather than participation which advances the full range of social and cultural, in addition to economic, rights. 'Healthy ageing' is a preferable concept for ensuring policies align with human rights given its focus on the health of the individual older person. However, even this approach ought not to be employed blindly, but should always be guided by the principles of the human rights framework outlined in Chap. 3. The following sections expand beyond these narrow conceptualisations of participation to explore how a human rights-based approach to elder law ought to understand and promote older persons' participation more broadly. It does this by considering a number of thematic areas related to economic, social and cultural rights, beginning with issues related to education and work, particularly in the context of new technologies and the emerging digital economy.

4.3 Participation in Education and Employment

The ICESCR guarantees to all people, without discrimination, the right to education and the right to work, including favourable working conditions.[44] For older persons, these two rights are closely related and can directly affect the extent and nature of

---

[40] Katz (2000), 148, in Kesby (2017).

[41] Minkler and Holstein (2008), 197.

[42] Biggs (2001).

[43] Baltes and Carstensen (1996).

[44] ICESCR, arts 6, 7 and 13.

their participation in their communities, as well as their enjoyment of the full range of other economic, social and cultural rights.

The causes and effects of ageism will be discussed in more detail in Chap. 5, but it is important to note here that ageism can have a significant impact on older persons' experiences of work. Negative assumptions about older persons' capabilities, preferences, technological literacy, agility and relational styles can disadvantage older persons across the full spectrum of work-related interactions: job advertisements, recruitment processes and appointments; opportunities for on-the-job training, development and promotion; recognition of achievements; and a range of other work-related factors.[45] Ageism also influences the opportunities for education outside the workplace, where assumptions are sometimes made about the kinds of education that older persons wish to pursue, the motivation they have for enrolling, their learning styles and preferences for teaching delivery.[46] Older persons may face other barriers to education based on the location of educational institutions, accessibility of physical and online learning spaces, design of courses and cost.[47]

Ageism can therefore have a snowballing effect leading to a range of negative human rights consequences. As explained, age discrimination can lead to redundancy, but also makes it harder for older individuals to find new work, especially where that requires further training or education. This can lead to poverty and other vulnerabilities, placing a wide range of rights at risk, and further limiting the ability of older persons to participate in the wider community. These problems are particularly acute in the context of the emerging digital economy, which is changing the nature of work and the skills required of workers. This section will explore the human rights implications of education and employment for older persons, with a particular focus on the impact of the digital economy and the need for a human rights-based approach to law and policy in this area.

4.3.1 Education

One of the core rights protected in the ICESCR is the right to education. While article 13 of the Covenant makes specific provision for primary and secondary school education, it is premised on a broader understanding of education as being fundamental to the 'full development of the human personality and the sense of its dignity' and which strengthens 'respect for human rights and fundamental freedoms'. By signing the Covenant, states parties 'agree that education shall enable all persons to participate effectively in a free society, promote understanding, tolerance and friendship among all nations and all racial, ethnic or religious groups, and further the activities of the United Nations for the maintenance of peace.'[48]

---

[45] Thornton and Luker (2010).

[46] Kornfeld-Matte (2018), paras 76–78.

[47] Ibid.

[48] *ICESCR*, art 13.

Education early in life has obvious implications for our human rights as we age, because higher levels of education tend to lead to higher incomes, providing a more stable platform for the enjoyment of human rights in retirement, as discussed in more detail in Chap. 8. However, supporting people to take up educational opportunities in older age is also an important means of advancing human rights. As well as being a human right in itself, education and lifelong learning are important preconditions for longer participation in the labour market, avoiding the negative human rights impacts of unemployment discussed in Chaps. 5 and 8. Education also has immense potential to promote social and intergenerational interaction and increasing inclusivity and diversity within education is an important way of countering ageist stereotypes which undermine the enjoyment of human rights.[49]

The importance of education and life-long learning for older persons has been identified in the work of the Open-ended Working Group on Ageing, where discussions and submissions have noted the various barriers which older persons can face in accessing education.[50] Where an older person experiences illness, disability or poverty, this can interfere with the accessibility of educational services. Physical barriers can exist if the location of education services is remote or if the older person's mobility is limited. Visual or hearing impairments can also limit older individuals' ability to access educational services, particularly where these are delivered via online platforms which do not adhere to accessibility standards. Ageist attitudes among institutions can limit the opportunities and support available to older learners, and institutions may lack qualified teachers to address the needs and preferences of older learners.[51] As will be explained in Chap. 5, ageism can also affect older persons themselves, who may take on stereotypical ideas of their capabilities and entitlements when it comes to education and lifelong learning. This can lower their participation, and consequently reduce demand and supply of educational opportunities for older persons.[52]

Once again, a reflection on the work of international bodies in this area can shed light on some of the considerations which need to be borne in mind when devising laws and policies in support of older persons' education. In 1982 the Vienna Plan of Action recommended the need for 'informal, community-based and recreation-oriented programmes for the elderly in order to develop their sense of self-reliance and the community's sense of responsibility.'[53] The idea of 'recreation-oriented' education programs is, however, outdated, and fails to appreciate that education for older persons is not merely about keeping them 'occupied' or 'entertained'.

Principle 16 of the UN Principles states that governments should provide access to suitable educational programs and training.[54] The Committee on Economic, Social

49 Kornfeld-Matte (2016), para 14; Office of the High Commissioner for Human Rights (2012).

50 Open-ended Working Group on Ageing (2019a); Open-ended Working Group on Ageing (2019b); AGE Platform Europe et al. (2019).

51 Open-ended Working Group on Ageing (2019b), paras 18–19.

52 Ibid., para 19.

53 VIPAA, Recommendation 47.

54 UN Principles, Principle 16.

and Cultural Rights elaborates that the right to education entails both the right of older persons to benefit from educational programs as well as the right of younger generations to benefit from older persons' experience.[55] The Madrid Plan of Action recognises education as a 'crucial basis for an active and fulfilling life', in addition to its importance in ensuring 'the productivity of both individuals and nations.'[56]

These approaches to education for older persons reveal a key challenge, which is itself influenced by the various narratives of ageing discussed in the previous section. To support those older persons who wish to continue working, it is essential that they are able to access affordable and appropriate education and training. This is particularly relevant in the context of rapid changes in the labour market due to the emergence of the digital economy (discussed in more detail below). However, it should not be assumed that vocational training or professional development are the only forms of education older persons wish to pursue, and care needs to be taken not to develop educational policies which perpetuate the view that older persons' activities are only of value when they make an economic contribution or are otherwise 'productive'. Policies around education and lifelong learning for older persons need to support those older persons who wish to continue working, while at the same time avoiding entrenching expectations of longer working lives or undervaluing non-economic or 'unproductive' activities.

The answer to addressing this challenge lies in a human rights-based approach, and particularly in the principle of participation. An approach which is genuinely guided by the principle of participation will take a broad understanding of the various activities that older persons will engage in and why they are doing so, and for which education and training might be instrumental, as well as appreciating the participatory value of education itself. It will also demand that older persons be involved in the design of policies and of educational opportunities to maximise their participation.

In terms of how educational programs should be designed for older persons, Kornfeld-Matte has argued that consideration must be had for the specific needs and interests of older persons and their different learning styles.[57] However, she has also stressed how crucial it is that a range of programs is offered which do not target older persons exclusively, as this significantly limits the opportunities for social and intergenerational interaction which education can offer. Education must not be designed in a way which creates 'social spaces of exclusion,' but should instead be developed in a way which maximises community integration, and the enjoyment of the full range of human rights by older and younger generations alike.[58] In this way, the participatory benefits of education can be maximised for older persons, with positive consequences for a wider range of human rights.

As will be discussed below in relation to the emerging digital economy, the expansion of new technologies creates particular challenges for older persons and highlights the need for policies which support them to acquire competencies with digital and

[55]Committee on Economic, Social and Cultural Rights (1995), para 8.

[56]Madrid Plan of Action, paras 35–39.

[57]Kornfeld-Matte (2016), para 14.

[58]Ibid.

other new technologies. New technologies offer enormous potential to benefit older persons in terms of reducing marginalisation through enhanced communication and mobility, improving health outcomes through new diagnostic technologies and treatments, and improving care and physical environments through the use of assistive technologies and robotics. However, these new technologies can have an alienating effect where older persons are not provided with sufficient education, training and support to enable effective adoption and use.[59] They also present new risks relating to privacy, autonomy and liberty, and potential for exploitation and abuse.[60] It is therefore crucial that older persons understand these risks and are able to give informed consent to the use of new technologies. Appropriate education is essential to ensure that older persons can take advantage of the benefits of new technologies across a range of areas.[61]

As noted above, a human rights-based approach demands that barriers are removed which would prevent older persons from accessing educational services, and educational opportunities should be made available that fit their needs, capabilities, preferences, and diverse identities.[62] Technology offers great opportunities to expand the availability of education to older persons, particularly those who live in remote or regional areas or who have limited mobility. However, the effectiveness of technology-based education and training depends on affordability and accessibility, reliability of infrastructure, and inclusive design that supports users of all ages and abilities. The Declaration adopted at the meeting of experts on the human rights of older persons, held in Vienna in 2018, stated that:

> The use of technologies, including in the field of education and lifelong learning, must be geared towards enabling older persons to live autonomous and independent lives, fulfil their aspirations, build their skills and capacities, develop their full human potential and sense of dignity and self-worth and participate fully in society, and must not deprive older persons of their liberty, exclude them from decision-making, or stigmatize and objectify them.[63]

To achieve this, and ensure that the right to education can be fulfilled for older persons, it is essential that older persons participate in the design of policies and educational opportunities. Regulation of new technologies (in education and other fields) should support co-design and consultation in both development and deployment. Protections should also be in place to protect older persons' privacy, dignity and autonomy. Further human rights implications of new technologies are explored below in relation to the emerging digital economy.

[59] Madrid Plan of Action, Political Declaration, para 38.

[60] Kornfeld-Matte (2017); Lewis et al. (2018).

[61] International Expert-Conference on the Human Rights of Older Persons (2018), p. 2; Open-ended Working Group on Ageing (2019b), para 51.

[62] Vienna Declaration, 3.

[63] Ibid., 2.

4.3.2 Employment and the Digital Economy

As noted above, the right to education for older persons has in the past been championed as a means of enabling older persons to continue in the workforce. While it is problematic to limit educational participation to this context for the reasons outlined above, vocational education and training are essential components of fulfilling both the right to education and the right to work for older persons. Ageism in the provision of education and training is a key barrier to older persons' ability to secure new jobs or to advance in their careers, compounding the negative impact of ageism in the workplace more broadly (discussed in more detail in the next chapter). These impacts can be seen clearly in the context of the emerging digital economy, in which traditional jobs and ways of working are giving way to new, often technology-based, professions, business models and working relationships.[64] This transformation presents challenges for older workers, who are often assumed to be 'too old' to adapt to new processes or to contribute value to new industries. To fulfil the right to work for older persons, laws and policies relating to the digital economy must combat these ageist attitudes and support older workers where it is necessary to transition to new modes of work.

As discussed in Chap. 5, many older persons have reported experiencing ageist attitudes in the workplace where it is assumed that they are resistant to innovation or unable to adapt to new technologies.[65] A report commissioned by the Australian Human Rights Commission, for example, found that older persons were commonly thought to display 'stale thinking', where they were considered unable to contribute 'fresh' ideas. This research also found a prevalent belief that 'older workers … simply knew less about technology than their younger peers.'[66] Another Australian study found evidence of a belief on the part of employers that they would get better value for money training younger employees rather than investing in developing the skills of their older workers.[67] These sorts of attitudes perpetuate the devaluing of older workers' skills and experience. Thornton and Luker have linked ageism to the postmodern, neoliberal context in which the digital economy has emerged. They argue that this incorrectly privileges younger, more flexible workers at the expense of older workers.[68]

In fact, other data reveal that older workers are equally able to contribute to emerging industries and professions. The assumptions about older workers' inability to adapt to technological innovates are unfounded, and many are successfully able to learn new skills and work within new business models.[69] Further, older workers bring

---

[64] World Economic Forum (2016).

[65] Thornton and Luker (2010), 141.

[66] Australian Human Rights Commission (2016), 47.

[67] The Benevolent Society (2017), 19.

[68] Thornton and Luker (2010), 143.

[69] Bersin and Chamorro-Premuzic (2019).

other experiences and abilities to the workplace which make a valuable contribution, even in new industries.[70]

One of the facets of the digital economy which can impact on older workers is the changing nature of work practices. In many sectors, work is transitioning away from traditional 9–5, permanent positions and towards more short-term, casual or freelance work. This is often referred to as the 'gig economy', and it raises a number of concerns more broadly relating to labour rights such as wage rates and overtime, leave entitlements, insurance, contributions to pension funds, collective bargaining and dispute resolution processes.[71] In many cases, employers seek to classify staff as 'independent contractors' rather than 'employees' in order to avoid the usual obligations which would exist in an employer-employee relationship.[72]

While the human rights implications of the gig economy are numerous and require appropriate regulatory responses, they have the potential to impact on older persons in particular ways. As noted above, ageist attitudes may create barriers for older persons obtaining work in these industries. While the flexibility of this work may appeal to some older persons, the lack of legally protected rights and entitlements may have a negative impact on older workers with family responsibilities or greater financial commitments, and therefore may operate as a form of indirect discrimination against older persons.

Ageist assumptions about older workers' capacity to adapt to the digital economy represent discrimination which needs to be addressed within legislative frameworks. However, as Thornton and Luker have argued, the emergence of the digital economy is paralleled by economic changes which challenge the effectiveness of current models of anti-discrimination law. In the aftermath of the global economic crisis 'restructuring and downsizing have become everyday occurrences, which mean that a respondent is invariably going to be able to rationalise its actions, including dismissal.'[73] This creates a situation of competing rationalities, where the priority given to productivity, consumerism and growth in the neo-liberal employment landscape may mean that arguments based around profitability, transformation and workforce renewal receive the benefit of assumed truth and are given relatively more weight than claims based on individual workers' rights. Thornton and Luker argue that anti-discrimination legislation itself has limited effectiveness because it too is predicated on out-dated ideas of the modernist, linear career path and has failed to keep pace with the newer, post-modern work ethic.[74]

Laws and policies need to adapt to the digital economy to guarantee older persons' employment rights. Policies are required to combat ageist assumptions about older workers' capabilities in relation to new technologies. As will be explained in the

---

[70]Baum (2018); Thornton and Luker (2010), 151.

[71]Business and Human Rights Resource Centre (2019a).

[72]Business and Human Rights Resource Centre (2019b); *Dynamex Operations West, Inc. v. Superior Court of Los Angeles County*, No. S222732 (Supreme Court of California 30 April 2018); *Aslam* et al. *v Uber*, No. 2202550/2015 (Employment Tribunals 28 October 2016); Frias (2018).

[73]Thornton and Luker (2010), 151.

[74]Ibid., 167.

next chapter, ageist attitudes are socially and culturally constructed and will take time to shift. In the meantime, law reforms which encourage or mandate affirmative action strategies may be beneficial to support older workers. Within the gig economy, stronger regulation is required to ensure that all workers—young and old—are protected from exploitation and afforded basic employee entitlements.

In relation to anti-discrimination laws in particular, reforms may be required to ensure that forms of discrimination that are currently obscured within the digital economy can be identified and redressed. In a context where transformation of the workforce is largely an assumed and unchallenged good outcome, greater protections are required to ensure that older workers' rights are protected against discrimination in favour of younger workers who are perceived to be more flexible, technically proficient or able to work for a longer period of time. As explained in Chaps. 1 and 3, access to justice is a key pillar of a human rights-based approach, thus ensuring that anti-discrimination laws are still fit for purpose in the age of the digital economy ought to be a priority of the law reform agenda.

4.4 Social and Cultural Participation

A fundamental aspect of ensuring the economic, social and cultural rights of older persons and respecting their dignity and autonomy is advancing their ability to participate in the social and cultural life of their community. There are individual and collective aspects to this. Individuals must be able to pursue cultural, recreational and other pursuits of their own choosing, but must also be supported to enjoy activities in conjunction with others. Not only is social and cultural participation a human right, it also brings numerous benefits associated with inclusion and engagement, and avoids the negative impacts which can flow from social isolation.

For older persons who are members of cultural, ethnic or religious minorities, the right to participate in and pass on cultural, spiritual or linguistic practices must be safeguarded,[75] and the harm caused when this participation is impeded must be recognised. Related to this is the need for culturally appropriate services and programs for older persons who are members of particular cultural or ethnic groups, which demands co-design or at least meaningful participation of older persons in developing and evaluating such programs and services.[76]

The Madrid Plan of Action recognises the need for holistic approaches which draw on the interconnectedness of specific human rights with principles of dignity, participation and respect for cultural and religious traditions. The commentary to the Plan of Action states that:

> Implementation (of the Plan of Action) … requires, inter alia, a political, economic, ethical and spiritual vision for social development of older persons based on human dignity, human

[75] *International Covenant on Civil and Political Rights* (1966), art 27 ('*ICCPR*').

[76] See, for example: Australian Department of Health (2017); Cain (2007); Australian Human Rights Commission (2019).

> rights, equality, respect, peace, democracy, mutual responsibility and cooperation and full respect for the various religious and ethical values and cultural backgrounds of people.[77]

This section explores a number of more specific elements of social and cultural participation to highlight key priorities for a human rights-based approach to elder law.

4.4.1 Social Inclusion

As a starting point, older persons are entitled to social interaction on an equal basis with all other members of society. Social interaction is important for older persons' physical and mental well-being, allows them to express their individual identities, and helps to build strong and cohesive communities.[78] Unfortunately, older persons can often experience social exclusion, as a consequence of any number of financial, physical, social and structural factors.[79] Often it is a result of ageism, which leads to a devaluing of older persons' contributions to social activities and society more generally, and limits their opportunities to participate.[80] While there is no settled definition of social exclusion, a working definition proposed by Walsh et al. (based on the work of Levitas et al.), indicates its multidimensional nature and the potential to impact on a range of human rights:

> Social exclusion of older persons is a complex process that involves the lack or denial of resources, rights, goods and services as people age, and the inability to participate in the normal relationships and activities, available to the majority of people across the varied and multiple domains of society. It affects both the quality of life of older individuals and the equity and cohesion of an ageing society as a whole.[81]

Social exclusion is a key risk factor for elder abuse (discussed in Chap. 7), with research demonstrating that many older people who experience abuse lack social connection and interaction, often resulting in isolation.[82] A number of factors contribute to this. When older persons are isolated and lack interaction with the broader community, abuse may go unnoticed and the older person may have fewer opportunities to report their experiences. At a deeper level, social exclusion is closely connected to the systemic ageism which is at the root of elder abuse. In an ageist society older people are made invisible, their views are overlooked and their contributions devalued, and these same attitudes and behaviours create fertile ground for elder abuse. At the same time, social exclusion of older persons limits our potential to combat ageism through intergenerational interaction, which can help to embed respect for older persons' dignity and worth.

---

77 Madrid Plan of Action, Paragraph 115.

78 Barbosa Neves et al. (2019).

79 Nicholson (2012); Ellwardt et al. (2015).

80 Kornfeld-Matte (2018), para 25.

81 Walsh et al. (2017), 83, adapted from Levitas et al. (2007).

82 Dow and Johnson (2012).

There are numerous negative human rights implications associated with social exclusion. Kornfeld-Matte has recently explored the relationship between human rights and social exclusion and highlighted the close connection between social exclusion and the rights to an adequate standard of living, the right to work, and the right to the highest attainable standard of health. She articulates how social exclusion can flow from violations of other rights, including rights to housing and work. For instance, older persons who lack financial resources due to breaches of their right to work may find themselves forced to move to cheaper accommodation outside their communities, leading to isolation.[83] This in turn compounds interferences with their right to an adequate standard of living, which encompasses the social and family dimensions of the right to housing, and the right to health, including mental health.[84]

As a result of these relationships between social exclusion and negative human rights impacts, Kornfeld-Matte and others have argued that older persons have a right to independent living and social inclusion. As Kornfeld-Matte defines it, inclusion as part of the right to independent living requires equal access to all services that are offered to society generally, with necessary support to enable older persons to fully participate in all spheres of social life, including intergenerational activities. 'Those services encompass transport, health care, public spaces and buildings, shopping, volunteering, leisure, events of a political nature, sports, technology and any other activity in which the older person wishes to participate.'[85]

Along similar lines, Brownlee has argued that the importance of social inclusion ought to be recognised in the form of a dedicated human right to be protected from social deprivation. For Brownlee, social deprivation is not merely associated with poverty, but is 'a persisting lack of minimally adequate opportunities for decent or supportive human contact including interpersonal interaction, associative inclusion, and interdependent care.'[86] A right to inclusion, or a right to be protected from social deprivation, would recognise the multiple ways in which inclusion contributes to the well-being of older persons and supports their full enjoyment of human rights.

To date, a human right along these lines has not been recognised in any multilateral human rights treaty. However, as noted above, the Inter-American Convention specifically recognises that: 'Older persons have the right to active, productive, full, and effective participation in the family, community, and society with a view to their integration. States Parties shall adopt measures to enable older persons to participate actively and productively in their community and to develop their capacities and potentialities.'[87]

To facilitate social integration, article 8 of the Inter-American Convention further outlines a number of actions which states are expected to take. These are to:

---

[83] Kornfeld-Matte (2018).

[84] Committee on Economic, Social and Cultural Rights (1991); Committee on Economic, Social and Cultural Rights (2000).

[85] Kornfeld-Matte (2018), para 70.

[86] Brownlee (2013).

[87] IACROP, art 8.

i. create and strengthen mechanisms for the participation and social inclusion of older persons in an environment of equality that serves to eradicate the prejudices and stereotypes that prevent them from fully enjoying those rights;
ii. promote the participation of older persons in intergenerational activities to strengthen solidarity and mutual support as key components of social development; and
iii. ensure that facilities and community services for the general population are available to older persons on an equal basis and that they take account of their needs.

While the Inter-American Convention is obviously only legally binding on the small number of states that have signed on to its obligatory framework, it provides a useful and appropriate template for how social and cultural participation should be approached within a human rights-based approach to elder law. Key elements include removing barriers to older persons' social inclusion by combatting ageism and promoting intergenerational activities which enhance interaction. States must also address discrimination, including indirect discrimination, and ensure that all social and other services are equally accessible to older persons. These steps are necessary to ensure that older persons are meaningfully able to participate in society, and thereby reap the many benefits that this can bring.

4.4.2 *Age-Friendly Urban Design*

Social participation and inclusion are also affected by physical factors, including the design of cities and other urban spaces. Physical design of public and private spaces, accessibility and affordability of transportation, and availability of spaces, programs and services for social interaction can all contribute to the ease with which older persons can take part in the social and cultural life of their communities.[88] Architectural and landscape design and planning of urban spaces can have a significant impact on older persons' ability to interact with their communities on an equal basis with other people. While it should not be assumed that older persons will experience disability, as people age they are more likely to have difficulty with mobility or experience other impairments which make it difficult to access, for example, buildings or public spaces with steps or inaccessible footpaths.

Accessibility and affordability of transportation are also relevant factors for older persons' social participation, as is the availability of information about public transport timetables and fares, and the ease of operation of ticketing systems. The importance of accessible transportation becomes all the more significant if older persons are required to relocate away from home, either because of increased care needs or economic changes which force them to move to more affordable accommodation. In her report on human rights and social exclusion, the UN Independent Expert paid particular attention to the impact of gentrification of urban neighbourhoods on older

[88] Tinker and Ginn (2016); Kornfeld-Matte (2016), para 76.

persons. Through the process of gentrification, older persons may be priced out of neighbourhoods as rents increase, dislocating them and increasing social exclusion, as noted above. For those who can afford to stay, gentrification may lead to the closure or privatisation of community facilities as economic priorities and market interests often take precedence over social goods and the interests of long-term residents. Entrenched ageist attitudes can also mean that older persons are not included in consultation and planning discussions, and their interests can be easily overlooked. Consequently, as neighbourhoods gentrify older residents can lose opportunities to participate in the community, exacerbating the fragmentation that occurs through the urban renewal process.[89]

To avoid these problems and protect older persons' rights to social inclusion and participation, city plans and associated regulations must ensure that older persons' interests are adequately taken into account. This can be achieved through a combination of meaningful consultation and implementation of age-friendly design principles. The WHO has identified the importance of age-friendly urban design for facilitating good health in older age, as part of its 'active ageing' approach, discussed above.[90] Such design is guided by the objective of 'optimizing opportunities for health, participation and security in order to enhance quality of life as people age.'[91] Key to this is recognising that older persons are not a homogenous group, and that cities and urban spaces need to be flexible to their abilities, resources, choices, interests and needs. Urban planning regulations need to facilitate this flexibility, but must ultimately guarantee that older persons are equally able to access and participate in the life of their city or neighbourhood, and benefit equally from the process of gentrification or urban renewal.[92] To achieve this it is essential to mainstream any consultation with older persons into the planning process and provide adequate opportunities for both consultation and review of decisions.

Age-friendly design helps to facilitate older persons' social interaction, but as a consequence it benefits entire communities. This is the key objective behind the "8 80 Cities" initiative, which seeks to encourage the design of 'safe and happy' cities that cater to people of all ages and improve the quality of life of whole communities. Its central principle is that if cities are designed to cater for eight-year olds as well as eighty-year olds then they will be more vibrant, healthy and equitable cities for all.[93]

---

[89] Kornfeld-Matte (2018), paras 75–80.

[90] World Health Organization (2007).

[91] Ibid., 1.

[92] Kornfeld-Matte (2018), para 75.

[93] 8 80 Cities (n.d.).

4.4.3 *Cultural Participation and Cultural Safety*

In addition to requiring that older persons be protected from social exclusion, a human rights-based approach also demands that specific forms of cultural participation be safeguarded. Article 15 of the ICESCR guarantees to all persons the right to take part in cultural life. Further, article 27 of the ICCPR recognises the right of all persons who are members of religious, ethnic and linguistic minorities 'in community with other members of their group, to enjoy their own culture, to profess and practise their own religion, or to use their own language.' This is also reflected in the Inter-American Convention at article 21, which provides that:

> Older persons have the right to their cultural identity, to participate in the cultural and artistic life of the community, to enjoy the benefits of scientific and technological progress and those resulting from cultural diversity, and to share their knowledge and experience with other generations in any of the contexts in which they participate.

Laws and policies which affect older persons must therefore work to support, rather than hinder, their ongoing participation in cultural and religious community to which they belong.

This right can be particularly difficult to guarantee in aged care settings, where people from minority backgrounds may have limited choices and culturally-specific facilities may be unavailable.[94] The right to cultural participation can be assisted through the provision of culturally-safe aged care. Cultural safety originated in the health professions, and has been defined by the Nursing Council of New Zealand as:

> The effective nursing practice of a person or family from another culture, and is determined by that person or family. Culture includes, but is not restricted to, age or generation; gender; sexual orientation; occupation and socioeconomic status; ethnic origin or migrant experience; religious or spiritual belief; and disability.... Unsafe cultural practice comprises any action which diminishes, demeans or disempowers the cultural identity and well-being of an individual.[95]

Cultural safety is closely linked to cultural participation as it is understood within a human rights framework because it ensures that opportunities for participation are not blocked by culturally inappropriate practices. It also requires participation as part of its implementation, to ensure that culturally safe practices and policies are developed with the consultation of the cultural community. A further alignment with human rights is that, as Ahmad puts it, cultural safety: 'repositions the care recipient from a passive individual to an influential partner who has both the power and the responsibility to determine whether the care experience has been positive.'[96] Cultural safety not only helps to ensure that the individual older person is able to participate in cultural or religious practices, but it is also 'rooted in the fundamental "human rights of respect, dignity, empowerment, safety, and autonomy"'[97] and therefore helps

---

[94] Ahmad (2018); Khan and Ahmad (2014).

[95] Nursing Council of New Zealand (2011), 7 in Ahmad (2018), 26.

[96] Ahmad (2018), 26.

[97] Phiri et al. (2010), p. 107 cited in in Ahmad (2018), 27.

to achieve exactly the kind of paradigm shift envisioned by a human rights-based approach to elder law.

4.4.4 Leisure

A final component of social and cultural participation that should be facilitated within a human rights-based approach is the right to leisure. This right is found in the ICESCR, which includes leisure as part of the right to satisfactory conditions of work in article 7. States must ensure that all people are entitled to: 'rest, leisure and reasonable limitation of working hours and periodic holidays with pay, as well as remuneration for public holidays.'[98] In this sense, the right to leisure is often thought of as a counter-point to the right to work, but it should be understood as more than just the 'right to not-work', especially in the context of older persons who may be retired or semi-retired.[99] Karev and Doron have argued instead that there should be a broader understanding of the right to leisure which encompasses the qualitative nature of leisure, and that this is a more meaningful construction for older persons in particular.

Karev and Doron defined three distinct conceptual aspects to leisure. These include a 'negative-temporal' aspect, where leisure is essentially the absence of work or other obligation (the sense conveyed by article 7 of the ICESCR), and a 'positive-active' dimension, where leisure is defined by the activity which the individual engages in.[100] This sort of positive formulation can be found in other international rights instruments. For instance, the *Convention on the Rights of Persons with Disabilities* provides that, 'with a view to enabling persons with disabilities to participate on an equal basis with others in recreational, leisure, and sporting activities, State Parties shall take appropriate measures … [t]o ensure that persons with disabilities have access to services from those involved in the organization of recreational, tourism, leisure, and sporting activities.'[101]

The Inter-American Convention provides a similar protection for older persons:

> Older persons are entitled to recreation, physical activity, leisure, and sports.
>
> States Parties shall promote the development of recreational services and programs, including tourism, as well as leisure and sports activities, taking into account the interests and needs of older persons, particularly those receiving long-term care, in order to improve their health and quality of life in all respects and to promote their self-fulfilment, independence, autonomy, and inclusion in the community.[102]

---

[98] ICESCR, art 7(d).

[99] Karev and Doron (2017).

[100] Ibid., 278.

[101] *Convention on the Rights of Persons with Disabilities* (2006), art 30(5).

[102] *IACROP*, art 22.

The *Chicago Declaration*, which puts forward a proposed text for a new international treaty on the rights of older persons, also lists the right to leisure among other participation rights.[103] These positive-active formulations go beyond simply limiting permissible working hours, and focus on the right to partake in activities of the individual's choosing.

However, Karev and Doron's third dimension is most relevant to older persons because it focuses on the quality of the activity and seeks to identify the intrinsic nature of leisure. In their words: 'leisure is a choice made out of a sense of freedom and internal motivation … there is more to "leisure" than simply time off work or time free from obligation. Recognition of this sophisticated qualitative aspect of leisure is essential to the discussion at hand, as it dismisses the idea that retirement is leisure by default.'[104]

In terms of what governments must do to safeguard older persons' right to leisure, Karev and Doron stress that it goes beyond simply ensuring sufficient financial resources so that people are free not to work. As they say, 'pensions alone cannot guarantee that the right to leisure is protected'[105] and some of the strategies discussed above to protect social inclusion and participation more broadly are also necessary to help facilitate meaningful leisure for older persons. Karev and Doron propose a definition of the right to leisure which fleshes out the content of these obligations. They say that,

> As part of their right to leisure older persons have the right
>
> (a) to retire—fully or partially—from work;
>
> (b) to engage and participate in social, cultural, economic, recreation, and sports activities; and
>
> (c) to choose and exercise individual and meaningful free time according to their preferences and beliefs.[106]

Like all forms of social participation, there are broader societal benefits to be gained from safeguarding the right to leisure for older persons. Certain leisure activities in particular (Karev and Doron give the examples of surfing or competitive sports) can challenge the dominant ageist discourses about older persons and ageing.[107] This can help to break down the ageist stereotypes which otherwise hinder older persons' participation in the social and cultural life of their communities, and which can lead to multiple other human rights violations.

---

[103] *Chicago Declaration on the Rights of Older Persons* (2014). Adopted at the International Elder Law and Policy Conference, John Marshall Law School, Chicago (11 July 2014), art 2(r).

[104] Karev and Doron (2017), 279.

[105] Ibid., 289.

[106] Ibid., 291–292.

[107] Ibid., 291; Dionigi (2006); Wheaton (2017).

4.5 Conclusion

Protection and fulfilment of economic, social and cultural rights is a cornerstone of ensuring a human rights-based approach to elder law. Rights to an adequate standard of living including adequate housing and nutrition, the highest attainable standard of health, fair working conditions, social security and participation in social and cultural life are crucial to ensuring that older persons can enjoy lives of dignity and agency equally with other members of society.

Participation is a key guiding principle for delivering these rights. Practically speaking, it should inform the design of all law and policy by facilitating meaningful consultation and co-design in partnership with older people. In this way, laws and policies can best take account of older persons' needs, interests and experiences, helping to improve human rights outcomes. Participation also has a substantive dimension in terms of fulfilling economic, social and cultural rights. In addition to explicit rights to cultural participation and social inclusion, economic, social and cultural rights are all oriented towards enhancing older persons' interaction with their communities by removing barriers related to income, health, geography or education. Participation can therefore act as an overarching concept which gives content and meaning to specific economic, social and cultural rights. It demands that the determinants of participation be provided in relation to specific rights, for example through the provision of social security and through age-friendly design principles. A broad understanding of participation is required which goes beyond merely involvement in the workforce and encompasses participation in all aspects of society. It must respect older persons' choices as to the degree and nature of participation they wish to engage in. This facilitates a greater respect for older persons' varied contributions, regardless of their perceived 'productivity'.

This chapter has explored a number of dimensions of participation in relation to economic, social and cultural rights. The following chapters will look in more detail at particular legal areas which impact not only on economic, social and cultural rights but also on civil and political rights. The analysis will identify legal and policy mechanisms which can hinder or promote participation and equal respect for the dignity and autonomy of all older persons.

References

8 80 Cities (n.d.) Creating cities for all. https://www.880cities.org/. Accessed 17 Feb 2020

AGE Platform Europe et al (2019) Social security and social protection, education and lifelong learning: a discussion paper for the 10th open-ended working group on ageing. https://social.un.org/ageing-working-group/documents/tenth/Social%2520protection%2520and%2520social%2520security,%2520education%2520and%2520lifelong%2520learning%2520A%2520discussion%2520paper.pdf. Accessed 7 Feb 2020

Ahmad M (2018) Cultural safety in muslim aged care: taking the bull by the horns. J Relig Spirituality Aging 30(1):25–47

Australian Department of Health (2017) Aged care diversity framework. https://www.health.gov.au/initiatives-and-programs/aged-care-diversity-framework-initiative. Accessed 14 July 2020

Australian Government (2015) Intergenerational report 2015: overview. https://static.treasury.gov.au/uploads/sites/1/2017/06/2015IGR_Overview.pdf. Accessed 6 Feb 2020

Australian Human Rights Commission (2016) A qualitative study of employment discrimination against older Australians. https://www.humanrights.gov.au/our-work/age-discrimination/publications/qualitative-study-employment-discrimination-against-older. Accessed 7 Feb 2020

Australian Human Rights Commission (2019) A human rights perspective on aged care (submission to the Royal commission into aged care quality and safety). https://agedcare.royalcommission.gov.au/system/files/2020-06/AWF.500.00267.0002.pdf. Accessed 14 July 2020

Baltes MM, Carstensen LL (1996) The process of successful ageing. Ageing Soc 16(4):397–422

Barbosa Neves B et al (2019) 'It's the worst bloody feeling in the world': experiences of loneliness and social isolation among older people living in care homes. J Aging Stud 49:74–84

Baum C (2018) The ugly truth about ageism. The guardian online (14 Sep 2018). https://www.theguardian.com/lifeandstyle/2018/sep/14/the-ugly-truth-about-ageism-its-a-prejudice-targeting-our-future-selves. Accessed 10 Feb 2020

Bersin J, Chamorro-Premuzic T (2019) The case for hiring older workers. Harvard business review. https://hbr.org/2019/09/the-case-for-hiring-older-workers. Accessed 9 Dec 2019

Biggs S (2001) Toward critical narrativity: stories of aging in contemporary social policy. J Aging Stud 15(4):303–316

Brownlee K (2013) A human right against social deprivation. Philos Q 63(251):199–222

Business and Human Rights Resource Centre (2019a) The future of work: litigating labour relationships in the gig economy. https://www.business-humanrights.org/en/the-future-of-work-litigating-labour-relationships-in-the-gig-economy. Accessed 1 Oct 2019

Business and Human Rights Resource Centre (2019b) Corporate legal accountability quarterly bulletin. https://us3.campaign-archive.com/?u=bdd1a6a40fffad39c8719632f&id=96a2a661a7. Accessed 1 Oct 2019

Butler RN (1975) Why survive? Being old in America, Harper and Row, New York

Butler RN (1989) Dispelling ageism: the cross-cutting intervention. Ann Am Acad Polit Soc Sci 503:138–147

Cain L (2007) Providing culturally appropriate and inclusive aged care. United Care Ageing

Chicago Declaration on the Rights of Older Persons (2014) Adopted at the international elder law and policy conference. John Marshall Law School, Chicago (11 July 2014)

Dionigi R (2006) Competitive sport and aging: the need for qualitative sociological research. J Aging Phys Act 14(4):365–379

Doron I, Georgantzi N (eds) (2018) Ageing. Ageism and the law, European perspectives on the rights of older persons, Edward Elgar

Dow B, Johnson M (2012) Understanding elder abuse: a social rights perspective. Int Psychogeriatr 24(6):853–855

Ellwardt L et al (2015) Personal networks and mortality risk in older adults: a twenty-year longitudinal study. PLoS ONE 10(3):30116731

Foster L (2018) Active ageing, pensions and retirement in the UK. J Popul Ageing 11(2):117–132

Frias MC (2018) 'Uber drivers are employees, court decides in Brazil'. Folha de S. Paulo, 28 Aug 2018. https://www1.folha.uol.com.br/internacional/en/business/2018/08/uber-drivers-are-employees-court-decides-in-brazil.shtml. Accessed 7 Feb 2020

Friedan B (1985) The Mystique of age. In: Butler RN, Gleason HP (eds) Productive aging: enhancing vitality in later life. Springer, New York, pp 37–45

Grover G (2011) Thematic study on the realization of the right to health of older persons by the special rapporteur on the right of everyone to the enjoyment of the highest attainable standard of physical and mental health. UN Doc A/HRC/18/37

International Expert-Conference on the Human Rights of Older Persons (2018) Vienna declaration on the human rights of older persons. https://ageing.at/site/pub_html/sites/default/files/ICHRoP%2520Conference%2520Declaration%2520Final_0.pdf. Accessed 19 Dec 2019

Karev I, Doron I (2017) The human right to leisure in old age: reinforcement of the rights of an aging population. J Aging Soc Policy 29(3):276–295
Katz S (2000) Busy bodies: activity, aging, and the management of everyday life. J Aging Stud 14(2):135–152
Kesby A (2017) Narratives of aging and the human rights of older persons. Hum Rights Rev 18(4):371–393
Khan S, Ahmad M (2014) The case for muslim aged care in the west. J Relig Spirituality Aging 26(4):281–299
Kornfeld-Matte R (2016) Report of the independent expert on the enjoyment of all human rights by older persons. UN Doc A/HRC/3344
Kornfeld-Matte R (2017) Report of the independent expert on the enjoyment of all human rights by older persons. UN Doc A/HRC/36/48
Kornfeld-Matte R (2018) Report of the independent expert on the enjoyment of all human rights by older persons. UN Doc A/HRC/39/50
Levitas R et al (2007) The multidimensional analysis of social exclusion. Department of sociology and school for social policy townsend centre for the international study of poverty, bristol institute for public affairs
Lewis B et al (2018) Submission by the Australian centre for health law research to the Australian human rights commission human rights and technology issues paper. https://eprints.qut.edu.au/122938/. Accessed 7 Feb 2020
Minkler M, Holstein MB (2008) From civil rights to … civic engagement? Concerns of two older critical gerontologists about a "new social movement" and what it portends. J Aging Stud 22(2):196–204
Nicholson NR (2012) A review of social isolation: an important but underassessed condition in older adults. J Primary Prevent 33(2):137–152
Nursing Council of New Zealand (2011) Guidelines for cultural safety, the treaty of waitangi and maori health in nursing education and practice. https://pro.healthmentoronline.com/assets/Uploads/refract/pdf/Nursing_Council_cultural-safety11.pdf. Accessed 9 Dec 2019
Office of the High Commissioner for Human Rights (2012) Normative standards in international human rights law in relation to older persons: analytical outcomes paper. https://social.un.org/ageing-working-group/documents/OHCHRAnalyticalOutcomePaperonOldePersonsAugust2012.doc. Accessed 9 Dec 2019
Open-ended Working Group on Ageing (2019a) Substantive inputs on the focus area 'education, training and life-long learning and capacity-building'. Working document submitted to department of economic and social affairs. UN Doc A/AC.278/2019/CRP.2
Open-ended Working Group on Ageing (2019b) Report of the open-ended working group on ageing on its tenth working session. UN Doc A/AC.278/2019/2
Phiri J, Dietsch E, Bonner A (2010) Cultural safety and its importance for Australian midwifery practice. Collegian 17:105–111
Rowe JW, Kahn RL (1997) Successful aging. The Gerontologist 37(4):433–440
The Benevolent Society (2017) The drivers of ageism. https://www.benevolent.org.au/ArticleDocuments/340/Ageism_SummaryReport_Final.pdf.aspx. Accessed 7 Feb 2020
Thornton M, Luker T (2010) Age discrimination in turbulent times. Griffith Law Rev 19(2):141–171
Tinker A, Ginn J (2016) An age friendly city: how far has London come? King's College, London. https://www.housinglin.org.uk/_assets/Resources/Housing/OtherOrganisation/Age-Friendly-London-Report-final.pdf. Accessed 9 December 2019
Townsend P (2007) Using human rights to defeat ageism: dealing with policy-induced 'structured dependency.' In: Bernard M, Scharf T (eds) Critical perspectives on ageing societies. Policy Press, Bristol, pp 27–44
Walsh K, Scharf T, Keating N (2017) Social exclusion of older persons: a scoping review and conceptual framework. Eur J Ageing 14(1):81–98
Weicht B (2013) The making of 'the elderly': constructing the subject of care. J Aging Stud 27:188–197

Wheaton B (2017) Surfing through the life-course: silver surfers negotiation of ageing. Ann Leisure Res 20(1):96–116
World Economic Forum (2016) The future of jobs. https://wef.ch/1RkF7w5. Accessed 7 Feb 2020
World Health Organization (2002) Active ageing: a policy framework. https://www.who.int/ageing/publications/active_ageing/en/. Accessed 7 Feb 2020
World Health Organization (2007) Global age-friendly cities: a guide. https://www.who.int/ageing/publications/Global_age_friendly_cities_Guide_English.pdf. Accessed 3 December 2019
World Health Organization (2015) World report on ageing and health. https://www.who.int/ageing/events/world-report-2015-launch/en/. Accessed 19 December 2019

International Instruments

Committee on Economic, Social and Cultural Rights (1990) *General Comment No. 3: The nature of States parties' obligations (Art 2 par 1)* (14 December 1990)
Committee on Economic, Social and Cultural Rights (1991) *General Comment No. 4: The right to adequate housing (Art 11 par 1)* (13 December 1991)
Committee on Economic, Social and Cultural Rights (1995) *General Comment No. 6: The Economic, Social and Cultural Rights of Older Persons* (8 December 1995)
Committee on Economic, Social and Cultural Rights (2000) *General Comment No. 14: The right to the highest attainable standard of health (Art 12)* UN Doc E/C.12/2000/4 (11 August 2000)
Convention on the Rights of Persons with Disabilities (2006) Opened for signature 13 December 2006, entered into force 3 May 2008. 2515 UNTS 3
Inter-American Convention on Protecting the Rights of Older Persons (2015) Opened for signature 15 June 2015, entered into force 11 January 2017. 55 ILM 985
International Covenant on Civil and Political Rights (1966) Opened for signature 16 December 1966, entered into force 23 March 1976. 999 UNTS 171
International Covenant on Economic, Social and Cultural Rights (1966) Opened for signature 16 December 1966, entered into force 3 January 1976. 993 UNTS 3
Madrid International Plan of Action on Ageing (2002) Adopted by Second World Conference on Ageing, Madrid, 8–12 April 2002
Parliamentary Assembly of the Council of Europe (2011) *Promoting Active Ageing – Capitalising on Older People"s Working Potential*. Resolution 1793 (28 January 2011)
United Nations Principles for Older Persons (1991) Adopted by United Nations General Assembly Resolution 46/91. UN Doc A/RES/46/91 (16 December 1991)
Vienna International Plan of Action on Ageing (1982) Adopted by World Assembly on Ageing, Vienna, 26 July–6 August 1982. Endorsed by UN General Assembly Resolution 37/51. UN Doc A/RES/37/51 (3 December 1982)

Cases

Aslam et al v Uber, No. 2202550/2015 (Employment Tribunals 28 October 2016)
Dynamex Operations West, Inc. v. Superior Court of Los Angeles County, No. S222732 (Supreme Court of California 30 April 2018)

Chapter 5
Ageism and Equality

Ageism is a significant barrier to older persons' enjoyment of their human rights. Ageism is socially constructed and flows from various, usually negative, assumptions about older persons' abilities, preferences, opinions, vulnerabilities and needs. Even where these attitudes are relatively benign, they can nevertheless result in discrimination and can lead to both direct and indirect human rights violations. At its worst, ageism can lead to elder abuse and be internalised by older individuals themselves, making them less likely to report human rights violations when they occur. Ageist assumptions about older people filter down into the way law and policy are framed, preventing a meaningful implementation of a human rights-based approach to elder law. While many jurisdictions have made age-based discrimination unlawful, an effective response to ageism requires a much broader engagement with the social and cultural factors which drive it, and a detailed analysis of elder law to identify its effects in their various forms.

5.1 Introduction

Discrimination against older persons—ageism—is a fundamental cause of many of the human rights violations and restrictions which older persons experience. Ageism flows from the various, usually negative, assumptions that are regularly made about older persons' abilities, preferences, opinions, vulnerabilities and needs. Sometimes these are subtle and relatively benign, but they can nevertheless result in discrimination and even elder abuse. The assumptions and judgments inherent in ageism have led to the creation of laws and practices which both engender and perpetuate human rights concerns and, as such, it is a problem which cuts across the range of elder law issues addressed in this book.

Ageism represents a major barrier to the implementation of an effective human rights-based approach to elder law. Thinking about the framework developed in

© Springer Nature Singapore Pte Ltd. 2020

B. Lewis et al., *The Human Rights of Older Persons*,
https://doi.org/10.1007/978-981-15-6735-3_5

Chap. 3, we can see how pervasive the human rights impacts of ageism can be. Age-discrimination, like all forms of discrimination, is fundamentally inconsistent with the core human rights values of equality, universality, dignity and autonomy because it is based on assumptions which devalue older individuals and diminish their agency and opportunities. Depending on how it manifests, ageism can amount to a violation of specific human rights recognised by law, including the right to freedom from discrimination and equality before the law, the right to work, to choose a place of residence and to social and cultural participation. To effectively combat ageism, laws and policies must be guided by human rights principles of non-discrimination and participation—the voice of older persons must be included in the development and implementation of laws and policies in order to genuinely understand and begin to address the repercussions of ageism. There must also be an increased focus on access to justice in order to be able to remedy the negative outcomes for those whose rights have been affected by ageist behaviours, laws, policies and/or systems.

Ageism has personal, social and structural dimensions. As individuals we acquire attitudes and beliefs about older people that can settle as stereotypes. These stereotypes are often caused or reinforced by social and cultural norms about age and ageing, as well as by how older persons are portrayed in the media and popular culture. Even where individual attitudes towards older persons may be enlightened, long-term ageist assumptions have infiltrated laws and systems such that many aspects of elder law still display and perpetuate structural ageism. Further, as ageism is socially constructed, people are often not conscious of it, and come to accept 'casual' ageism as commonplace and unproblematic. When we are attuned to it, however, we can see it in all dimensions of our society. It is present in our physical environment, for instance in the design of transportation, housing and urban spaces, and in our economies, where older people are frequently discriminated against in relation to employment, insurance and other financial services. It also plagues the cultural and social aspects of our communities, where older people's participation and contribution is often undervalued, blocked or measured in fiscal terms only.

Ageism operates in society to make acceptable treatment of older people that would not be tolerated for younger generations. Expectations are lowered about what it means to live a 'full life' as people age, particularly for those in the 'old, old' category, and this results in allowances being made for poorer treatment, restricted opportunities and social marginalisation. Instead of recognising these as human rights concerns, we view them as inevitable consequences of ageing, and as an acceptable trade-off for the security and care that we assume older people require. Further, these sorts of ideas about ageing become internalised, so that older people themselves can come to believe or at least tolerate more limited, and limiting, views about their capabilities and entitlements. This can mean that older persons are less likely to report elder abuse or other human rights violations when they occur because they are more willing to accept such treatment as being 'a part of life'.[1]

Ageism, even when seemingly innocent or incidental, can come to have significant ramifications for how we approach the economic and social challenges associated

[1] Applewhite (2016), in Lagacé et al. (2012), 337.

with the ageing population. Generalisations or assumptions about older people filter down into the way we frame policy. Consequently, ageism operates not only to limit older people's enjoyment of their individual human rights, but also as a barrier to the sort of paradigm shift towards a rights-based approach that is needed in elder law (and was introduced in Chaps. 1 and 3).

The implications of ageism for human rights are profound. Given its pervasive and variable forms, ageism impacts the enjoyment of human rights in numerous interrelated and often self-perpetuating ways. In its most egregious form, ageism leads to elder abuse. This is because ageism causes us to strip away the humanity and dignity of older people, casting them as 'the other' and opening the way for them to be victimised, exploited and abused.[2] This othering allows for older people to become objects within the various legal, health, housing and other social systems, which undermines their agency and devalues their contributions.[3]

In terms of the specific human rights guaranteed under law, detailed in Chap. 2, ageism leads to both direct and indirect human rights violations. Age-based discrimination is a direct violation of the right to equality before the law and to equal enjoyment of other rights.[4] Indirectly, ageism also leads to violations of a range of other rights by limiting older persons' freedoms with respect to employment, accommodation, health and financial services, and participation in political, social and cultural life, as explained in Chap. 4. It undermines older persons' options and quality of life by assuming that they do not want or need the same richness of life that 'younger people' enjoy.

Both the Human Rights Committee, which monitors the *International Covenant on Civil and Political Rights* (ICCPR) and the Committee on Economic, Social and Cultural Rights, which monitors the *International Covenant on Economic, Social and Cultural Rights* (ICESCR), have stressed the need for governments to address both formal and substantive discrimination.[5] They note that even implicit discrimination can have damaging impacts on the enjoyment of the full range of human rights, and explain that governments may need to take positive action to address the effects of discrimination and disadvantage. However, even in implementing such corrective measures, we must be cautious that ageist stereotypes do not lead us to presume an automatic connection between age and vulnerability requiring protection—a truly effective human rights-based approach must instead foreground the dignity and autonomy of older individuals.

As noted in the previous chapters, specific protections of older persons' human rights are uncommon in international and domestic laws. In many places, anti-discrimination legislation is the pinnacle of legal protections for older persons.[6] However, often discrimination is only unlawful in certain contexts, for example,

---

[2]Sethi et al. (2011); Pillemer et al. (2016), S200.

[3]See Footnote 1.

[4]*International Covenant on Civil and Political Rights* (1966), arts 2, 26 ('*ICCPR*').

[5]Human Rights Committee (1989), para 10; Committee on Economic, Social and Cultural Rights (2009), para 8.

[6]Mégret (2010), 6.

in relation to employment or delivery of public services. Further, despite the existence of relevant laws, age-discrimination still occurs, even in areas where it has been expressly outlawed. This raises questions about the effectiveness of complaints-based anti-discrimination models to address ageism and its human rights implications.[7]

Further, 'hidden ageism' remains pervasive. It is inherent in society's attitudes, behaviours and cultural norms, and has become built into laws and policies. Addressing ageism is therefore key to improving elder law and giving effect to a human rights-based approach. Simultaneously, a greater emphasis on the human rights of older persons may start to challenge many of our assumptions about older persons and what it is to age, and thus contribute to undoing long held ageist stereotypes.

Ultimately, as Mégret has explained, '[a]lthough the right to equality is a distinct right, the problem of discrimination against the aged is also a much broader conceptual and social problem that deserves to be analyzed from multiple angles.'[8] This chapter contributes to such an analysis by exploring the causes and effects of ageism through a human rights lens. Given that ageism cuts across all areas of elder law, the discussion here connects with the analysis presented in the other chapters. This was seen in the previous chapter, where it was explained that ageism operates as a significant barrier to older persons' participation in economic, social and cultural rights. This chapter goes further, however, to analyse the ways in which a human rights-based approach might help address ageism and lead to greater equality for older persons. It also foreshadows the need for wider engagement with the social and cultural drivers of ageism.

5.2 Definition of Ageism

The first use of the term 'ageism' is attributed to Robert Butler, a psychiatrist and researcher, who wrote extensively on the phenomenon. Butler used the term in an interview with the *Washington Post* in 1968 to explain the community resistance he had observed to a public housing project in Chevy Chase, Maryland, which would provide low-cost housing to poor older people in the Washington DC area.[9] Speaking of that experience and others subsequent, Butler explained that, 'I saw ageism manifested in a wide range of phenomena, on both individual and institutional levels—stereotypes and myths, outright disdain and dislike, simple subtle avoidance of contact, and discriminatory practices in housing, employment, and services of all kinds.'[10] In his words, '[a]geism can be seen as a systematic stereotyping of and

[7] Kornfeld-Matte (2016), para 12.

[8] Mégret (2010), 14.

[9] Bernstein (1969).

[10] Butler (1989), 139.

discrimination against people because they are old, just as racism and sexism accomplish this with skin color and gender. Old people are categorized as senile, rigid in thought and manner, old-fashioned in morality and skills.'[11]

Ageist stereotypes typically focus on older persons' physical and cognitive abilities.[12] In reviewing scholarly literature on different manifestations of ageism, Lagacé et al. identified ageism in depictions of older persons as incapable and childlike,[13] weak and unhappy,[14] depressed,[15] and self-centred.[16] These assumptions about older persons' capabilities and vulnerabilities are translated into discourse and behaviours, including laws, regulations and policies, which marginalise and stigmatise older persons, deny their agency and undervalue their contributions.[17]

Butler identified three separate, though related, dimensions of ageism. First are 'prejudicial attitudes toward the aged, toward old age, and toward the aging process, including attitudes held by the elderly themselves.' The second aspect is discriminatory behaviour against older persons in various fields, but especially in employment. Third, institutional policies and systems which, 'often without malice, perpetuate stereotypic beliefs about the elderly, reduce their opportunities for a satisfactory life and undermine their personal dignity.'[18] Significantly, Butler noted that these attitudes, behaviours and institutional norms are interrelated and mutually reinforcing. 'All three have contributed to the transformation of aging from a natural process into a social problem in which the elderly individual bears the detrimental consequences.'[19]

As alluded to above, ageist attitudes are not always explicitly negative—they can also manifest in feelings of sympathy, respect and protectiveness.[20] However, these views are still grounded in stereotypical notions of older persons' capabilities and vulnerabilities. As Harpur explains, they can operate so that 'ageists erroneously believe they are protecting the human rights of the old rather than denying rights.'[21] Whatever the motivations, ageism is fundamentally at odds with a human rights-based approach to elder law. Therefore, to address it successfully, we must first understand more about its origins, changeable nature and effects.

---

[11] Ibid., quoting Butler (1969a).

[12] Lagacé et al. (2012), 336.

[13] Levy et al. (2000), in Lagacé et al. (2012), 336.

[14] Montepare and Zebrowitz-McArthur (1988), in Lagacé et al. (2012), 336.

[15] Palmore (1999), in Lagacé et al. (2012), 336.

[16] Hummert et al. (1994), in Lagacé et al. (2012), 336.

[17] Lagacé et al. (2012), 336; Butler (1969a, b); Palmore (2001).

[18] Butler (1980).

[19] Ibid., 8; see also Iversen et al. (2009), who identify cognitive, emotional and behavioural dimensions of ageism.

[20] Harpur (2015), 1055.

[21] Ibid.

5.3 Drivers of Ageism

Butler's three dimensions of ageism demonstrate that it is a socially constructed phenomenon, stemming from long-held cultural norms and reinforced by behaviours and practices. In his early work, Butler identified that ageism 'reflects a deep seated uneasiness on the part of the young and middle-aged—a personal revulsion to and distaste for growing old, disease, disability; and fear of powerlessness, 'uselessness,' and death.'[22] It is a response to culturally-ingrained fears of frailty and death, and a tendency to associate youth with beauty, strength and independence.[23] This translates into a privileging of youth (sometimes called 'youthism') and a 'desire to expel the aged "other".'[24]

These cultural norms and associated attitudes towards older persons have links to a traditional 'medicalisation' of age and perspectives within the medical field in the twentieth century have contributed to a structural ageism which persists today.[25] Butler recounted the ageism that he observed in medical school in the 1960s, noting that 'on the whole, physicians do not invest the same amount of time in dealing with elderly patients as they do in their younger counterparts. Doctors question why they should even bother treating certain problems of the aged; after all, the patients are old. Is it worth treating them? Their problems are irreversible, unexciting, and unprofitable.'[26]

Kesby has explained that this medicalisation has continued as one of the dominant narratives of ageing seen in representations in the mainstream media, government and intergovernmental reports and policies, and even in scholarly literature. She refers to this as the 'narrative of decline, dependency and pathology,' which foregrounds the physical decline of the ageing body thus leading to a construction of older persons as being in need of care.[27] Such stereotypes feed into and reinforce a related narrative of ageing, which Kesby calls the 'narrative of crisis'. This narrative focuses on demographic change and the increasing proportion of older persons within the population. The data pointing to the ageing population are interpreted through a neo-liberal lens which emphasises the cost of such a shift, drawing on assumptions about older persons as welfare beneficiaries who are dependent on support, and ultimately perpetuates the stereotype of their being a burden.[28]

Other scholars, along with Kesby, have highlighted the links between increasing ageism and neo-liberalism. Mégret, Townsend, Thornton and Luker, and Lagacé et al. have all pointed to the way that older people are considered to be of less value to

---

[22]Butler (1969b), 243; Baum (2018); Lagacé et al. (2012).

[23]Greenberg et al. (2006).

[24]Macnicol (2006), 11; Thornton and Luker (2010), 143; Kesby (2017), 371; Butler (1989).

[25]Ng et al. (2015).

[26]Butler (1989), 140. The notion that older people may not be worth the cost of care can be seen in the rationing of care which is discussed in Chap. 10 in relation to health and aged care.

[27]Kesby (2017), 377.

[28]Ibid.

the community as they make less of an economic contribution.[29] As Mégret says, 'ageism may be accentuated by societies' prevalent individualism and consumerism, and a tendency to discount the value that its older members can make to society given their perceived non-productive status.'[30]

The influence of neo-liberalism on contemporary ageism can be explained by Wolfensberger's 'social valorisation theory', according to which the value attributed to a person's or group's social role shapes how they are treated by others.[31] The emphasis within Western, neo-liberal societies on individualism, consumerism and economic productivity both engenders and reinforces ageist stereotypes.[32] The value of older people is measured according to their ability to contribute to economic prosperity, while at the same time assuming that, due to their age, they are unable to make such a contribution. Instead, commercial success and productivity are equated with youth, and older persons are viewed as an economic burden because they are presumed to be dependent on care and support. Consequently, older persons are devalued and their status as 'other' is further entrenched.

The role of neo-liberal market domination contributes to many of the human rights issues that are discussed in this book, most notably in relation to the provision of health and aged care (discussed in Chap. 10) as well as housing (Chap. 9), where privatisation and protection of commercial interests have hindered implementation of rights-based models. These same influences can be seen in the design of pension schemes (Chap. 8) where the economic viability and sustainability of non-contributory pension schemes can be questioned by policy-makers.

Another factor which may be perpetuating ageist stereotypes is the rapid rise of new technologies, and the consequential devaluing of 'older' ways of knowing and doing. Mégret refers to the 'difficulty of some among the senior population to keep up with changes (for example, the increasing digital divide between generations)',[33] though even this notion needs to be treated with circumspection to avoid assumptions about older persons' technological capabilities. However, Mégret is correct to point to the negative consequences that can flow if 'elderlies' know-how' is dismissed as outdated, particularly when such knowledge had, in the past, provided older persons with a 'strong sense of social relevance'.[34] There is also a loss to the whole of society if older people's knowledge and experience is dismissed or ignored. The impact of the emerging digital economy on ageism, particularly in the workforce, will be explored in more detail below.

An important factor in understanding the prevalence and longevity of ageist stereotypes is that, because we are socially habituated to ageist views, ageism is internalised by older people themselves. Older persons' own ideas about what they are entitled to expect in terms of participation, choices and treatment have become aligned with

[29] Mégret (2011); Townsend (2007); Thornton and Luker (2010); Lagacé et al. (2012), 336.

[30] Mégret (2010), 14, citing Townsend (1986).

[31] Wolfensberger (2000); Kesby (2017), 381.

[32] Lagacé et al. (2012), 340.

[33] Mégret (2010), 14; Kiel (2005).

[34] See Footnote 33.

ageist ideas. Research demonstrates that where negative perceptions about older people prevail, older people themselves behave in ways that conform to those stereotypes.[35] Lagacé et al. describe this process of internalization as '[o]ne of the most insidious and subtle impacts of negative stereotypes.'[36]

The potential for ageism to intersect with other forms of discrimination is significant. Butler's early work described the intersectionality of race and age discrimination that he observed in public housing debates in the United States of America in the 1960s.[37] The aggravating effect of age- and race-based discrimination combining persists for many older persons today, and can be particularly profound for Indigenous peoples or people of culturally and linguistically diverse backgrounds.[38]

The compounding effect of age and gender has been recognised as a significant barrier to older women's full enjoyment of their human rights. Systemic and structural gender-based discrimination which affects women during their lives frequently operates to leave them vulnerable to human rights abuses later in life. The Committee on Economic, Social and Cultural Rights has explained that older women, 'because they have spent all or part of their lives caring for their families without engaging in a remunerated activity entitling them to an old-age pension, and who are also not entitled to a widow's pension, are often in critical situations.'[39] Women who find themselves out of work at an older age face considerable barriers obtaining employment, but are simultaneously less financially secure due to structural factors which have affected their earning capacity in earlier life (discussed further in Chap. 8). As these circumstances compound, women become vulnerable to a range of human rights abuses, and data indicate a troubling prevalence of poverty and homelessness among older women.[40]

The intersectionality of different forms of discrimination must be borne in mind in the search for legal responses to ageism. Age discrimination will be experienced differently by different people, and nuanced approaches are therefore required. As well as understanding the drivers of ageism, it is necessary to appreciate the common areas in which it manifests and the various ways it impacts on human rights. The next section identifies some of the areas in which ageism commonly manifests, noting the impacts it has on human rights.

---

[35] Applewhite (2017).

[36] Lagacé et al. (2012), 337.

[37] Butler (1969a, b).

[38] Australian Human Rights Commission (2016a), 72–74; Kesby (2017), 385.

[39] Committee on Economic, Social and Cultural Rights (1995), para. 20; see also Committee on the Elimination of All Forms of Discrimination Against Women (2010), paras 11, 13; Kesby (2017), 385.

[40] Australian Human Rights Commission (2015, 2019).

5.4 Impact of Ageism on Human Rights

As will be discussed below, anti-discrimination laws, including those prohibiting age discrimination, typically focus on discrimination in certain contexts, most notably in employment, education, accommodation or the delivery of other public services. Ageism, particularly cultural or 'casual' ageism, is much more widespread than these narrow contexts defined by legislation however, and older people report regularly experiencing ageism on public transport, on social media, or while in public spaces.[41] The following sections examine selected areas where ageism is prevalent and consider the legal responses which have been, or should be, introduced in response. Some of these areas are the subject of more developed fields of elder law and are consequently the subject of later chapters.

5.4.1 Employment

One of the most commonly reported spheres in which ageism occurs is the workforce. Despite the fact that age-discrimination in employment is widely prohibited, numerous studies demonstrate that it still occurs across many societies, though some of its manifestations have diminished over time while other new forms and causes have emerged.[42] It represents a clear violation of the right to work, which encompasses rights to fair remuneration and conditions, and equal opportunities for training, promotion and other recognition.[43] It is also obviously inconsistent with the right to be free from discrimination and undermines fundamental human rights values of equality and respect for dignity.

It is worth noting that age discrimination in the workforce can occur for both young and old. People as young as forty have experienced discriminatory treatment where potential employers assume they are 'too old' for a particular position.[44] Age discrimination also works in the opposite direction where employers make assumptions based on a person's youth, for example determining that they would be unreliable, irresponsible or disloyal.[45] However, studies indicate that age discrimination is most commonly experienced by people aged fifty-sixty years.[46] These people would

---

[41] SunLife (2019); Hill (2019).

[42] Australian Human Rights Commission (2016a, b); Macnicol (2006); The Benevolent Society (2017).

[43] *International Covenant on Economic, Social and Cultural Rights*, (1966), art 6.

[44] Thornton and Luker (2010), 153, discuss a number of complaints against the Australian Defence Force where people aged as young as thirty-five were successful in establishing unlawful age-discrimination against them in securing employment or promotion eg *Commonwealth of Australia v Human Rights and Equal Opportunity Commission* (1998) 158 ALR 468. See also, cases of Bradley, Barty, Peterson and Van Den Heuvel reported in Human Rights and Equal Opportunity Commission (2000).

[45] Snape and Redman (2003).

[46] Thornton and Luker (2010), 149; Glover and Branine (2001); Australian Human Rights Commission (2016a).

typically fall outside the definition of 'older persons' which, as discussed in Chap. 1, is normally understood to be people over the age of sixty or sixty-five. The realities of ageism in the workplace highlight the problem of defining 'older persons' too prescriptively. Definitional issues aside, the degree and frequency of age discrimination in the workplace typically worsen for older cohorts so that people who fall within the classical definition of 'older persons' are more likely to experience the negative human rights impacts of ageism in relation to employment.[47]

Ageism can occur across the full life cycle of employment feeding into the heightened risk of experiencing financial insecurity post-paid employment (discussed further in Chap. 8). It can begin in relation to education, training and job-seeker support services, which are frequently tailored to younger users and may exclude or ignore the particular needs of older persons. This is especially problematic for older persons wishing to upskill to find work in emerging industries or professions, or in relation to technological advances. Older workers can be further disadvantaged by ageist assumptions and practices present in recruitment processes including, for example, in the way that job advertisements and selection criteria are drafted.[48] Interview panels may also have ageist pre-conceptions or unconscious biases that privilege younger applicants. Studies have also demonstrated that older workers frequently experience discrimination in accessing professional development or other workplace training, and in securing promotion or wage increases.[49]

Ageism is frequently reported in relation to retirement and redundancy. While mandatory retirement ages have gradually been eradicated, in some parts of the world it is still lawful for employers to stipulate a mandatory retirement age in employment contracts.[50] This is usually justified on the grounds of ensuring workforce renewal and freeing up employment opportunities for younger generations, but it remains problematic from a human rights perspective.[51] In the absence of individualised assessments of a person's capabilities, forced retirement rests on ageist generalisations about older people. In some professions there may be stronger arguments for mandatory retirement than in others, but making assumptions about a person's capabilities based on their age is inherently ageist, and justifications for mandatory or contracted retirement ages therefore need careful scrutiny.[52]

Even without contractual terms, older persons may still feel pressure to take up redundancy processes when they are offered. While some older persons no doubt welcome retirement, where it is coerced or imposed against a person's choice it

[47] Glover and Branine (2001), 5; Thornton and Luker (2010), 142.

[48] Bennington (2001), 128; Thornton and Luker (2010), 161.

[49] Kornfeld-Matte (2016), para 7.

[50] For instance, both Oxford and Cambridge Universities mandate an 'Employer Justified Retirement Age' which applies to staff over a specified employment level, see https://hr.admin.ox.ac.uk/the-ejra and https://www.hr.admin.cam.ac.uk/policies-procedures/1-retirement-policy/2-statement-policy. See also Blackham (2018).

[51] Kornfeld-Matte (2016), para 16.

[52] Mégret (2010), 16. For example, in Australia, the Constitution (1900, s 72) mandates that all federal judges must retire at age seventy and Australian Defence Force personnel must generally retire at age sixty.

represents a violation of their agency and dignity. Retirement can be a traumatic experience, and is sometimes linked to feelings of uselessness and other impacts on mental health.[53] Retirement can also have a disproportionately negative impact on the poor, especially women, who may lack the financial security post-paid employment to ensure their human rights during a longer retirement period. Once an older person is made redundant, the ageist practices mentioned above will make finding alternative employment difficult, and studies demonstrate that the longer a person has been out of work the more difficult it can become to secure appropriate employment.[54]

This demonstrates the pervasive and reinforcing effect that ageism has in the workforce, the consequences of which are of serious concern for human rights. Not only is ageism a breach of the right to be free from discrimination and the right to work, but it also has implications for a range of other human rights. A person who finds themselves out of work as a result of age discrimination is likely to experience financial insecurity, poverty and even homelessness—placing a wider range of human rights at risk, most notably healthcare (discussed in Chaps. 8 and 10).

Thornton and Luker have considered the impact on older workers of the transformation to the knowledge or digital economy.[55] Many older professions are transforming or disappearing with the introduction of robotics, artificial intelligence and machine learning, and new jobs are emerging. As discussed in Chap. 4, the nature of work in many sectors is also transforming away from traditional 9–5, permanent positions and towards short-term, casual and freelance work—commonly known as the 'gig economy'. It is widely stated that many young people today will end up working in positions and fields that do not yet exist, such is the rate of change that is occurring.[56] For various reasons, older persons may be excluded from this transformation, with potentially significant impacts on their human rights.

Thornton and Luker have identified cultural dimensions of the shift to the digital economy which have particular implications for older workers, and which may engender and/or exacerbate ageist attitudes. They point to a culture of 'youthism', 'characterised by technological know-how, flexibility and choice' in which 'traditional values such as maturity, experience and loyalty have become passé.'[57] Such a shift may privilege younger over older workers, though the idea that older workers bring 'traditional values' may itself have ageist connotations worthy of analysis. The gig economy may also be more suited to younger workers, who are likely to have fewer financial liabilities or family responsibilities, and be less attractive to older workers seeking more stable, reliable, long-term work, especially as they may be increasingly considering life post-paid employment.

New technologies can also promote ageism in other ways. For instance, research indicates that artificial intelligence is being used to target job advertisements to

[53]Mégret (2010), 16.

[54]Ranzijn et al. (2006) in Thornton and Luker (2010), 65.

[55]Thornton and Luker (2010).

[56]World Economic Forum (2016).

[57]Thornton and Luker (2010), 141.

younger rather than older candidates.[58] This indicates either ageism on the part of employers in deciding who they think is suited for the job, or ageist biases built into the algorithms (something that occurs with both race and gender[59]), or both. The existence of ageist biases in algorithm design is likely to be a result of ageism within the technology sector itself, since biases are frequently attributed to a lack of diversity among those working to design them.[60]

The neo-liberal attitudes which appear to privilege younger workers need to be challenged. Research has shown that assumptions about older workers' inability to adapt to new technologies are unfounded, and that they can make a valuable contribution to new industries, especially the creative industries.[61] Even if we accept the neo-liberal calculus wherein economic contribution is a means of determining the 'value' of older workers, the evidence confirms that there is no basis for assuming that older workers are less capable, or that they are not suited to newer forms of work.

5.4.2 *Housing*

As will be discussed in Chap. 9, the availability of suitable accommodation in older age is essential to the enjoyment of a range of human rights, as well as being a right in itself.[62] Inappropriate housing can undermine the dignity and autonomy of the individual, and create barriers for social inclusion. Yet many societies struggle to ensure that older persons can access appropriate, affordable housing and other associated services.[63] There is a strong correlation between inequality, poverty and housing deficits, and in many societies existing vulnerabilities are compounded by the unavailability of adequate housing for older persons. Ageism is a key reason underpinning the struggles to provide appropriate housing for older people, and the failure to address associated human rights concerns.[64]

To understand the scale of the problem of ageism in relation to older people's housing, we must appreciate the centrality of home to an individual's sense of identity, independence, and emotional and physical well-being. As Adams has said: 'the quality of a home shapes the quality of life … Quality and suitability of place are key determinants of the experience of growing older.'[65] Housing is also closely linked to general well-being. It goes beyond the physical need for shelter, and is fundamental to a range of other rights, including the right to health, freedom to choose a place of

58 Australian Human Rights Commission (2018).

59 Lambrecht and Tucker (2019); Turner Lee (2018); Danks and London (2017).

60 Turner Lee (2018).

61 Baum (2018).

62 ICESCR, art 11.

63 See, for example: Australia's Royal Commission inquiry into Aged Care Quality and Safety (2018).

64 Adams (2009), 77.

65 Ibid.

residence, and the right to participate in social and cultural life. Moreover, a person's ability to choose where and how they live is important for dignity and autonomy.

Conversely, poor quality or unsuitable housing presents numerous risks to human rights, and this has been acknowledged by international human rights bodies.[66] For example, inappropriate housing increases the risk of falls, which are one of the major causes of death and health decline among older people. Improperly insulated or climate-controlled housing also increases health risks associated with exposure to extremes of heat and cold—an increasing concern in the context of climate change.[67] Modifications to housing may be required to address these risks, presenting the question of how these works are to be funded when older individuals may lack the means themselves.

Ageism has had a profound impact on the way we approach housing for older persons, with potentially significant human rights implications. It affects the way we conceive of housing options and living arrangements for older people, the styles of housing that are supported by public policies and funding, and the way that specialised housing and neighbourhoods are designed. To begin with, ageist assumptions have narrowed our ideas about what constitutes appropriate housing for older people, leading to a narrowing of available offerings. As discussed in Chap. 9, a recent trend has emerged of favouring 'ageing in place', in part as a means of delaying the social and financial cost of supporting older persons to move into higher-care facilities. Despite this growing emphasis, thinking of housing for older persons typically conjures images of retirement villages and nursing homes as the paradigms of accommodation options in older age.[68] This focus is based on ageist assumptions about the kind of lives that older people want to lead and the level of support they require, and gives little regard to individual preferences. Fisk has argued that the design of specialist accommodation for older people is 'the physical expression of a benevolent paternalism.'[69] Hanson et al. have similarly characterised it as conforming to an 'alms house' stereotype, conjuring notions of dependence and charity which undermine older persons' autonomy and dignity.[70] Research conducted by the Housing and Older People Development Group, based in the United Kingdom, explains the ramifications of such ageist assumptions: 'the options offered to older people are not what they want. Those involved in planning or providing house are making false assumptions about older people, their needs, and the way they choose to live their lives. The result? Older people are devalued, ignored, dismissed or badly served.'[71]

Ageism has also influenced the physical design of specialist housing for older people, where spaces are typically small, formulaic, medicalised or functional.[72] In response to a survey on accommodation for older people conducted in the United

[66]See, for example: Committee on Economic, Social and Cultural Rights (2000).

[67]Adams (2009), 91; Smith et al. (2014) in Field et al. (2014), 709–754.

[68]Neuberger (2009); Adams (2009), 91.

[69]Fisk (1999), cited in Adams (2009), 87.

[70]Hanson et al. (2001), cited in Adams (2009), 87.

[71]Harding (2006), 3.

[72]Adams (2009), 88.

Kingdom in the early 2000s, one interviewee explained: 'Because you don't cook anymore, do you, and you never entertain! So what do you want a kitchen for?'[73] Space matters in design for both emotional and practical reasons. As Adams explains, 'the acquisition of possessions over a lifetime means that the availability of space for furniture, storage of valued items and space for hobbies and activities are important in designing housing for older age.'[74] Adams further notes that this has been a feature of social housing generally, where space is limited and designs basic. Consequently, social housing often leaves limited room for modifications as people age, creating barriers for older people who find themselves in housing which is no longer adequate for their needs.[75] The same applies to private rentals, which are often unsuitable for ageing tenants. As Power has said, '[i]t's rare to find a rental property you can age in.'[76] She has stressed the need for regulations which can require or encourage landlords to support mobility-oriented modifications so that older people can remain in rented accommodation longer.[77] This is particularly important given the added emotional and financial stress that comes with finding new rental accommodation if older persons are required to move.

As noted, ageism has restricted housing options to those which fit the model of the residential aged care and the archetypal nursing home. Even the 'resort style' retirement accommodation discussed in Chap. 9 conforms to particular perceived ideals of what older people want in terms of housing and community living. In addition to the design deficits discussed above, this sector rightly deserves attention and requires reform for various reasons discussed further in Chap. 9. However, the vast majority of older persons worldwide reside in private accommodation—either with family or independently—and numerous reports have confirmed that most older people wish to remain at home as long as possible.[78] Consequently, far more attention must be paid to the way their human rights can be protected in this context.

Despite the growing emphasis on ageing in place and the fact that most older people wish to remain in their own homes, and despite the increasing proportion of older people within our societies, housing policies and strategies have typically focussed on supporting home ownership for younger people, especially young families as a central objective. This has left older people's needs relatively obscured. In Western societies it is commonly assumed that older people will own their own home outright. This in itself is an ageist assumption which ignores the reality of many older persons' lives. In fact, an increasing number of older people rent or rely on social housing. Ageism affects the reliability and suitability of these accommodation options as well. Older renters, particularly those relying on government pensions, may find it difficult to secure properties because landlords prefer to rent to younger

[73] Mrs. B, Research participant in EPSRC EQUAL study 2001, cited in Adams (2009), 88.

[74] Adams (2009), 88.

[75] Ibid., 87.

[76] As quoted in Stewart and Stein (2019); Power (2016).

[77] See also Stewart and Stein (2019); Power (2016).

[78] Age UK (2019); Administration for Community Living (2017), 14.

tenants, partly because they may be able to offer higher rents, but sometimes for purely ageist ideas about the 'ideal' tenant.[79]

Enabling older people to remain living at home can help to maintain their social connections and engagement with society, with positive impacts on their well-being. However, we must be cautious not to make ageist assumptions about older people's social engagement. The level of participation in society is not simply a factor of older people's individual mobility, capacity or preference—it is in large part a function of the way our neighbourhoods are designed. Thoroughly addressing the impacts of ageism requires that we look beyond just the individual living arrangements of older people to consider how we design our neighbourhoods and communities to ensure they are not excluded and can continue to lead a full life of their choosing.

In recent years we have seen some new and innovative ideas emerging in the design of aged accommodation and communities. For example, the '8 80 Cities' initiative (mentioned in the previous chapter) aims to change the way cities and public spaces are designed to ensure they are accessible and enriching for people of all ages, with a particular focus on physical spaces and removing barriers to mobility.[80] Throughout Europe, particularly in Belgium and the Netherlands, communities are working collaboratively to design more dementia-friendly accommodation, community spaces and programs, recognising that a whole-of-community approach can deliver profound outcomes in advancing the human rights of people with dementia. These human rights-based approaches to accommodation and aged care will be explored in more detail in Chaps. 9 and 10 respectively. Ensuring suitable housing is therefore fundamental to human rights but requires that we address ingrained ageism and move towards more inclusive and creative planning and design, funding models, and provision of services.

5.4.3 Healthcare

As noted above, one of the drivers of ageism can be linked to the medicalisation of old age, and feelings of anxiety about future physical and cognitive decline. Studies have shown that even healthcare professionals are not immune to these ageist attitudes. Early scholars, like Butler, identified ageist attitudes present within the medical profession, and recent studies have shown that ageism is alive and well within the healthcare sector, both at a personal level, where it manifests in provider-patient communication and clinical practice, as well as at a structural level, where it can be seen in funding arrangements and institutional policies.[81] Chapter 10 explores the need for a human rights-based approach to healthcare in more detail, but some

---

[79] Stewart and Stein (2019).

[80] 8 80 Cities (n.d.); see also Tinker and Ginn (2016); World Health Organization, Global Network for Age-Friendly Cities and Communities; Kornfeld-Matte (2016), para 76.

[81] Ambady et al. (2002); Caris-Verhallen et al. (1999); Adelman et al. (2000); Wyman et al. (2018), 202.

brief examples are given below to illustrate the challenge of addressing ageism in healthcare settings.

Wyman et al. have hypothesised that our tendency to want to 'deny and distance ourselves from the negative aspects of old age' may be intensified for healthcare professionals, who are exposed to illness, senility and death among older patients on a regular basis.[82] Numerous studies have identified negative age-based attitudes among different healthcare professionals, and these can manifest in poorer communication with older patients. For example, communication may be patronising or disrespectful, or may be less optimistic than equivalent discussions with younger patients.[83] Studies have shown that healthcare professionals can be less inclusive of older patients' views about their diagnosis and treatment, and may 'bypass' the patient and go directly to family members, or make clinical decisions without any input at all.[84] Ben-Harush et al. explored the reasons for excluding older patients from these discussions, and found that some professionals found it easier or simpler not to include the patients, or found it difficult to relate to them.[85] These explanations indicate underlying ageist assumptions about older patients' ability to understand information about their health and/or their preferences in terms of treatment options. These communication issues go beyond simply 'poor bedside manner' and can have real negative consequences for older patients' health, both in terms of the impact on clinical decisions, and the emotional impact of dismissive or disrespectful interactions.[86]

Discrepancies are also evident in terms of diagnostic procedures and treatment options for older patients, pointing to the influence of both personal attitudes among healthcare professionals and institutional policies which entrench ageist assumptions. For example, studies have revealed differential screening and treatment of patients with breast cancer,[87] lung cancer,[88] and heart disease[89] attributable to purely age-based factors rather than other clinical indicators.[90] Structural factors also contribute to lower rates of access to necessary medical care for older patients. These issues include a lack of geriatric-trained professionals, a shortage of preventive or rehabilitative-focused care options for older patients, and geographic or transportation barriers for older patients to access services.[91]

Healthcare funding and reimbursement models also frequently affect older persons in a disproportionately negative way, given that medical costs typically increase with old age, but financial resources become more constrained.[92] In some

[82] Wyman et al. (2018), 195.

[83] Ambady et al. (2002); Caris-Verhallen et al. (1999); Adelman et al. (2000).

[84] Ben-Harush et al. (2017); Caris-Verhallen et al. (1999).

[85] Ben-Harush et al. (2017).

[86] Nussbaum et al. (2005).

[87] Haigney et al. (1997); Madan et al. (2001).

[88] Peake et al. (2003).

[89] Wenger (1997).

[90] Wyman et al. (2018), 196.

[91] Ibid., 202.

[92] Ibid.

systems, availability of certain screening programs, treatment options or clinical trials is cut-off beyond a certain age as funding models struggle to cope with ageing populations.[93] These policies are based on ageist assumptions about older persons' treatment needs and outdated ideas about what constitutes good clinical care for older patients. They demonstrate that personal beliefs and attitudes have become formalised in healthcare policies and clinical practice, making it more challenging to combat ageism in the healthcare sector.

The existence of ageism in the healthcare sector presents numerous concerns from a human rights perspective. Most notably, it has potential to interfere directly with the right to the highest attainable standard of health.[94] As introduced in Chap. 2 and further explained in Chap. 10, this entails the right to timely and appropriate healthcare, access to health-related information, and the protection of the underlying determinants of health.[95] Further, healthcare services need to be provided in a manner which is sensitive to culture, gender and individual preferences. To fully address ageism in healthcare, a human rights-based approach needs to be adopted which embeds these principles and values.

5.5 Countering Ageism

As has been shown, ageism is widespread and has significant ramifications for the enjoyment of human rights. Given that ageism is a socially constructed phenomenon, steps need to be taken to shift attitudes and combat negative stereotypes if policy reforms are to endure. Legal measures are also essential to prohibit discrimination and protect the rights of older persons in areas where ageism typically manifests. This section will therefore consider a number of the steps which can, and should, be taken to address ageism as part of embedding a human rights-based approach to elder law, beginning with a discussion of relevant international and domestic legal frameworks.

5.5.1 Legal Protections Against Ageism

As noted in Chap. 2, the major human rights treaties do not mention age as a specific head of prohibited discrimination, though most include some 'catch-all' language, for example, by prohibiting discrimination on grounds of 'other status', which arguably brings ageism within the scope of anti-discrimination provisions. The omission of age as a specific ground of discrimination has been explained by the Committee on Economic, Social and Cultural Rights: 'rather than being seen as an intentional

---

[93] Ibid.

[94] ICESCR, art 12.

[95] Committee on Economic, Social and Cultural Rights (2000), para. 11.

exclusion, this omission is probably best explained by the fact that, when these instruments were adopted, the problem of demographic ageing was not as evident or as pressing as it is now.'[96]

Other international instruments have extended protection against discrimination specifically on the basis of age. In Europe, the *European Communities Treaty* of 1997 had empowered the European institutions to 'take appropriate action to combat discrimination based on sex, racial or ethnic origin, religion or belief, disability, age or sexual orientation.'[97] The European Union's *Charter of Fundamental Rights* goes a step further and demands from member states that '[a]ny discrimination based on any ground such as sex, race, colour, ethnic or social origin, genetic features, language, religion or belief, political or any other opinion, membership of a national minority, property, birth, disability, age or sexual orientation shall be prohibited.'[98] This is arguably the most comprehensive list of heads of prohibited discrimination, although interestingly does not include a 'catch-all' term to cover any bases not explicitly mentioned. The *Convention on the Rights of Persons with Disabilities* (CRPD) recognises the potential for intersectional discrimination and obliges governments to take measures 'to combat stereotypes, prejudices and harmful practices relating to persons with disabilities, including those based on sex and age, in all areas of life.'[99]

There are numerous pieces of domestic legislation which prohibit discrimination on the basis of age though, as stated above, these are frequently restricted to certain spheres of activity, such as work, education, accommodation, access to goods and services and so on.[100] The general approach amongst these legal regimes is that differential treatment with respect to older persons can be justified as long as it is proportionate and rationally connected to a legitimate policy goal. This means that positive discrimination or affirmative action which supports older persons is usually accepted. Age-based rules relating to social security or self-funded/contributory retirement schemes, for example Australia's superannuation system, can also be justified where they promote efficiency and fairness within the system and are in the best interests of the community as a whole. However, as will be discussed in more detail in later parts of this book, rules such as these must still adhere to basic human rights standards for each individual affected, and should avoid ageist assumptions about people's capabilities and preferences.

Mégret identifies a paradox in legislating a particular category of age-based discrimination—'the very creation of a category of elderly rights might reinforce

---

[96]Committee on Economic, Social and Cultural Rights (2009), para 11; Mégret (2010), 15.

[97]*Treaty of Amsterdam* (1997), art 6(1).

[98]Charter of Fundamental Rights of the European Union (2007), art 21.

[99]*Convention on the Rights of Persons with Disabilities* (2006), art 8(1)(b).

[100]See, for example: Age Discrimination Act 2004 (Cth); *Equality Act* (2010) (UK); *Age Discrimination Act of 1975* 42 USC Sects. 6101–6107; *Age Discrimination in Employment* USC Title 29 Chap. 14 Sections 621–634; *Canadian Charter of Rights and Freedoms* (1982), art 15(1). Specific age-discrimination legislation is also in place in many sub-national jurisdictions in Canada, the United States of America and Australia.

some of the problems it is supposed to alleviate.'[101] A similar concern was raised by some disability advocates during the negotiation of the CRPD, who were concerned that the adoption of a specific treaty might reinforce stereotypical assumptions about persons with disabilities, and arguments along these lines have been made in relation to a proposed convention on the rights of older persons (CROP).[102] On balance though, it was concluded that a specific treaty was necessary to articulate the obligations of governments with respect to the human rights of persons with disabilities.[103] Given the prevalence of age-based discrimination in certain socio-economic settings, and the pervasiveness of ageism in society generally, recognition of age as a specific head of prohibited discrimination is appropriate. Further, the damaging impacts of ageism for other human rights justify the adoption of a dedicated human rights instrument, as has been argued throughout this book.

Most anti-discrimination legislation adopts a complaints-based model where the individual bears the burden of bringing the complaint, progressing the claim, and proving that they have suffered harm which is caused by a prohibited form of discrimination. This is an important part of guaranteeing access to justice for people affected by discrimination, and recognises the obligation of governments to provide remedies for human rights violations.[104] These sorts of models do not necessitate that public or private entities take positive steps to address the causes of discrimination, however. Thornton and Luker argue that '[i]n the absence of significant damages awards, incentives or penalties, the present legislative schema is unlikely to do very much to effect social change.'[105]

Consequently, these claims-based models are not effective in preventing discrimination. They can result in a remedy for those individuals who choose to pursue a complaint, but often people who have experienced discrimination choose not to take formal legal action or, if they commence an action, may decide not to see it through, often due to the financial and emotional costs involved.[106] In relation to employment, Thornton and Luker further argue that the emphasis on downsizing and restructuring which is part of the neo-liberal workplace and the shift to the digital economy is likely to allow employers to rationalise actions which negatively affect older workers, including dismissal or forced redundancy.[107] In such cases, it is possible that an affected individual will be unsuccessful in proving their claim in the absence of clear evidence of direct discrimination on the basis of their age.

Broader measures are therefore needed to address both direct and indirect ageism and to properly embed the human rights principles of equality and non-discrimination. The United Nation's Independent Expert on the Enjoyment of All

---

[101] Mégret (2010), 15.

[102] See discussion in Chap. 1. See also Williams (2003); Avers et al. (2011), 153–154; Greengross (2019).

[103] Mégret (2010), 15.

[104] *ICCPR*, art 2.

[105] Thornton and Luker (2010), 151.

[106] Ibid.

[107] Ibid.

Human Rights by Older Persons, Rosa Kornfeld-Matte, has suggested that, where they do not already exist, equality bodies, such as dedicated ombudspersons or commissioners, should be established to monitor discrimination and advocate for its elimination across all sectors. These bodies should be empowered with the ability to hear complaints to ensure enforcement, but their responsibilities would go beyond individualised matters to address systemic and cultural factors.[108] While protections against discrimination are essential, they are not sufficient for ensuring that all older persons are able to enjoy their human rights without discrimination and on an equal footing with the rest of the population. Given the potential for ageism to cause or perpetuate a range of human rights interferences, anti-discrimination legislation ought therefore comprise just one pillar of a comprehensive structure of laws and policies aimed at achieving substantive equality for older persons.

5.5.2 *Culture Shift*

Major shifts are clearly required in the way that society thinks about ageing: recognition is needed that ageing is an ongoing process that is universally experienced throughout the life course; it is not a binary division between young and old which renders 'old' people less capable or valuable once they cross a certain pre-defined, often chronological, threshold. Applewhite argues that even the act of defining 'older people' as a group can propagate or perpetuate ageist attitudes, as it defies the reality of age as a spectrum.[109] While some categorisation of 'older people' is useful to enable meaningful discussion of the issues which confront them and a thorough analysis of elder law (discussed in Chap. 1), recognising ageing as a continuum that we are all a part of can help to combat stereotypes about older people and to create a more positive and more accurate narrative about what it is to be older.

The fundamental argument of this book is that human rights can lead the way in achieving this paradigm shift. Through their foundation in respect for individual dignity and autonomy, their insistence on equality and non-discrimination, and their emphasis on empowering the enjoyment of a wide variety of interconnected and indivisible rights, human rights-based approaches to ageing and elder law offer the potential to radically shift expectations and redefine responsibilities. Not only is ageism antithetical to human rights—human rights can form the basis of the antidote to ageism and elder abuse.

Attempts to combat ageism can be found in narratives of 'successful ageing', 'productive ageing' and 'active ageing' (discussed further in Chap. 4).[110] Kesby has identified these as a counter to the narratives of crisis and dependency (noted above) which have come to shape so many of our responses to the demographics of

---

[108] Kornfeld-Matte (2016), para 12.

[109] Applewhite (2017).

[110] Kesby (2017), 379.

ageing.[111] They aim to highlight the abilities and independence of older people and promote greater integration and active engagement. However, as Kesby argues, these concepts themselves can be susceptible to ageist manipulation. As was explained in Chap. 4, we commonly define 'successful ageing' by the contribution an older person makes to the community, usually in terms of either paid or unpaid work. Even 'active ageing', which is often defined to encompass activity across a range of spheres including economic, social, cultural and civic affairs, often emphasises productivity or economic contribution. This can have a positive impact in terms of countering ageism (particularly within the workforce) but it risks reinforcing the notion that an older person's value is attached to their productivity, with obvious negative implications for those older persons who are unwell or have a disability, and are therefore 'unproductive'.[112] Productivity is aligned to neo-liberal values of economic growth, and 'successful' ageing is achieved where older persons contribute to defraying the economic 'burden' of the ageing population. Those older persons who are 'unproductive' are easily undervalued, marginalised or made invisible.

While programs that fit within the 'active ageing' or 'successful ageing' paradigm can generate multiple benefits for older people, they might offer little to those older individuals who do not fit the ideal of a 'successful' older person. Further, they create the risk that, rather than truly combatting age discrimination, they create a new morph of ageism, where some older individuals are 'excused' from the usual negative assessments because they have proved to be 'productive'. Those who fall outside this ideal of productive, successful or active ageing remain marginalised and their rights obscured, while the way we ascribe value to some social roles and not others remains unchallenged.[113] The concept of 'healthy ageing' advanced by the World Health Organization is less at risk of ageist distortions, and is more closely aligned to a human rights-based approach, but is still dependent on many structural and systemic factors which themselves may be influenced by ageist assumptions. More is clearly required to truly combat ageism and advance the human rights of all older individuals.

A key way to achieve this is to ensure that programs and policies aimed at addressing ageism and supporting older persons' rights more generally are designed in partnership with older people themselves.[114] A commitment to co-design in elder law can contribute to addressing ageism because at its core it demonstrates a respect for older persons' experiences and knowledge, as well as their fundamental dignity and autonomy. It also promotes participation, integration and inclusivity by directly involving them in the policy-design and decision-making processes, challenging ageist stereotypes about older people's capabilities. In practical terms it also helps to realise policies and laws which are more likely to be successful in addressing ageism because they can be informed by older persons' lived experiences.

---

[111] Ibid., 377.

[112] Kesby (2017), 382.

[113] Kesby (2017), 385.

[114] Kornfeld-Matte (2016), para 34.

A specific measure which would help to address ageism both legally and culturally, and which presents the opportunity for co-design in action, would be a dedicated international instrument along the lines of a CROP. Progress towards this objective was discussed in more detail in Chap. 2. In terms of combatting ageism, a dedicated treaty could include specific provisions relating to age discrimination, rather than perpetuate the invisibility of older persons within the 'other status' category of discrimination used in existing laws. It would also explicitly recognise the rights of older persons, their particular needs and vulnerabilities, and may therefore help to change attitudes and challenge negative stereotypes. A new treaty would challenge governments to translate new thinking about older persons into actual, practical strategies that advance their rights, combatting the 'structural dependency' of older persons which has become entrenched in our laws and policies through decades of ageist thinking.[115]

Some of these objectives could potentially be achieved by a human rights-based approach more generally—it may not be essential to have a dedicated treaty. However, we need to avoid the risk, identified by various scholars, that we are addressing age discrimination for the wrong reasons.[116] Ageism needs to be addressed because it is intrinsically incompatible with respect for human rights, and not just because it prevents us from harnessing the economic and productive potential of older persons. As Chung has said, '[w]hile it is important to take care of the elderly on economic and social grounds, such efforts must be grounded on the belief that the elderly have the inherent and inalienable right to enjoy a life of dignity, security, and independence, and free from discrimination based on their age.'[117]

5.6 Conclusion

The discrimination, 'othering' and dehumanisation inherent in ageism and its socially constructed origins point to a need for a paradigm shift in the way society thinks about ageing and older people. The legal protections currently in place focus on addressing age-discrimination, usually in certain settings like employment or the provision of goods and services. Yet the effects of ageism are pervasive and varied, going beyond explicit or even indirect discrimination, and impacting widely on older persons' enjoyment of their human rights. This chapter has argued that a human rights-based approach to elder law—one which embeds respect for dignity, autonomy and the full range of human rights throughout all legal and policy areas affecting older people's lives—would not only combat the effects of ageism, but also help drive a shift in culture to one which properly values older people and eliminates negative stereotypes.

Achieving this sort of paradigm shift requires a recalibration of the way we appreciate older persons' participation in the community. Models which focus on

---

[115] Kesby (2017), 385.

[116] Ibid., 377; Minkler and Holstein (2008); Chung (2009).

[117] Chung (2009).

enhancing older persons' 'contributions' are susceptible to manipulation driven by neo-liberal objectives of increasing economic growth and lowering the financial burden of ageing demographics. Concepts of 'successful' or 'productive' ageing must be scrutinised to ensure they are truly geared towards enhancing older persons' opportunities, choices and experiences. A human rights-based approach can help here, because it places the older individual at the centre of policy and inherently values their dignity and autonomy.

Even with a human rights-based approach to elder law, addressing ageism in our societies will be a long-term undertaking, and will require a multidimensional strategy. This chapter has identified a number of areas where ageism has typically been common, with notably negative consequences. Chapters 8–10 will focus on some of these specific areas in more detail, analysing current law and policy and offering recommendations for how a human rights-based approach could be better implemented. First however, Chaps. 6 and 7 will explore two more cross-cutting issues of capacity and elder abuse.

References

8 80 Cities (n.d.) Creating cities for all. https://www.880cities.org. Accessed 17 Feb 2020

Adams S (2009) No place like home? Housing inequality in later life. In: Cann P, Dean M (eds) Unequal ageing: the untold story of exclusion in old age, chap 4. Policy Press, University Press Scholarship Online

Adelman RD, Greene MG, Ory MG (2000) Communication between older patients and their physicians. Clin Geriatr Med 16(1):1–24

Administration for Community Living (2017) Profile of older Americans. https://www.acl.gov/sites/default/files/Aging%2520and%2520Disability%2520in%2520America/2017OlderAmericansProfile.pdf. Accessed 19 Sept 2019

Age UK (2019) Later life in the United Kingdom. https://www.ageuk.org.uk/globalassets/age-uk/documents/reports-and-publications/later_life_uk_factsheet.pdf. Accessed 19 Sept 2019

Ambady N et al (2002) Physical therapists' nonverbal communication predicts geriatric patients' health outcomes. Psychol Aging 17(3):443–452

Applewhite A (2016) Who me, ageist? https://thischairrocks.com/wp-content/uploads/2016/02/ConciousnessRaisingBooklet.pdf. Accessed 19 Jan 2020

Applewhite A (2017) Let's end ageism. TED talk. https://www.ted.com/talks/ashton_applewhite_let_s_end_ageism?language=en. Accessed 19 Jan 2020

Australian Human Rights Commission (2015) Willing to work. National prevalence survey of age discrimination in the workplace. https://www.humanrights.gov.au/our-work/age-discrimination/publications/national-prevalence-survey-age-discrimination-workplace

Australian Human Rights Commission (2016a) Willing to work: national inquiry into employment discrimination against older Australians and Australians with disability. https://www.humanrights.gov.au/our-work/disability-rights/publications/willing-work-national-inquiry-employment-discrimination. Accessed 9 Dec 2019

Australian Human Rights Commission (2016b) Qualitative study of employment discrimination against older Australians. https://www.humanrights.gov.au/our-work/age-discrimination/publications/qualitative-study-employment-discrimination-against-older. Accessed 19 Sept 2019

Australian Human Rights Commission (2018) Human rights and technology issues paper. https://www.humanrights.gov.au/sites/default/files/document/publication/AHRC-Human-Rights-Tech-IP.pdf. Accessed 19 Aug 2019

Australian Human Rights Commission (2019) Risk of homelessness in older women: background paper. https://www.humanrights.gov.au/our-work/age-discrimination/projects/risk-homelessness-older-women. Accessed 21 Jan 2020
Avers D et al (2011) Use of the term "elderly". J Geriatr Phys Ther (34)(4):153–154
Baum C (2018) The ugly truth about ageism. The guardian online (14 Sept 2018) https://www.theguardian.com/lifeandstyle/2018/sep/14/the-ugly-truth-about-ageism-its-a-prejudice-targeting-our-future-selves. Accessed 10 Feb 2020
Ben-Harush A et al (2017) Ageism among physicians, nurses, and social workers: findings from a qualitative study. Eur J Ageing 14(1):39–48
Bennington L (2001) Age discrimination: converging evidence from four Australian studies. Empl Responsibilities Rights J 13(3):125–134
Bernstein C (1969) Age and race fears seen in housing opposition. Washington Post (7 Mar 1969)
Blackham A (2018) Why mandatory retirement ages should be a thing of the past. The conversation (9 Apr 2018). https://theconversation.com/why-mandatory-retirement-ages-should-be-a-thing-of-the-past-94484. Accessed 10 Feb 2020
Butler RN (1969a) The effects of medical and health progress on the social and economic aspects of the life cycle. Paper delivered at the National Institute of Industrial Gerontology, Washington DC, 13 Mar 1969
Butler RN (1969b) Age-ism: another form of bigotry. The Gerontologist 9(4):243 246
Butler RN (1980) Ageism: a foreword. J Soc Issues 36(2):8–11
Butler RN (1989) Dispelling ageism: the cross-cutting intervention. Ann Am Acad Polit Soc Sci 503:138–147
Caris-Verhallen WM et al (1999) Factors related to nurse communication with elderly people. J Adv Nurs 30(5):1106–1117
Chung CS (2009) The necessity of a human rights approach and effective United Nations mechanism for the human rights of the older person. Human Rights Council Advisory Committee, Working Paper. UN Doc A/HRC/AC/4/CRP
Danks D, London AJ (2017) Algorithmic bias in autonomous systems. In: Proceedings of the twenty-sixth international joint conference on artificial intelligence, Melbourne, Australia, pp 4691–4697
Field CB et al (2014) Climate change 2014: impacts, adaptation, and vulnerability. Part A: global and sectoral Aspects. Contribution of working group II to the fifth assessment report of the intergovernmental panel on climate change
Fisk M (1999) Our future home: housing and the inclusion of older people in 2025. Help the Aged, London
Glover I, Branine M (eds) (2001) Ageism in work and employment. Ashgate, Adershot
Greenberg J, Schimel J, Martens A (2006) Ageism: denying the face of the future. In: Nelson TD (ed) Ageism: stereotyping and prejudice against older persons. MIT Press, Cambridge, p 27
Greengross S (2019) Human rights and ageing: are we doing it right? Keynote address to Australian association of Gerontology, Brisbane, Australia, 15 Nov 2019
Haigney E et al (1997) Breast examinations in older women: questionnaire survey of attitudes of patients and doctors. BMJ 315(7115):1058–1059
Hanson J, Kellaher L, Rowlands M (2001) Profiling the housing stock for older people: the transition from domesticity to caring. Final report of EPSRC EQUAL Research, London
Harding B (2006) Age equality in housing: a guide to tackling age discrimination for housing providers, commissioners, planners and builders. Housing and Older People Development Group. https://hoop.eac.org.uk/downloads/kbase/Age-Equality-LeafletOld.pdf. Accessed 10 Feb 2020
Harpur P (2015) Old age is not just impairment: the CRPD and the need for a convention on older persons. Univ Pennsylvania J Int Law 37:1027–1047
Hill A (2019) Over a third of britons admit ageist behaviour in new study. The guardian (18 Aug 2019) https://www.theguardian.com/society/2019/aug/19/over-a-third-of-britons-admit-ageist-behaviour-in-new-study. Accessed 19 Aug 2019

Human Rights and Equal Opportunity Commission (2000) Report of an inquiry into complaints of discrimination in employment and occupation: age discrimination in the Australian defence force (HRC Report No 8)

Hummert ML et al (1994) Stereotypes of the elderly held by young, middle-aged, and elderly adults. J Gerontol 49(5):240–249

Iversen TN, Larsen L, Solem PE (2009) A conceptual analysis of ageism. Nord Psychol 61(3):4–22

Kesby A (2017) Narratives of aging and the human rights of older persons. Hum Rights Rev 18(4):371–393

Kiel JM (2005) The digital divide: internet and e-mail use by the elderly. Med Inf Internet Med 30(1):19–23

Kornfeld-Matte R (2016) Report of the independent expert on the enjoyment of all human rights by older persons. UN Doc A/HRC/3344

Lagacé M et al (2012) The silent impact of ageist communication in long term care facilities: elders' perspectives on quality of life and coping strategies. J Aging Stud, Spec Sect Innov Approaches Int Comparisons 26(3):335–342

Lambrecht A, Tucker C (2019) Algorithmic bias? An empirical study of apparent gender-based discrimination in the display of STEM career ads. Manage Sci 65(7):2966–2981

Levy BR et al (2000) Reducing cardiovascular stress with positive self-stereotypes of aging. J Gerontol Ser B, Psychol Sci Soc Sci 55(4):205–213

Macnicol J (2006) Age discrimination: an historical and contemporary analysis. Cambridge University Press, Cambridge

Madan AK, Aliabadi-Wahle S, Beech DJ (2001) Ageism in medical students' treatment recommendations: the example of breast-conserving procedures. Acad Med J Assoc Am Med Coll 76(3):282–284

Mégret F (2010) The human rights of the elderly: an emerging challenge. SSRN Papers https://ssrn.com/abstract=1584303. Accessed 10 Feb 2020

Mégret F (2011) The human rights of older persons: a growing challenge. Hum Rights Law Rev 11(1):37–66

Minkler M, Holstein MB (2008) From civil rights to … civic engagement? Concerns of two older critical gerontologists about a "new social movement" and what it portends'. J Aging Stud 22(2):196–204

Montepare JM, Zebrowitz-McArthur L (1988) Impressions of people created by age-related qualities of their gaits. J Pers Soc Psychol 55(4):547–556

Neuberger J (2009) Not dead yet: a manifesto for old age. HarperCollins, London

Ng R et al (2015) Increasing negativity of age stereotypes across 200 Years: evidence from a database of 400 million words. PLoS ONE 10(2):0117086

Nussbaum JF et al (2005) Ageism and ageist language across the life span: intimate relationships and non-intimate interactions. J Soc Issues 61(2):287–305

Palmore E (1999) Ageism: negative and positive, 2nd edn. Springer Publishing Company, New York

Palmore E (2001) The ageism survey first findings. The Gerontologist 41(5):572–575

Peake MD et al (2003) Ageism in the management of lung cancer. Age Ageing 32(2):171–177

Pillemer K et al (2016) Elder abuse: global situation, risk factors, and prevention strategies. The Gerontologist 56(S2):S194-204

Power E (2016) Older women and home. Transforming Cities for Care (13 Sept 2016) https://dremmapower.com/older-women-and-home/. Accessed 17 Feb 2020

Ranzijn R et al (2006) On the scrap-heap at 45: the human impact of mature-aged unemployment. J Occup Organ Psychol 79:467–469

Sethi D et al (2011) European report on preventing elder maltreatment. World Health Organization, Geneva

Smith K et al (2014) Human health: impacts, adaptation, and co-benefits. In: Field CB et al (2014) Climate change 2014: impacts, adaptation, and vulnerability. Part A: global and sectoral aspects.

In: Contribution of working group II to the fifth assessment report of the intergovernmental panel on climate change, pp 709–754
Snape E, Redman T (2003) Too old or too young? The impact of perceived age discrimination. Hum Resour Manag J 13:78–89
Stewart E, Stein L (2019) One whole pension goes on Helen and Peter's rent. This is Australia's rental crisis. ABC News (28 Aug 2019). https://www.abc.net.au/news/2019-08-28/what-can-be-done-about-rental-stress/11452486. Accessed 19 Sept 2019
SunLife (2019) Ageist Britain? https://www.sunlife.co.uk/life-cover/over-50-life-insurance/over-50-data-centre/ageism/. Accessed 9 Dec 2019
The Benevolent Society (2017) The drivers of ageism. https://www.benevolent.org.au/ArticleDocuments/340/Ageism_SummaryReport_Final.pdf.aspx. Accessed 7 Feb 2020
Thornton M, Luker T (2010) Age discrimination in turbulent times. Griffith Law Rev 19(2):141–171
Tinker A, Ginn J (2016) An age friendly city: how far has London come? King's College, London. https://www.housinglin.org.uk/_assets/Resources/Housing/OtherOrganisation/Age-Friendly-London-Report-final.pdf. Accessed 9 Dec 2019
Townsend P (1986) Ageism and social policy. In: Phillipson C, Walker A (eds) Ageing and social policy: a critical assessment. Gower, Aldershot, pp 15–44
Townsend P (2007) Using human rights to defeat ageism: dealing with policy-induced 'structured dependency.' In: Bernard M, Scharf T (eds) Critical perspectives on ageing societies. Policy Press, Bristol, pp 27–44
Turner Lee N (2018) Detecting racial bias in algorithms and machine learning. J Inf Commun Ethics Soc 16(3):252–260
Wenger NK (1997) Coronary heart disease: an older woman's major health risk. BMJ 315(7115):1085–1090
Williams J (2003) When i'm sixty-four: lawyers, law and old age. Cambrian Law Rev 34:103–119
Wolfensberger W (2000) A brief overview of social role valorization. Ment Retard 38(2):105–123
World Economic Forum (2016) The future of jobs. http://wef.ch/1RkF7w5. Accessed 7 February 2020
World Health Organization, global network for age-friendly cities and communities. https://extranet.who.int/agefriendlyworld/. Accessed 10 Feb 2020
Wyman MF, Shiovitz-Ezra S, Bengel J (2018) Ageism in the health care system: providers, patients, and systems. Contemporary perspectives on ageism. In: Ayalon L, Tesch-Römer C (eds) International perspectives on aging. Springer International Publishing, Cham, pp 202

International Instruments

Charter of Fundamental Rights of the European Union (2007) Opened for signature 12 Dec 2007, entered into force 1 December 2009. OJ C326/391
Committee on Economic, Social and Cultural Rights (1995) *General Comment No. 6: The Economic, Social and Cultural Rights of Older Persons* (8 December 1995)
Committee on Economic, Social and Cultural Rights (2000) *General Comment No. 14: The right to the highest attainable standard of health (Art 12)* UN Doc E/C.12/2000/4 (11 August 2000)
Committee on Economic, Social and Cultural Rights (2009) *General Comment No. 20: Non-discrimination in Economic, Social and Cultural Rights (Art 2 par 2)* UN Doc E/C.12/GC/20 (2 July 2009)
Committee on the Elimination of All Forms of Discrimination Against Women (2010) *General Recommendation No. 27 on Older Women and the Protection of Their Human Rights.* UN Doc CEDAW/C/GC/27 (19 October 2010)
Convention on the Rights of Persons with Disabilities (2006) Opened for signature 13 December 2006, entered into force 3 May 2008. 2515 UNTS 3

Human Rights Committee (1989) *General Comment No. 18: Non-Discrimination* (10 November 1989)
International Covenant on Civil and Political Rights (1966) Opened for signature 16 December 1966, entered into force 23 March 1976. 999 UNTS 171
International Covenant on Economic, Social and Cultural Rights (1966) Opened for signature 16 December 1966, entered into force 3 January 1976. 993 UNTS 3
Treaty of Amsterdam amending the Treaty on European Union, the Treaties establishing the European Communities and certain related acts (1997) Opened for signature 2 October 1997, entered into force 1 May 1999. OJ C340/01

Cases

Commonwealth of Australia v Human Rights and Equal Opportunity Commission (1998) 158 ALR 468

Legislation

Age Discrimination Act 2004 (Cth)
Age Discrimination Act of 1975 42 USC Sections 6101-6107
Age Discrimination in Employment USC Title 29 Ch 14 Sections 621-634
Canadian Charter of Rights and Freedoms (1982)
Equality Act 2010 (UK)

Chapter 6
Legal Capacity and Decision-Making

The ability to make legally recognised decisions is fundamental to the exercise of human rights and is reflected in the core values of dignity, autonomy, participation and liberty. Respect for human rights requires that capacity be presumed absent evidence establishing incapacity. The process of capacity assessment also raises human rights issues as a determination of incapacity can have significant ramifications for the enjoyment of a person's human rights. This is particularly the case where the assessment is triggered by ageist assumptions or fails to respect the person's dignity or autonomy, or where it does not maximise their participation in the process as much as possible. This chapter argues that capacity and its assessment are fundamental to respecting individual human rights. It will explore the question of capacity and its assessment within the human rights framework established in Chap. 3. It will discuss the nature of capacity generally including the impact of age, particularly in the 'old, old' grouping. The question of legal capacity will then be explored including its assessment. Examples have been drawn from select areas relevant to persons as they age—wills, enduring documents and driving. Final comments will then be made demonstrating the inherent role that a human rights framework has in capacity and its assessment, and what this should look like.

6.1 Introduction

The presumption of capacity, that is the notion that everyone over the age of 18 has the cognitive and functional ability to make valid decisions recognised at law, is a fundamental tenet of modern law.[1] Questions of impaired and lost capacity are

---

[1] *Re Caldwell* [1999] QSC 182, 12 (Mackenzie J); Law Reform Committee, Parliament of Victoria (2010), 109–10.

© Springer Nature Singapore Pte Ltd. 2020

B. Lewis et al., *The Human Rights of Older Persons*,
https://doi.org/10.1007/978-981-15-6735-3_6

therefore fundamental to legal issues affecting older persons, particularly in relation to the continued ability to be able to make their own, legally recognised decisions about, for example, health, financial and accommodation matters. Capacity, and its assessment, also have the potential to significantly impact older persons' human rights. Respect for human rights dictates that capacity ought to be presumed in the absence of evidence to the contrary.

The law requires that people have a certain standard of capacity in order to make legally recognised decisions. Failing to meet that standard means that the decision will not be recognised at law. The standard required will depend upon the decision being made, for example, health, personal or financial decisions. The ability to make legally recognised decisions for oneself, particularly financial, health-related or personal decisions, is fundamental to the exercise of human rights and is reflected in the core values of dignity, autonomy and liberty. It is also inherent in the framework principles of the human rights-based approach outlined in Chap. 3 in the form of participation and respect for will and preferences.

The process of having one's capacity *assessed* also raises human rights concerns, particularly where the assessment is triggered by ageist assumptions or conducted in other ways which fail to respect the individual's dignity or autonomy. This is because an assessment that a person lacks capacity can have significant ramifications for the full enjoyment of that person's human rights, including rights to, for example, housing, freedom of movement, and to take part in the civil, economic, social and cultural life of their community.[2] It curtails older persons' opportunities to participate meaningfully in decisions which affect them and can restrict their liberty as well as disregard their dignity. For these reasons, a human rights-based approach requires that, wherever possible, capacity should not be viewed in binary terms of either having or not having it, but instead, and in line with the *Convention on the Rights of Persons with Disabilities* (CRPD), autonomous decision-making should be supported, noting that the diminution of both physical and cognitive capacity presents differently for all persons.[3] This can help to both maximise an older person's potential to participate as well as to minimise the negative human rights consequences of impaired or lost capacity.

An assessment of legal capacity, or more correctly legal *in*capacity given the presumption, generally occurs as a result of a trigger or triggers indicating that capacity may be impaired or lost in relation to the legal decision in question. The prevalence of ageism in our society, coupled with a lack of experience with capacity issues, frequently leads to erroneous assumptions that older persons lack capacity, something which is fundamentally contrary to a human rights-based approach. To counter this, a deeper understanding of capacity is needed by all within our communities. In particular, professions that are either involved in assessing capacity (particularly the health, allied health and legal professions) or which regularly encounter capacity issues with their clients (for example, financial institutions and financial

---

[2]See, for example: *International Covenant on Civil and Political Rights* (1966) arts 7, 9 12 ('*ICCPR*'); *International Covenant on Economic, Social and Cultural Rights* (1966), arts 11, 15.

[3]World Health Organization (2015), 7.

planners) need a greater awareness of the triggers indicating a loss of capacity and the significance of this. It is ultimately the practical issues associated with identifying, and then satisfactorily assessing, capacity within the applicable legal framework that can be the most problematic, and which therefore demand attention in implementing a human rights-based approach.

Assessments of capacity are currently undertaken on an ad hoc basis, dependent upon the experience and knowledge of the individual practitioner(s) involved in the assessment.[4] This variability in the assessment process can heighten the risk of abuse for already vulnerable individuals and infringe upon their human rights. Appropriate definitions, including of legal capacity, are imperative and the legal standards need to be clearly identified. This is fundamental for not only elucidating the level of understanding required on the part of the individual in question, but also so that assessors appreciate what is required of them to satisfactorily conduct an assessment. It is also necessary to consider the triggers which raise legal capacity as an issue, for example: cognitive, emotional, or behavioural signals; emotional distress; delusions; hallucinations; very poor hygiene; or displaying inappropriate behaviour.[5] These triggers can also be signifiers of an individual's vulnerability to abuse and other potential violations of their human rights. Consequently, the ability of assessors to identify and understand them is an important part of embedding a human rights-based approach within professions which regularly interact with older persons. Furthermore, it is also important to understand that lost or impaired capacity can be underpinned by the cumulative effects of the health and financial inequities discussed in Chaps. 8 and 10 respectively.[6]

With legal capacity assessments growing in complexity and frequency, it is generally accepted that assessment best practice guidelines are necessary that are responsive to discrepancies in the knowledge, skills and expertise of potential assessors.[7] More detailed guidance may be needed, for example, by lay people who are unfamiliar with either the legal or the health context in which the assessment is being conducted but who find themselves in the position of having to be involved in an assessment of capacity for another person. However, the persistent challenge is actually bringing this to fruition given the number of stakeholders who are involved, associated costs and the need for user testing, which adds ethical complexities when considering that the ultimate 'end-users' have diminished or lost capacity. Further exacerbating this

[4]Purser and Rosenfeld (2014).

[5]American Bar Association Commission on Law and Aging/American Psychological Association Assessment of Capacity in Older Adults Project Working Group (2008) 15–6.

[6]World Health Organization (2015), 8. See Box 1.2 on page 9 for a discussion of the result of health and pension inequities in relation to capacity. For example, a study in England demonstrated that life expectancy and the time lived free of disability varied, being conditional upon where someone resided. The life expectancy of those in wealthier neighbourhoods in England was 6 years more than those living in poorer neighbourhoods with the difference in the number of those who lived disability-free being 13 years.

[7]See, for example: Australian Law Reform Commission (2017), 286; Purser (2017); Law Society of New South Wales (2016); Queensland Law Society, Allens Linklaters and Queensland Advocacy Incorporated (2014).

problem is the fact that health and legal organisations, including insurers, continue to develop their own, discipline specific, guidelines around assessing capacity. Consequently, particularly when the legal and health professions intersect, there is confusion about what is required to be assessed, what the relevant legal framework consists of, how the assessment ought to be conducted, what format the assessment should take, and whose assessment should be preferred in cases of conflicting opinions.[8]

A multidisciplinary approach through the continued and better inclusion of other disciplines, such as health and allied health professionals as well as those in financial institutions, in the preparation of best practice guidelines will expose the process to wider scrutiny ideally resulting in better practice. Health and allied health professionals in particular are more 'frontline' and accessible to people. Consequently, they may be in a better position to identify changes in a person's capacity—the question then being how to map individual cognitive and functional change to the relevant legal standard for the specific decision. Further, there is often an element of trust that exists between patients and their health professionals that is not always present in a lawyer-client relationship given that people may only deal with lawyers infrequently and/or in negative or stressful circumstances. Such external investigation will strengthen the development and application of any guidelines. Interdisciplinary guidelines have been produced cooperatively in, for example, the United States of America by the American Psychological Association and the American Bar Association,[9] and in the United Kingdom by the British Medical Society and the Law Society.[10]

Developing best practice capacity assessment paradigms may help to address not only the complexity of the capacity assessment process but to also preserve individual human rights and ensure respect for autonomy and dignity. This chapter therefore explores the implications of human rights principles in the area of legal capacity, and the role of stakeholder involvement in best practice design and implementation. It argues that the human rights principles of individual autonomy and respect for dignity require that legal capacity ought to be presumed until it can be satisfactorily proven to be lacking or lost, and that all persons should be given the opportunity to participate in decisions which affect them as far as is possible. To this end, the general concept of capacity will be explored before looking at the intersection between capacity and ageing, focusing on those in the 'old, old' cohort. Capacity in the legal context ('legal capacity') will then be discussed. This discussion will include select areas where issues of legal capacity are commonly encountered by older persons such as wills, enduring documents (that is, those facilitating the appointment of substitute

[8]See, for example, Purser (2015); Purser and Rosenfeld (2014, 2016).

[9]American Bar Association Commission on Law and Aging/American Psychological Association Assessment of Capacity in Older Adults Project Working Group (2008); American Bar Association Commission on Law and Aging, American Psychological Association and National College of Probate Judges (2006).

[10]British Medical Association and the Law Society (2015).

decision-makers) and driving, as well as what is required to evidence the loss of legal capacity. The discussion in this chapter will conclude with a more detailed examination of the applicable human rights principles and how they can inform the realisation of best practice capacity assessment in the ageing context.

6.2 The General Concept of Capacity

Generally, capacity is the cognitive and functional ability of a person to be able to make a decision. The World Health Organization has defined healthy ageing in terms of promoting the functional ability of persons to be able to be and do what they value. This includes intrinsic capacity and pertinent environmental factors as well as the relationship between the person and these elements—all concepts existing in the capacity literature.[11] However, capacity can mean different things to different people, including, significantly, the various persons whose job it is to assess it. For example, legal professionals are concerned with notions of legal capacity—can the individual in question make the specific decision at the relevant time when taking into account the relevant legal test and standard? By comparison, medical and health practitioners assess variations in physical and mental capability. What is clear is that capacity is ultimately a legal determination.[12] However, the legal assessment of capacity cannot occur in isolation from the clinical notion, and it is at this juncture that the differing ideas as to what constitutes capacity and how it should be assessed can become problematic. The difficulties in assessment are therefore only reinforced by the terminological inconsistencies and unpredictability between those involved in assessments.[13]

This section will discuss the cognitive and functional nature of capacity generally before discussing the mentally disabling conditions that can affect it. First, however, a brief note on terminology. The terms 'capacity' and 'competency' are often used interchangeably. Traditionally the literature tended to try to separate them according to discipline, with capacity being the medical concept and competency the legal.[14] However, attempting to unambiguously distinguish between clinical capacity and

---

[11]World Health Organization (2015), 28–9, 32. See pages 31–2 and 100–3 for the trajectories of healthy ageing involving functional ability and intrinsic capacity. The trajectories can be divided into three main categories—that of high capacity, declining capacity and lost capacity, noting that the decline of capacity is not necessarily a steady downwards trend and that capacity differs for every person (that is, recognising its task and time specific nature). Figure 2.4 (page 33) sets out a public-health framework for healthy ageing identifying opportunities for action throughout the life course.

[12]Carney and Keyzer (2007).

[13]Purser (2017).

[14]Sullivan (2004), 131. See also Kitamura and Takahashi (2007), 578; Berg et al. (1996), 348–9; Kerridge et al. (2009), 244.

legal competency is practically futile. This is because, first, the existence, or otherwise, of 'capacity' is ultimately a legal determination; secondly, in order to satisfactorily make this assessment, legal questions of capacity (competency) cannot be considered distinct from the clinical construct, especially when considering mentally disabling conditions such as dementia; and, finally, as practitioners use the terms interchangeably, any attempted demarcation is, at best, likely to remain an exercise in linguistic futility.[15] Consequently, the terms capacity and competency will be used interchangeably here, being identified as legal or clinical where appropriate.

6.2.1 Cognitive and Functional Capacity

Capacity generally requires the ability to: understand the information being used when making the decision; appreciate the information; apply the information in a reasoned manner in order to make the decision, including evaluating the possible consequences (risks and benefits) arising from making the decision; and communicate consistent choices.[16] It is a highly fluid construct. It is also time and decision specific, and can fluctuate.[17] An individual who has the capacity to make one legally recognised decision may not have the capacity to make a different decision or potentially even the same decision at a different time. For example, a person who has an operative enduring power of attorney for the purposes of making financial decisions (such as selling a house to enter into a retirement village) because of a loss of capacity may nevertheless be able to execute a simple will. This variability highlights the importance of having nuanced guidelines for assessing capacity to ensure that a person's human rights are not unduly restricted.

In order to make a decision, an individual will employ cognitive and/or functional competencies.[18] These competencies can also be referred to in terms of decisional and executional capacity. Whereas decisional capacity is concerned with the person's cognitive ability to be able to make the specific decision, executional capacity deals with the person's functional ability to then be able to action the decision.[19] The combination of cognitive and functional capacity required in order to make a decision depends upon the specific decision in question.[20] Take, for example, financial decision-making. This type of decision-making is primarily a cognitive exercise. However, financial decision-making can also involve a functional element, such as

---

[15]Purser (2017).

[16]O'Neill and Peisah (2011), 955; Wilen Berg et al. (1996), 351; Karlawish et al. (2005), 1514. See also Cairns et al. (2005), 373; MacArthur Research Network on Mental Health and the Law (2004); Gurrera et al. (2006), 1367; Kitamura and Takahashi (2007), 579; Sturman (2005), 955; Lai and Karlawish (2007), 105; Purser et al. (2009), 796.

[17]Setterlund et al. (2002), 128; Attorney General's Department of New South Wales (2008), 27.

[18]Cockerill et al. (2005), 8.

[19]Falk and Hoffman (2014).

[20]Dārziņš et al. (2000), 8–9.

the signing of a bank withdrawal slip.[21] It is the combination of cognitive *and* functional elements needed to exercise financial decision-making capacity that makes assessing it particularly difficult.[22]

Common features underpinning a person's legally recognised cognitive ability to exercise decisional capacity therefore include understanding, reasoning, evaluation and communication. Such decisions are also recognised to be made in the context of the individual's own experiences, morals and beliefs with the notion of undertaking a 'reasonable process' being distinguished from the rationality of the outcome ('reasonable conclusion').[23] That is, decision-makers can make prima facie eccentric and irrational decisions even if the process (understanding, reasoning, evaluation and communication) is sound. This is an important point when assessing capacity as it is easy to overlook, and instead impose external notions of what the decision should be (the rejected outcome approach discussed below). As will also be discussed in more detail below, a human rights-based approach requires an understanding of the way that individual circumstances, experiences, opinions and relationships will shape a person's decision-making, so that capacity is assessed with these contextual factors in mind.

6.2.2 Mentally Disabling Conditions

The cognitive and functional capacity necessary to make decisions can be impaired or eliminated by any one or more of a combination of conditions, collectively termed 'mentally disabling conditions'. These conditions can generally be grouped as: those resulting in cognitive decline, such as dementia; developmental and/or intellectual disability; acquired and/or organic brain injury; mental illness such as schizophrenia; and alcohol and/or drug-related diseases.[24] Therefore, the character of the mentally disabling condition may be mental, psychological, intellectual and/or physical. A diagnosis of a particular condition, for instance dementia, should also not be taken to equate to an automatic loss of capacity. In fact, impaired or lost capacity may potentially be addressed through the adoption of appropriate supports, such as a treatment plan. This includes not only recognising the possible effects of any prescription medications and/or co- or multi-morbidities on the capacity of the person, but also through heightening the positive environmental factors in which the assessment occurs for that person.[25] For example, conducting the assessment in an environment familiar to the person being assessed, as well as ensuring that they have any hearing and/or visual aids present.

---

[21] Ibid., 72–3.

[22] Ibid., 73.

[23] Arias (2013), 144.

[24] Dunn et al. (2006), 1323–4; Carney and Keyzer (2007), 255.

[25] Frost et al. (2015), 8; Dārziņš et al. (2000), 4; *Sargent & Anor v Brangwin* [2013] QSC 306; World Health Organization (2015), 26, 58–9.

There are a number of markers that may indicate the existence of a mentally disabling condition. In addition to the above identified general categories, these can include: delirium; withdrawal; confusion and/or anxiety; a lack of motivation; indecisiveness; poor attention span; inferior memory retention; and/or neurodegenerative diseases, such as dementia.[26] Depression is another potentially significant indicator of a mentally disabling condition. It is crucial that depression be identified in capacity assessments.[27] This is because depression can result in people with mild cognitive impairments exaggerating those impairments which can, in turn, lead to incorrect outcomes when assessing capacity.[28]

It is difficult not to discuss dementia specifically when considering mentally disabling conditions given the prevalence statistics as well as the attention dementing illnesses attract.[29] Dementia rates and associated costs are only expected to increase over the next two decades, particularly as the 'baby boomer' generation reaches 'retirement' age.[30] It is anticipated that the number of people diagnosed with dementia, currently estimated to be approximately 44 million people, will almost double by 2030.[31] It is one of the main causes of disability in people sixty to sixty-five years and older in, for example, Australia and the United Kingdom.[32] In the United States of America, one in nine people who are aged sixty five years and over have been diagnosed with dementia, with this figure anticipated to almost triple by 2050.[33] One in four Australians aged eighty-five years and over are estimated to have been diagnosed with dementia,[34] while one in three people over the age of sixty-five are expected to die from dementia in the United Kingdom.[35] Particularly concerning, although somewhat unsurprising, is the fact that over half of the residents in aged care facilities in Australia in 2015 had been diagnosed with dementia.[36] Although dementia does not only affect older people, the risks of dementia do increase as people age, particularly for those people in the eighty year and over ('old, old') cohort. Understanding the effects of dementia as people age is vital when considering mechanisms to protect against the possible impairment or loss of capacity. Such mechanisms will often highlight the importance of estate or future planning, which will be discussed in Chap. 8.[37]

---

[26]O'Neill and Peisah (2011), 3.

[27]World Health Organization (2015), 58.

[28]Okonkwo et al. (2008), 656.

[29]For more on this see, for example: World Health Organization (2015), 59.

[30]Lin and Lewis (2015), 237.

[31]UK Government (2015).

[32]Australian Bureau of Statistics (2009), 2–3; Access Economics (2009), i; and AgeUK (2016), 12–3.

[33]Population Reference Bureau (2015); Alzheimer's Association (2016).

[34]Australian Bureau of Statistics (2009), 2–3; Access Economics (2009), i.

[35]AgeUK (2016), 12–3.

[36]Australian Institute of Health and Welfare (2016).

[37]Peisah and Brodaty (1994), 382; Liptzin et al. (2010), 950.

'Dementia' is a large grouping of a number of different conditions, with the most notable being Alzheimer's disease followed in frequency by dementia with Lewy bodies.[38] It is generally defined as a 'an acquired global impairment of memory, intellect and personality without impairment of consciousness'.[39] It is a degenerative neurological disorder that results in the decline of a person's cognitive abilities. Symptoms include progressive deterioration of cognitive abilities, memory, reasoning, communication skills, and the capability to undertake daily living tasks.[40] Paranoid delusions and severe confusion are common in people with dementia. This can therefore have a significant negative bearing on the assessment of capacity, particularly financial capacity (which is discussed below) if the assessor(s) is unaware of the dementia diagnosis. This is because, for example, delusions can impede a person's understanding of who their potential beneficiaries are in terms of having the testamentary capacity necessary to make a valid will.[41] Whereas, if the condition is known, it gives the solicitor preparing the will, for instance, the ability to try to maximise the person's understanding of any potential beneficiaries in the particular circumstances in order to meet the testamentary capacity standard.

As can be seen, mentally disabling conditions can result in the impairment or loss of capacity, with potentially serious implications for the enjoyment of human rights extending beyond the legal context. If someone is capable, they possess the right to be able to make their own, legally recognised decisions. A determination of a loss of capacity eliminates that decision-making autonomy for that particular decision. The impairment or total loss of decision-making autonomy can be confronting for not only the individual but can also affect familial and societal attitudes.

This is significant because individual and communal perceptions of how autonomous behaviour *should* present can be a powerful factor in capacity determinations and beyond, especially when coupled with ageist attitudes.[42] That is, the label of incapacity for *one* decision may erroneously morph into a label of incapacity to make *all* decisions thus resulting in an unnecessarily interventionist, paternalistic and protectionist view being adopted by family, friends and/or caregivers (depending upon the specific circumstances and how vulnerable the person is). Members of familial and social networks can often be either too hesitant or too fervent in imposing what they view as being 'for the best' in relation to the person whose capacity is in question. This can often result from a lack of understanding as to the decision-specific nature of capacity and what 'capacity' actually is. Significantly, however, such attitudes are not always informed by a nefarious goal such as to benefit financially from the assets of the person with impaired or lost capacity (elder abuse is discussed in Chap. 7). Whether malevolent or benevolent in intent, such approaches can feed into how the individual sees themselves, thus giving rise to a symbiotic relationship highlighting how important the social and familial nexus is. The concept

[38] O'Neill and Peisah (2011), 3.

[39] Peisah and Brodaty (1994), 382.

[40] Alzheimer's Australia (2009), 5.

[41] *Banks v Goodfellow* (1870) LR 5 QB 549; Purser (2015).

[42] Purser and Sullivan (2019).

of relational autonomy is therefore important, especially in the context of supported decision-making as required by the CRPD.[43] Although important, a discussion of relational autonomy principles is outside the scope of this work which is confined to exploring a human rights framework for elder law.[44]

Accurate identification of the relevant mentally disabling condition(s), the assessment process and the skills of the assessor(s) are therefore vital to ensuring that capacity assessments do not unnecessarily encroach upon human rights by making an incorrect determination. This is particularly the case in mentally disabling conditions, such as dementia, where the individual may not clearly present as experiencing decline and in which sufferers may lack the insight necessary to understand both the changes they are going through, as well as the necessity of conveying the diagnosis to the assessor(s).[45] The choice not to reveal pertinent information during the assessment process can arise, in part, because of the stigma that can attach to a diagnosis of any mentally disabling condition, but particularly that of dementia. Assessors and assessment processes therefore need to accommodate these factors to ensure that relevant information is available to enable accurate assessments, while at the same time respecting individuals' rights to privacy and respect for dignity.

6.3 Capacity and the 'Old, Old'

Populations globally are ageing and although not the sole cause, ageing is one of the main contributing factors to the growth and importance of capacity assessments. Ageing, both normal and pathologic, can have a substantial, detrimental and inescapable influence on a person's mental and physical capabilities and, consequently, their capacity to be able to make their own, legally recognised decisions. In fact, it has been proposed that 'diseases of the aged' should be a distinct category when considering the mentally disabling conditions discussed above.[46] However, ageing, in and of itself, is not a determinative factor for the existence of capacity—or for the loss or impairment thereof. In fact, to assume that a person lacks capacity because they have attained a certain chronological number is not only to perpetuate discriminatory and ageist assumptions, but also to encroach upon an individual's human rights.[47]

[43] See, for example: O'Connor (2010); Hall (2012); Purser and Sullivan (2019).

[44] See, for example: O'Connor (2010); Hall (2012).

[45] Falk and Hoffman (2014), 856.

[46] Creyke (1995), 10–2.

[47] *Universal Declaration of Human Rights* (1948), art 2.

Nevertheless, the risk of cognitive impairment, as well as certain medical and neurological diseases does increase with age, particularly in the old, old cohort.[48] Consider again, for example, dementia which is often intrinsically enmeshed with 'ageing'. It is important to appreciate that dementia is a clinical condition distinct from normal cognitive ageing. Age does not automatically equate to a dementia diagnosis, although the risk factors for dementia and cognitive impairment do strongly increase with age.[49] A number of conditions resulting in cognitive decline are associated with 'normal' ageing which can, in turn, affect an individual's functional capacity, although they will not generally result in difficulties when undertaking activities of daily living (ADLs).[50] There can, however, be a gradual decline in a person's complex functional abilities which impacts the ability to perform at a higher level, for example, being able to drive.[51] Other factors, such as isolation, sensory impairment and dependence can also increase not only the risk of incapacity, but also abuse.[52] There are some cognitive abilities, such as a person's vocabulary, that are relatively unaffected by ageing, however other functions, for instance reasoning and memory, do experience a decline.[53] It should, however, be noted that decision-making is not always a rational undertaking. It is certainly not one free from emotion.[54] As people grow older, their decision-making can be influenced by deficits in memory, education, attention, cognition and risk-taking.[55] Older people are also more prone to relying on experience rather than making a decision based purely upon the available facts.[56]

An issue closely related to both cognitive impairment as well as the notion of 'ageing well' is the effect of isolation and loneliness on older people. For instance, 40% of people diagnosed with dementia have reported feeling lonely, with 34% indicating that they do not feel as though they are part of their local community.[57] When these factors are combined with a loss or lack of capacity, the potential issues of isolation and/or loneliness faced more broadly by people as they age is concerning. This is because a loss of capacity, when mixed with loneliness and social isolation, can result in increased vulnerability to abuse as well as being subject to undue influence. This therefore has significant implications for the protection and promotion of human rights particularly when considering the fundamental right to independence and autonomy.[58]

---

48 Moye and Marson (2007), 3; Harada et al. (2013), 737–8; Moye et al. (2013), 162; World Health Organization (2015), 25.

49 Moye et al. (2013), 162.

50 Ibid.

51 Ibid.

52 Ibid.

53 Harada et al. (2013), 738.

54 Moye et al. (2013), 167.

55 Ibid.

56 Ibid.

57 Ibid.

58 World Health Organization (2015), 74.

6.4 Legal Capacity

It is the law that provides the mechanism through which to recognise incapacity, rationalising third party intervention to ensure the protection of the individual.[59] Legal capacity focuses on a person's ability to be able to make a specific decision and/or undertake a particular activity at the specific time of the decision/activity. It is 'relative to the particular transaction which is being effected by means of the instrument, and may be described as the capacity to understand the nature of that transaction when it is explained.'[60] Further, 'for a juristic act to be valid the person performing it should have the mental capacity (with the assistance of such explanation as he may be given) to understand the effect of that particular act …'.[61] Thus, the legal requirements for capacity generally focus on the individual making the decision (including the individual's *ability* to make the decision), the context in which the decision is being made, as well as the nature and complexity of the decision after the relevant information has been communicated.[62] Five fundamental factors have therefore been suggested as underpinning legal capacity: functional, causal, interactive, judgmental and dispositional.[63] First, functional capacity is concerned with a person's capability to complete tasks and the knowledge this requires (thus identifying the contextual capacity).[64] Secondly, the causal component assesses a person's functional competence given the previously identified contextual capacity.[65] Thirdly, whether the person's ability is sufficient in light of the decision should be investigated (the interactive component).[66] Finally, the judgmental and dispositional elements require that a decision be reached, that the person has the requisite capacity (or not as the case may be) and that appropriate consequences occur as a result of this determination in the specific context.[67]

In addition to being time and decision specific, both the test for, and the standard of, capacity required can also vary between differing legal jurisdictions.[68] For example, the legal capacity necessary to enter into a contract to sell a house differs from that required to make health decisions, which differs again from that needed to execute a valid will.[69] In relation to the ability to enter into a contract, for example, the capacity

---

[59]Grisso (2003), 2.

[60]*Gibbons v Wright* (1954) 91 CLR 423, 438 (Dixon CJ, Kitto and Taylor JJ). Applied in, for example, *Re Beaney deceased* [1978] 2 All ER 595 and *In the estate of Park deceased* [1954] P 89.

[61]*Hoff v Atherton* [2005] WTLR 99, [33] citing *Re K (Enduring Power of Attorney)* [1988] Ch 310, 313.

[62]Frost et al. (2015), 6; *Masterman-Lister v Brutton* [2003] WTLR 259 CA.

[63]Grisso (2003), 23.

[64]Ibid., 23–4.

[65]Ibid., 29.

[66]Ibid., 32–3.

[67]Ibid., 36.

[68]Attorney General's Department of New South Wales (2008), 27; Setterlund et al. (2002), 28; Dārziņš et al. (2000), 4–6; Sullivan (2004), 132.

[69]Grisso (2003), 9.

required is defined in terms of being able to understand both the nature and effect of the specific contractual terms after they have been explained.[70] Thus, if there is a sale of a house it will be the person's capacity to understand this in relation to that transaction that is being assessed. Significantly, this also applies in the context of retirement villages (discussed further in in Chap. 9).

Capacity can be assessed either contemporaneously with the decision or retrospectively after the decision has been made.[71] Contemporaneous assessments are ideal although they raise practical issues of cost, as well as tactically indicating that there may be an issue with legal capacity—which can subsequently be used in relation to allegations of a lack of capacity in relation to the validity of, for example, the transmission of assets or a will. Another potential issue in obtaining a contemporaneous assessment is the requirement to obtain instructions from the individual. First, the person in question may not realise that there is an issue with their capacity, which can be confronting for them to face but, additionally, they may not have capacity to give instructions—remembering that the task and time specific nature of capacity requires capacity not only for the specific decision in question but also to consent to the assessment.

The remainder of this section will focus on some of the main issues arising in relation to legal capacity. It will first discuss legal 'terms of art' such as lucid intervals, insane delusions and undue influence. It will then explore financial capacity given that this is one of the main domains of capacity that can impact legal decision-making for older people. Select areas of substantive application of capacity which are common as people age, notably, wills, enduring documents and driving will then be discussed.

6.4.1 Legal Terms of Art

The phrases 'insane delusion', 'lucid interval' and undue influence have all been critiqued as 'legal terms of art' meaning little in a modern, multidisciplinary context.[72] Although these are terms that resonate with legal professionals, they often mean little to nothing to those in other disciplines, particularly the health and allied health professions.[73] For example, although capacity fluctuates, particularly in the early stages of dementia, there is no such thing as a 'lucid interval' for people diagnosed with moderate to severe dementia. An insane delusion, which is loosely defined at law as being 'a fixed and incorrigible false belief which the victim could not be reasoned out of',[74] arises in the testamentary context and is only relevant

---

[70] *Gibbons v Wright* (1954) 91 CLR 423.

[71] O'Neill and Peisah (2011).

[72] Sprehe and Kerr (1996), 255; Moye et al. (2013), 163. On each of these terms see: Purser (2017).

[73] Sprehe and Kerr (1996), 255; Moye et al. (2013), 163.

[74] *Bull v Fulton* (1942) 66 CLR 295, 337 (Williams J). See also Frost et al. (2015), 250–2 for a discussion on delirium.

when affecting the dispositions contained in a will.[75] However, although perfectly reasonable to those with legal training, the concept presents the question of whether a delusion is ever 'sane' and how appropriate is this approach to classifying one as (legally) 'insane'? These terms are therefore suggested to be legal fictions providing a means to an end to resolve difficult questions of law rather than having any authentic medical basis.[76]

In relation to capacity and undue influence, whereas capacity examines the question of whether a person is cognitively able to make a decision, or to exercise their will, undue influence focuses on the question of whether an individual's will is overborne.[77] Both capacity and undue influence deal with issues around potential vulnerability including isolation, dependency (physical and emotional) and loneliness. Therefore, although a lack of capacity and undue influence can occur in factually similar environments, they are distinct issues at law. In fact, in order for a will to be unduly influenced, it has to be recognised at law as existing.[78] That is, a person has to have capacity in order for them to be unduly influenced. Consequently, particularly in relation to substitute decision-making (discussed below), if a person lacks capacity and substituted decision-making is necessary then there can be no allegation of undue influence. This is distinct from supported decision-making regimes in which capacity, albeit impaired, does exist, resulting in a complex interplay between impaired capacity and undue influence.

The recognition of discipline specific language is therefore also valuable when considering the practical importance of adopting clear terminology when assessing capacity and legal 'terms of art'. This is because if the professions do not understand each other there is an increased likelihood that this could negatively influence the assessment outcome resulting in an incorrect determination as to capacity. Such a determination could, in turn, severely impact the independence of an individual to be able to continue to make autonomous and legally recognised decisions, thus infringing upon their human rights.

6.4.2 Financial Capacity

Financial capacity is a complex, multifaceted concept which has been defined as a person's ability to manage financial assets, including money, in ways that are consistent with their morals and self-interest.[79] It encompasses a variety of abilities

[75] Tippet v Moore (1911) 13 CLR 248, 250 (Griffith CJ).

[76] Marson et al. (2004), 78. On the link between lucid intervals and fluctuating cognition see, for example: Shulman et al. (2015), 287.

[77] Purser (2017).

[78] *Johnson v Buttress* (1936) 56 CLR 113; *Union Bank v Whitelaw* [1906] VLR 711; *Lloyds Bank v Bundy* [1974] 3 All ER 757; Wingrove v Wingrove (1885) 11 PD 81; and *Hall v Hall* (1868) LR 1 P & D 481; and *Nicholson v Knaggs* [2009] VSC 64.

[79] Marson and Hebert (2007); Marson and Sabatino (2012), 6.

and skills including conceptual, functional, pragmatic and judgment—all of which vary greatly depending on the individual in question and their experience, education and socio-economic status.[80] These abilities are crucial to the independent and autonomous functioning of individuals.[81]

Financial capacity is a fundamental question in the ageing context because it underpins a person's ability to independently and autonomously manage their financial and testamentary affairs in accordance with their own self-interests, beliefs and morals.[82] Consequently, it is a significant issue in a setting where people are living longer than ever before, but often in circumstances where they have lost the capacity to be able to make their own, legally recognised decisions.[83] Financial capacity is crucial to planning for a financially secure future (discussed in Chap. 8) given that absent financial capacity a person cannot manage their financial affairs including, potentially, in relation to estate planning, securing accommodation and accessing healthcare.[84] In fact, there is an immense and 'underappreciated' challenge posed by the diminution or loss of financial capacity in older people.[85] This is because older people, particularly in the old, old cohort, are most vulnerable to lost financial capacity resulting from both cognitive decline in normal ageing, as well as other mentally disabling conditions such as dementia.[86] This is especially significant given that older people tend to hold the most wealth, particularly when taking into account the value of real estate and voluntary pension contributions.[87]

To remain living independently, an individual must demonstrate that they can complete the ADLs and instrumental activities of daily living (IADLs).[88] ADLs can be either household tasks, that is doing chores around the house, or basic duties, which include tending to personal hygiene.[89] IADLs require advanced cognitive functioning, which includes, for example, exercising financial capacity.[90] Financial capacity is therefore highly susceptible to the mentally disabling conditions that result in cognitive decline.[91] In fact, one of the earliest indicators that financial capacity is declining is the inability to undertake basic calculations, which can then obviously impact a person's ability to manage their money and assets.[92] Other factors relevant to the existence of financial capacity include knowledge, judgment and

---

80 ABA/APA (2008); World Health Organization (2015), 55.

81 Marson et al. (2000).

82 Marson (2013), 382.

83 Marson and Hebert (2007). See also Pinsker et al. (2010), 333; Webber et al. (2002), 250.

84 Marson and Hebert (2007). See also Pinsker et al. (2010), 333; Webber et al. (2002), 250.

85 Marson and Sabatino (2012), 6–7.

86 Ibid.

87 Jourdan and Glickman (1991), 415.

88 On the role of ADLs and IADLs in intrinsic capacity across the life course as viewed by the World Health Organization see: World Health Organization (2015), 65.

89 Marson (2013), 383; Falk and Hoffman (2014), 860.

90 Marson (2013), 383.

91 Marson and Hebert (2007).

92 Marson (2013), 383; Lock (2016), 18.

performance, that is, the ability to explain relevant concepts and implement sound financial decisions.[93]

Consequently, the factors affecting financial capacity vary significantly between individuals and can be influenced by any number of variable constituent parts, including not only the decision being made but also the financial experience, education and socio-economic level of the person making the decision.[94] Additionally, it is also important to recognise the role that other factors external to financial considerations can play in financial capacity. These include, for example, familial, social and cultural influences.[95] The person's demonstrated belief system, morals and attitudes towards finances are another significant aspect informing financial capacity and its assessment. There is also no one tool or standardised test that can be used to determine either financial capacity or capacity generally.[96] This is despite the alarming and erroneous reliance on standardised tests such as the Mini-Mental State Examination as being determinative of capacity in specific legal contexts—they are not.[97]

Assessing financial capacity is difficult with scant literature in this area mapping the impact of mentally disabling conditions to the relevant legal framework (test and required standard).[98] Such lack of guidance as to how to satisfactorily assess financial capacity, particularly in light of the ageing population and potential human rights implications, is concerning. The fact that legal, health and allied health professionals may lack the understanding necessary to satisfactorily assess financial capacity is also cause for unease.[99] Further research is therefore required consolidating and building upon current knowledge.[100] Research focusing on the relationship between financial capacity and depression would be especially useful given the hidden impact that depression can have on capacity and its prevalence amongst older persons, particularly those living in aged care accommodation.[101] Wills and enduring powers of attorney (EPAs) both require (differing) levels of financial capacity. They will be discussed in the next sections.

[93]Earnst et al. (2001), 110–11; Pinsker et al. (2010), 333; Marson (2013), 384.

[94]Marson (2013), 384.

[95]Pinsker et al. (2010), 333, 336, 338.

[96]See also Collier et al. (2005).

[97]Lonie and Purser (2017), 44.

[98]Moye and Marson (2007), 7; Marson et al. (2000); and Cockerill et al. (2005), 49; Marson et al. (2000); Moye and Marson (2007), 7–8.

[99]Cockerill et al. (2005), 49.

[100]Kershaw and Webber (2004), 338.

[101]Pinsker et al. (2010), 336.

6.4.3 Testamentary Capacity

The question of capacity in the context of making a will and/or enduring document (which will be discussed next) is an issue which seems to arise more frequently when people are older. This can often be because people are more aware of the possibility of death and/or incapacity as they age—or as their family members and/or friends age. Wills and enduring documents are also common vehicles utilised in estate planning as people begin to plan for their future, including their financial security beyond paid employment (discussed in Chap. 8).

In order to validly execute a will, the document has to meet the necessary formal requirements. The testator also has to have the requisite mental elements, including testamentary capacity.[102] That is, a testator must be able to understand the nature and extent of her or his property (at least generally), the potential beneficiaries who may have a moral claim, and the effect of making a will.[103] The testator must also not suffer from any 'insane delusion' or cognitive impairment which is of such a degree and form so as to directly impact the dispositions in the will.[104] It is noted that a person may have capacity during a 'lucid interval'.[105]

The task and time specific nature of capacity means that a person's testamentary capacity can also depend upon both the complexity of the testator's estate and the testamentary instrument. For example, the capacity needed to execute a will establishing a discretionary trust is higher than a 'straightforward' will wherein the testator leaves all of their assets to their partner and then to any children.[106] Whether capacity existed at the relevant time is a question of fact, ultimately to be determined by the court.[107] The relevant time here is when the testator signed the will. However, if the testator loses capacity between the giving of instructions and the execution of the will, where the instructions were reasonably proximate, for instance the day before execution, then that will be taken as the relevant time.[108] The aim, where possible, is to give effect to the testamentary intentions of the testator.

It is important to note that often solicitors have a duty to act on any coherent instructions and thus must prepare the will, especially as the decision as to whether

---

[102] See generally *Banks v Goodfellow* (1870) LR 5 QB 549, 565–6 (Cockburn CJ) for common law countries such as Australia, the United Kingdom and the United States of America. The position in the United Kingdom is also informed by the Mental Capacity Act 2005 (UK). On this, see the British Medical Association and the Law Society (2015), 85–6; Myers (2014), 44; and Frost et al. (2015), 55–63. For the position in the United States of America see, for example, *Newman v Smith* 77 Fla. 633, 673–74, 82 So. 236, 247–48 (1918); Kapp (2015), 165; and Shulman et al. (2009), 436.

[103] *Banks v Goodfellow* (1870) LR 5 QB 549, 565–6 (Cockburn CJ).

[104] *Banks v Goodfellow* (1870) LR 5 QB 549, 565–6 (Cockburn CJ). Banks has been applied in, for example, *Read v Carmody* [1998] NSWCA 182 (23 July 1998); *Boughton v Knight* (1873) LR 3 P & D 64, 65 (Sir James Hannen); Posener and Jacoby (2008), 754; and Frost et al. (2015), 40–8.

[105] Timbury v Coffee (1941) 66 CLR 277 applied in *Challen v Pitt* [2004] QSC 365.

[106] Purser and Lonie (2019).

[107] Jacoby and Steer (2007), 155; Shulman et al. (2009), 434; Marson et al. (2004), 82.

[108] Parker v Felgate (1883) 8 PD 171; *Bailey v Bailey* (1924) 34 CLR 558, 567, 572.

the testator had the requisite capacity or not is ultimately that of the court.[109] This duty is not always well understood by health professionals, often causing confusion when legal and health professionals intersect throughout the assessment process. The onus is on the person propounding the will, the executor, to raise the presumption of capacity.[110] The burden then shifts to the person alleging a lack of capacity to evidence this. The courts will then make a determination as to testamentary capacity based on all of the evidence, and on a case by case basis.[111] This is where the evidence in relation to the assessment process, by both the legal and (any) health professionals as well as any lay witnesses (often family and friends), is particularly important. The assessment process will be discussed further in the next section.

Discussion has emerged in the literature questioning the currency of the test for testamentary capacity.[112] It should be noted, however, that this is predominantly occurring in the medical literature distinct from the legal academic and judicial discourse.[113] Further, the legal test for testamentary capacity is a general statement that was never intended to have the force or rigidity of a legislative provision.[114] That is not to say, however, that the scrutiny brought to bear on the traditional legal test by the adoption of a multidisciplinary approach is a bad thing, it is not. Indeed, it is essential. Opening the test to critique from a number of interested stakeholders will only serve to strengthen the assessment process. In fact, critical engagement with not only the legal test for capacity, but also the assessment process, is vital in developing best practice. Given the necessary flexibility of the legal test itself, the development of best practice is fundamental to ensuring the protection and promotion of human rights, notably the right, in the case of capacity, to engage in autonomous decision-making where possible.

Court ordered or statutory wills in particular pose an interesting conundrum when applying a human rights framework to testamentary capacity—are they interventionist and paternalistic or do they ensure equality before the law by giving effect to the testamentary intentions of persons who would otherwise have been unable to make a valid will?[115] Statutory wills enable a person to seek an order from a court to make, alter or revoke a will on behalf of another person allegedly lacking testamentary capacity but who is still alive.[116] They are not uncommon for older people in relation to estate planning purposes.[117] One of the questions the court will ask in deciding

---

[109] Jacoby and Steer (2007), 156.

[110] *Bull v Fulton* (1942) 66 CLR 295, 299 (Latham CJ); 341 (Williams J); *Shaw v Crichton* [1995] NSWCA BC 9,505,228 (23 August 1995) 2–4 (Handley, Cole & Powell JJ).

[111] *Bailey v Bailey* (1924) 34 CLR 558.

[112] For more on this see, for example: Purser K (2015. For a modern interpretation of the testamentary capacity test see, for example: *Re Loxston, Abbott v Richardson* [2006] WTLR 156, and *Read v Carmody* [1998] NSWCA 182 (23 July 1998).

[113] See, for example: Peisah (2005).

[114] Purser (2017).

[115] On this see: Croucher (2009).

[116] Purser (2017).

[117] *Re Matsis* [2012] QSC 349; *GAU v GAV* [2014] QCA 308.

whether to make the will or not is whether the will is one the testator would make if they had the necessary capacity to do so. Therefore, on the one hand, statutory wills are vehicles which enable the testamentary intentions of people lacking capacity to be carried into effect and thus support their right to dispose of their property as they would have wished had they had the capacity to be able to make a will themselves. On the other hand, there is significant difficulty in being able to determine what someone else would have wanted, absent clear instructions, particularly if the person never had the requisite capacity to be able to indicate what they would want done with their estate, that is a 'nil capacity' case. Consequently, while a statutory will can be an important means of ensuring that a person's estate is managed in a way which is consistent with their anticipated intention, a human rights-based approach demands that the court make such decisions with human rights principles firmly in mind, especially in the 'nil capacity' cases.

6.4.4 The Capacity to Make Enduring Documents

As people age, potentially without decision-making capacity, the ability to appoint someone, or multiple people, that they trust to make those decisions becomes increasingly important. This is the notion of substitute decision-making. Substitute decision-making tends to occur in two main areas, financial and healthcare/personal decisions. These documents demonstrate the traditional societal and government approaches to enabling people to have decisions made for them when they are no longer able. That is, by enabling the individuals to appoint person(s) they trust to make decisions on their behalf, this gives the person who loses capacity the greatest level of control (that is possible given the circumstances) in the decisions that will be made for them when they lose capacity. It is therefore an attempt to acknowledge the person's decision-making autonomy through the appointment of someone they trust to make financial and/or health/personal decisions on their behalf. The powers given to attorneys and guardians under enduring documents are immense. They can basically do anything the principal (the person who made the enduring document) could do if that person had capacity.[118] This therefore makes enduring documents effective vehicles for abuse, particularly financial abuse.[119]

Substitute decision-making therefore raises significant legal, ethical and human rights issues. In particular, the human right objectives of respecting decision-making autonomy and protecting vulnerable individuals against exploitation and abuse may come into conflict where the tool used to facilitate the exercise of autonomy is open to misuse. To address this potential conflict, the concept of supported decision-making as an alternative to substitute decision-making may be more appropriate within a

---

118 Standing Committee on Social Issues, Parliament of New South Wales (2010), 8.

119 Setterlund et al. (2002), 128. Australian Law Reform Commission (2017); Purser et al. (2018); World Health Organization (2015).

human rights framework.[120] Supported decision-making emphasises the ability of the individual to be able to make their own decisions for as long as possible with appropriate support. The support options employed vary, and often involve family and carer input where necessary. New technologies also provide innovative ways in which to provide support to people with diminished capacity.

A move to supported decision-making would be in line with the CRPD which recommends a paradigmatic shift away from substitute towards supported or assisted decision-making.[121] The CRPD has as its aim the idea that all individuals, including people with a disability, should experience the effects of legal capacity equally, with the ultimate goal being to give effect to the individual's preferences and thus promote their human rights.[122] To achieve this it draws a distinction between legal standing and legal agency as two different, though related, dimensions of capacity. Legal standing is the ability to have legal rights and duties, while legal agency is the ability to exercise those rights and duties.[123] By drawing this distinction the CRPD focuses on ability rather than capacity, with the understanding that even if a person has impaired capacity at law, they may still be able to exercise legal agency provided that appropriate supports are implemented to assist them in exercising their decision-making autonomy.[124] Supported decision-making is a method of promoting 'ability' rather than capacity, and is therefore more consistent with a human rights-based approach.[125]

Supported decision-making schemes have been introduced in a number of countries, for instance, Australia, Canada, Germany, the United Kingdom, Ireland, Sweden and the United States of America.[126] Many countries however have adopted a 'continuum approach'.[127] That is, rather than replacing substitute decision-making with supported decision-making, this approach recognises that some people will lose capacity (distinct from impaired capacity) and consequently will not have the cognitive ability to be able to make decisions, even with support. This approach is more representative of the practical problems that can arise and the need to recognise that at some point, for instance if a person is in a coma, capacity can be completely lost. This approach therefore promotes a system whereby substitute and supported decision-making co-exist.

Several challenges with implementing supported decision-making schemes exist, notably, the issue of finding appropriate support mechanisms, but also the risk of abuse, particularly through undue influence where the person has diminished but not lost capacity which can place the person with diminished capacity in a greater position

---

[120] Then et al. (2018); Carney et al. (2019).

[121] Notably articles 5 and 12. Perlin (2013), 1179.

[122] Australian Law Reform Commission (2013), 30, 36; Blanck and Martinis (2015), 26.

[123] Australian Law Reform Commission (2013), 28.

[124] Gooding (2013), 437.

[125] Perlin (2013), 1176–7; Australian Law Reform Commission (2014), 38.

[126] Blanck and Martinis (2015), 26.

[127] Browning et al. (2014).

of vulnerability.[128] Therefore, even though the aim of supported decision-making is to support the individual to be able to retain their independent decision-making ability for as long as possible, by doing so questions in relation to increased risk of both undue influence and abuse are raised which can be just as significant a threat to human rights. However, as noted above, if a human rights-based approach is implemented then principles of autonomy, dignity, equality before the law and access to justice can be better embedded to ensure safeguards are in place to minimise these risks in both supported and substitute decision-making regimes.

6.4.5 Driving

Another significant practical issue arising in relation to capacity assessments is driving. Older people may place heavy reliance on private vehicles for transportation, particularly in rural, regional and remote areas where public transport may not be readily accessible.[129] It seems likely that the number of people seeking medical advice in relation to their capacity to drive is only going to increase given the risk of age-related cognitive and physical decline, including with respect to visual and motor skills, on driving.[130] Driving is significant not only because it provides a means of transportation but it also fosters the ability to be socially connected, that is, being able to attend activities external to the person's house, particularly if they live alone. Such activities also extend beyond participation in social events to include other facets of life, such as promoting health and wellbeing, employment, and community involvement. In fact, isolation can increase in older people who are no longer able to drive.[131] As will be discussed in relation to elder abuse (Chap. 7), isolation is also one of the risk factors for a heightened vulnerability to elder abuse. Driving, therefore, can be an important factor in an older person's independence and freedom, and can be a crucial determinant of their ability to enjoy a wide range of interdependent human rights (as discussed in more detail in Chap. 4).

The questions around an older person's ability to drive are twofold. First, whether it is safe for the older person to continue to do so, both for themselves but also in relation to any risks posed to other drivers. Secondly, there may be legal restrictions and mandatory licence renewal processes imposed on older people in order to retain their licences once a certain age is reached.[132] For example, drivers must carry a current medical certificate upon attaining the age of seventy-five years or risk being fined in Queensland, Australia. The issue of age-based mandatory medical testing and licence regulation is obviously contentious. Those supporting the policy tend to use the safety argument as a justification. Conversely, however, this policy is also one

[128] Blanck and Martinis (2015), 27.

[129] Wong et al. (2016), 133.

[130] Horswill et al. (2013), 130; Wong et al. (2018), 495; Wong et al. (2016), 133.

[131] Qin et al. (2019).

[132] Wong et al. (2016), 133.

which is clearly based upon ageist and discriminatory assumptions that once a person reaches a certain age, they become a danger both to themselves as well as to other road users. Further, research suggests that older people can also face discrimination when they submit to the age-based mandatory driving assessments.[133]

Human rights are not absolute and it is lawful to restrict some rights in order to pursue a legitimate aim, such as public safety. However, any such restrictions need to be imposed under law (that is, they cannot be arbitrary) and they need to be necessary and proportionate to their stated objective.[134] It is therefore not incompatible with human rights to make provision for older people to lose their driving licences when they are no longer considered capable of driving safely. In order to comply with human rights principles the regulation of older drivers must avoid unreasonable assumptions about a person's abilities. Further, any processes for testing or medical assessments must be conducted in accordance with due process and afford older drivers due respect. Importantly, pathways need to be available to ensure that an individual can challenge any negative assessment against them and has an opportunity to have the decision over-turned if it is found to be contrary to law or an unjustified restriction on their human rights.

6.5 Evidencing the Loss of Legal Capacity

The loss of legal capacity results in the loss of decision-making autonomy and independence. This outcome not only impacts a person's human rights, it can also expose them to an increased risk of abuse as well as to the stigma that can attach to a label of 'incapable'. There have traditionally been three main legal paradigms for assessing capacity. First, there is the *status* approach whereby a lack of capacity is determined by the status of the individual, for example, their age and/or whether the person has a disability. This approach is useful in circumstances where the person clearly lacks capacity, such as if they are in a coma. It can also clearly result in discriminatory outcomes when adopted in terms of people with a disability and consequently is no longer favoured. A *functional* approach can also be adopted whereby the person's ability to make a context specific decision is assessed, for example, can the person make the decision to enter into aged care? The third approach to assessing capacity is the *outcome* approach. This model dictates that a person will lack capacity if they do not make the decision which is representative of what people think the decision should be, that is, whether the decision objectively appears to be in the best interests of the individual.[135] Significantly, both the status and outcome approaches would appear to contravene article 12 of the CRPD which, as noted above, requires that rules around capacity promote respect for individual will and preferences and are

[133] See, for example, Victorian Equal Opportunity and Human Rights Commission (2012).

[134] See, for example: *ICCPR*, art 12 for permitted restrictions on freedom of movement.

[135] Queensland Law Reform Commission (2008), 106; Queensland Law Reform Commission (2010), 243.

tailored to the person's circumstances. Assessment processes which are discriminatory or which disproportionately affect persons with disabilities are likely to be in violation of this provision, as would be assessments which determine a lack of capacity purely on the grounds that the decision is unusual or unexpected.[136]

Tests for assessing capacity in specific contexts have increasingly been developed since the early 1990s ranging from formal standardised tests to semi-structured interviews and observational methodologies.[137] Tests can be separated into two main types, general ability tests assessing cognitive abilities and purpose-built tests.[138] Task specific models were developed in response to the realisation that capacity is task specific rather than global or domain centred.[139] Most recently, a task specific model has been developed in relation to testamentary capacity.[140] However, there is currently no universally accepted standardised test through which to objectively determine decision-specific capacity.[141] This is also unlikely to occur given the difficulties in being able to accurately 'measure' capacity.[142] Further, such a test would also need to surmount the hurdles presented by accurately assessing capacity not only in relation to the nature and stage of the specific mentally disabling condition, but also with respect to the specific legal framework.

What it is possible to develop, however, is a unifying conceptual assessment paradigm that promotes best practice. The flexibility of such a model would make it more adaptable to both specific mentally disabling condition as well as jurisdictional peculiarities.[143] However, a 'gold standard' in assessment process (distinct from standardised testing) is also lacking in many countries. On the one hand this is somewhat surprising, especially given what is at risk—an individual's ability to be able to make their own, legally recognised decisions and, consequently, function independently within their own communities.[144] However, on the other hand, it is also unsurprising. The number of stakeholders necessary for the required multidisciplinary input, the subsequent intersection of discrete disciplines, as well as the sheer complexity of developing best practice guidelines makes for a very difficult task to successfully implement.

The consequence, however, of not having any generally accepted assessment methodology is the myriad of different approaches that are then adopted in order to assess capacity in the different contexts.[145] Each of the different methods is informed

[136] Australian Law Reform Commission (2013), 37.

[137] Sullivan (2004), 137. See also Berg et al. (1996), 349–51.

[138] Sullivan (2004), 135.

[139] Ibid., 137.

[140] Papageorgiou et al. (2018).

[141] Sullivan (2004), 135; Cairns et al. (2005), 377; Kitamura and Takahashi (2007), 579; Ganzini et al. (2003), 237.

[142] Dārziņš et al. (2000), 7.

[143] Wadley et al. (2003), 1621. See also Pinsker et al. (2010), 333. See also Suto et al. (2005), 202; Kershaw and Webber (2008), 40.

[144] Marson et al. (2006), 81.

[145] For more on the different approaches to assessing capacity see, for example: Purser (2017).

by the skill, experiences, interests and abilities of the individual assessor(s) which makes for ad hoc assessments that can result in unsatisfactory and/or erroneous outcomes.[146] The number of assessment paradigms also calls into question the quality and validity of each—with so many of them, how do practitioners, law-makers and judges determine which ones are rigorous and reliable? Further, do the assessor(s) understand both the nature and stage of the mentally disabling condition and how that particular condition maps to the relevant legal framework, including the standard of capacity required to be able to make the specific decision? Who are the assessor(s) and what are their discipline backgrounds? Are they talking about the same concept of capacity bearing in mind, for example, the legal terms of art discussed above? All of these factors will determine the accuracy of each individual assessment, as well as complicating broader policy decisions about which approaches ought to be preferred within a relevant area of law.

A constituent part of the 'ideal' assessment is education about the process and what is occurring. This should ideally engender trust in the system so that the person who is being assessed is able to 'perform' as best as possible in order to achieve an optimal assessment experience and outcome (even if the outcome is a determination of a lack of capacity).[147] The idea is that the person should in no way be harmed by having their capacity assessed. The lack of a best practice paradigm fails to create educational opportunities for the individuals to be fully informed about the process they are to undergo—which can then negatively impact the assessment and can impinge on their human rights in the process.

People will experience impaired and lost capacity, but the aim should be to ensure that any assessment process determining this maximises the person's chances of having an optimal assessment process and outcome. Therefore, the capacity assessment process is fundamental to ensuring the protection and promotion of human rights when considering that without capacity an individual is no longer able to make their own legally recognised decisions. Participation in the process not only heightens the chances of retaining capacity for as long as possible, it also demonstrates respect for the inherent dignity of older persons, as well as their independence and individual autonomy, including the freedom to make their own choices.[148] As seen, establishing standardised tests remains improbable if not impossible. In fact, the protracted use of assessment paradigms that lack veracity actually serves to threaten individual autonomy because they can more readily lead to erroneous outcomes than assessments informed by best practice.[149] The focus should therefore be on developing guidelines drawing upon multidisciplinary expertise that detail what amounts to sufficient cognitive performance of tasks relevant to the determining if the person has the legal capacity in question.[150] Both national and international discussions are therefore needed to consider what best practice paradigms should include, but also

[146] Purser and Rosenfeld (2014).

[147] Purser and Sullivan (2019).

[148] *Convention on the Rights of Persons with Disabilities* (2006).

[149] Dārziņš et al. (2000), 139.

[150] Moye et al. (2013), 167.

to best address the human rights issues that can arise as a result of an erroneous capacity assessment outcome. Fundamental to this is an examination of the relevant principles that should feature in such guidelines. The following section explores in more detail the way in which a human rights-based approach can be embedded in relation to capacity assessments.

6.6 Capacity and Human Rights Principles

Assessing capacity is a position of immense power which can both protect and threaten an individual's human rights if capacity is found to be impaired or lacking. Therefore, assessors need to be aware of their legal and ethical obligations. A human rights-based approach ought to be fundamental to any set of guidelines developed for assessing capacity, be it either a context- or discipline-specific guide or, ideally, a best practice paradigm. Guidelines informed by human rights serve three main functions in this context. First, they can ensure that assessment processes are aligned to the core values of respect for dignity and autonomy, and that they promote specific rights and guiding principles relating to individuals' freedom to make their own choices and participate fully in society. Secondly, guiding principles can function as an internationally accepted base from which to anchor and develop best practice. Finally, where human rights principles are translated into specific practices and policies they can offer much needed guidance to assessors.[151] This section will consider some of the more specific ways that a human rights-based approach can be embedded into practices for assessing capacity, with a view to informing the developing of such guidelines.

Human rights principles can currently be seen to varying degrees throughout international assessment models. As with the assessments, however, there is again inconsistency in their initial adoption and then in their implementation. This calls into question both the reliability and the transparency of the various assessment processes.[152] Some of the most common principles adopted include respect for the presumption of capacity,[153] as well as the acknowledgment of freedom of decision-making. That is, irrational, immoral or unwise decisions as judged by a third party do not determine incapacity. However, there are a number of other areas where human rights principles could be better embedded into assessment guidelines.

Fundamentally, guidelines for assessing capacity should strive to avoid negative human rights impacts and promote the greatest enjoyment of human rights, shaped by fundamental principles of respect for dignity, autonomy and equality of all individuals. Assessment guidelines should ensure that assessments are conducted free from the influence of ageist stereotypes. Respect should also be had for the welfare and best interests of the individual incorporating safeguards against abuse, neglect

---

[151] Queensland Law Reform Commission (2010), 61.

[152] On this, see: Purser (2017).

[153] See also British Medical Association and the Law Society (2015), 11.

and exploitation, while also striving for the option which is the least restrictive for the individual in question.[154] The least restrictive option here is premised upon the belief that an individual should be able to make their own decisions, where possible, and their participation in the decision-making process should be maximised as much as practicable with minimal limitations given the specific circumstances.[155] It promotes freedom of decision—even where the decision is prima facie irrational—and freedom of action, focusing on the welfare and interests of the individual. It has, as its core, respect for human dignity.

Supporting familial and social connections is particularly important in assessing and managing capacity through a human rights-based approach. As noted above, decisions about capacity (and particularly determinations of incapacity) can have significant ramifications for interpersonal and familial relationships. At the same time, family and social networks provide crucial support to older persons, including through taking on roles with regard to supported decision-making, and through helping older persons who have lost capacity in certain areas (for example driving). Family members and friends may also be called on to provide evidence when decisions about capacity are being made. A human rights approach recognises the importance of family and social connections, both in relation to specific rights like rights to housing, social and cultural inclusion, but also through key principles of respect for dignity and participation. Guidelines for assessing capacity should therefore prioritise, where possible, maintaining *healthy* relationships with family and community that are free from abuse and undue influence.

Another aspect of community participation which is relevant to capacity assessment is respect for the cultural and linguistic background of the individual being assessed. If not identified and understood, cultural and/or linguistic differences can have a negative impact on the assessment process and potentially the outcome. An appreciation of these factors is necessary for the assessor in order to understand how a particular individual might be understanding or interacting with the process, but also to ensure that they are able to adequately inform the individual about the process and its implications. This is supported by human rights principles of non-discrimination and respect for dignity, and the rights of minorities protected under human rights law. It suggests that professionals involved in capacity assessment need to appreciate the value of cultural safety and the implications of cultural rights discussed in Chap. 4, and that these factors ought to be included in the development of assessment guidelines.

A human rights-based system demands an inclusive approach which respects an individual's wishes and value as a human being. Self-reliance and supported decision-making should be promoted where possible, with intervention only to occur where necessary in order to protect individual safety. Maintenance of the individual's environment and values is key to a human rights approach to assessments, as is maintaining confidentiality. Notwithstanding the importance of preserving confidentiality

[154]Mental Capacity Act 2005 (UK) c 9 s 1.

[155]*Powers of Attorney Act 1998* (Qld) schedule 1, part 1; Guardianship and Administration Act 2000 (Qld) schedule 1, part 1 principle 7(1)-(2).

as part of the right to privacy, this needs to be handled cautiously as confidentiality is one of the elements existing in relationships of undue influence. For example, if a niece is managing her aunt's financial affairs and no one is aware of this—perhaps at the aunt's request—the aunt is at a heightened risk of abuse and undue influence. This raises the key challenge of how to balance respect for individual autonomy and dignity against the need for protection when impaired or lost capacity might create risks for a person's physical or financial security.

The competing concepts of autonomy and dignity on the one hand, and protection of the vulnerable on the other, create a persistent tension in a human rights approach. Autonomy is particularly problematic, as it goes to the heart of capacity but is a nebulous and dynamic concept. It has been described as 'self-determination within a sphere of personal sovereignty', and only permits another person to intervene when the individual is deemed to have lost the ability (capacity) to exercise their own sovereignty.[156] In this sense, accurate assessments of capacity are essential to ensuring that autonomy is respected to the greatest degree possible. This is also fundamentally linked to respect for the dignity of each person, which extends to the manner in which capacity assessments are conducted.

Crucially, respect for autonomy and dignity dictate that an individual's choices should only be overridden where it can be clearly established that they lack capacity to make those decisions. As noted above, this includes situations where they might choose a path which others consider to be less preferable, irrational, or even risky. The dignity of risk is a pertinent concept in the context of older persons and capacity assessments because it recognises that the act of taking a risk can be a powerful exercise of autonomy and is not necessarily evidence of incapacity. While human rights emphasise the importance of protecting the physical and emotional well-being and security of the individual, autonomy and dignity remain fundamental core values. This points to the need to ensure that older persons are provided with the necessary information and support to make informed and reasonable decisions, but emphasises that, absent a clear lack of capacity, they are entitled to exercise their individual sovereignty in the form of autonomous decision-making.

Arguably, rising awareness of and respect for individual autonomy and the importance of human rights has resulted in a deterioration of the traditional trust placed in the legal and health professionals. This includes in the capacity assessment process.[157] This has both negative and positive repercussions. On the one hand, people no longer have the security of belief that the system 'works', or faith that the people conducting the assessments are qualified and will act in the best interests of the person being assessed. It is, however, positive in the sense that knowledge is power. People should be reluctant to place unquestioning trust in others absent appropriate education. Subsequently, the emergence of a more rights-aware society is a constructive development when considering that assessor(s) are now on notice that they need to conduct assessments in accordance with best practice principles

[156] Schopp (1996), 727–8.

[157] Purser and Sullivan (2019).

(even absent guidelines clearly establishing what this is).[158] As argued here, the development of best practice guidelines needs to be informed by human rights, and ideally would entrench a human rights-based approach to capacity assessment. The implementation of best practice based on human rights would then have the effect of generating trust in the system, which is an important element in ensuring both optimal assessment processes as well as outcomes.[159]

6.7 Conclusion

Although age is not an automatic indicator of diminished or lost capacity, cognition does deteriorate with age, particularly advanced age. With this reality, capacity is fundamental to the notion of 'healthy ageing' because it underpins legally recognised decision-making. Absent legal capacity, an individual is not recognised at law as being able to make a particular decision. This has obvious implications for individual autonomy and dignity, as a finding that capacity has been lost or impaired will invariably curtail a person's ability to make decisions for themselves, including in relation to some of the most significant aspects of their lives. It can also significantly restrict their participation in the community and exercise of other rights. The concept of capacity is therefore a crucial component of a human rights-based approach to elder law.

The process of assessing capacity also has significant human rights implications. The time when concerns about capacity are first identified can be confronting for the individual and can coincide with issues relating to their rights to health, housing and an adequate standard of living. Assessments of capacity are currently undertaken on a haphazard basis dependent upon the skill and experience of the assessor(s) involved. Where multiple professionals are involved there is potential for inconsistent information, particularly given that health and legal professionals may be applying different standards or attaching weight to different factors. This can result in a confusing, unduly lengthy and potentially flawed process for the individual involved. Given the significance of a negative determination for the individual's autonomy and other human rights, it is therefore important to ensure that both the assessment process and the outcome are optimal. Optimal in this case does not necessarily mean a finding of capacity. It means that respect was shown for the person's inherent dignity throughout the process and that they were given maximum opportunity for effective participation in that process, regardless of the eventual determination.

As seen above, standardised tests for capacity do not exist. They are also unlikely to be developed given the complexity involved in establishing a 'number' or threshold for individual capacities subject to the nature and the stage of the mentally disabling condition, as well as the need to then map this against the requisite legal framework

[158] Schopp (1996), 472.

[159] Purser and Sullivan (2019).

whilst taking into account the specificities of an individual's circumstances. Consequently, best practice guidelines are required. Guidelines offer the flexibility to apply a case-by-case approach adapted to the particular mentally disabling condition, as well as to the relevant legal context. Significantly, guidelines can—and should—be informed by human rights values and principles to ensure the most positive experience for the individual.

A key challenge for the development of guidelines has traditionally been the siloed manner in which capacity issues are dealt with. That is, legal professionals assess legal notions of capacity while health professionals assess clinical notions. This must change. Legal professionals are not trained to, for example, assess the effect of mid-stage dementia upon the legal test for testamentary capacity. Conversely, health professionals are not trained to assess the clinical notions of capacity within the requisite legal framework, and often may also not understand the effects of the mentally disabling conditions upon the particular decision-making capacity being assessed.[160] There are also a number of stakeholders in addition to the legal and health professionals who have a vital role to play in developing a best practice paradigm, for instance governments and policy makers, as well as allied health professionals, including social workers. Given the importance of financial capacity, banks and financial institutions also have a significant role to play in identifying capacity issues and working with older persons. This is especially the case when considering their potential role in addressing problems of elder abuse arising from the misuse of EPAs. All of these professionals have a role to play and a responsibility to respect and uphold the human rights of the individual. Best practice guidelines must therefore address the respective responsibilities of all relevant stakeholders. One other significant stakeholder who is often overlooked is the individual themselves. Stress, anxiety, fear, a lack of trust and confusion can all have deleterious effects upon the assessment process—ultimately also potentially infringing on the person's decision-making autonomy. The individual's voice in the assessment process is therefore incredibly powerful when considering the impact of not only the outcome of the assessment, but also the assessment process itself.

Guidelines informed by human rights therefore offer possibly the most achievable way of implementing best practice in capacity assessment. They should promote an approach which prioritises listening to and respecting the views of various stakeholders, especially the individuals who are having their capacity assessed. This will help to promote the full and effective participation of the older person in the assessment process. It will also demonstrate respect for the dignity of older persons. A human rights-based approach to capacity assessment, embedded in best practice guidelines, offers the best chance to protect the autonomy, independence and dignity of older people undergoing capacity assessments and can promote the fuller enjoyment of all human rights.

[160] Purser (2017).

References

Access Economics (2009) Keeping dementia front of mind: incidence and prevalence 2009–2050. https://www.dementia.org.au/sites/default/files/20090800_Nat__AE_FullKeepDemFrontMind.pdf Accessed 20 Feb 2020

AgeUK (2016) Later in life in the United Kingdom. https://www.ageuk.org.uk/Documents/EN-GB/Factsheets/Later_Life_UK_factsheet.pdf?dtrk=true. Accessed 10 February 2020

Alzheimer's Association (2016) 2016 Alzheimer's disease facts and figures. https://www.alz.org/facts/. Accessed 10 Feb 2020

Alzheimer's Australia (2009) Dementia: facing the epidemic. a vision for a world class dementia care system. https://www.dementia.org.au/files/20090901_Nat_Sub_DemFacingEpidemic.pdf. Accessed 10 Feb 2020

American Bar Association Commission on Law and Aging/American Psychological Association Assessment of Capacity in Older Adults Project Working Group (2008) Assessment of older adults with diminished capacity: a handbook for psychologists. https://www.apa.org/pi/aging/programs/assessment/capacity-psychologist-handbook.pdf. Accessed 10 Feb 2020

American Bar Association Commission on Law and Aging, American Psychological Association and National College of Probate Judges (2006) Judicial determination of capacity of older adults in guardianship proceedings: a handbook for judges. https://www.nawj.org/uploads/files/annual_conference/session_materials/aging/judicial_determination_of_capacity_of_older_adults.pdf. Accessed 10 Feb 2020

Arias JJ (2013) A time to step in: legal mechanisms for protecting those with declining capacity. Am J Law Med 39(1):134–159

Attorney General's Department of New South Wales (2008) Capacity toolkit https://www.justice.nsw.gov.au/diversityservices/Documents/capacity_toolkit0609.pdf. Accessed 17 Feb 2020

Australian Bureau of Statistics (2009) Australian social trends 4102.0—future population growth and ageing. https://www.ausstats.abs.gov.au/ausstats/subscriber.nsf/0/4FABEA5D1AA59548CA2575830015E7B0/$File/41020_populationprojections.pdf. Accessed 10 Feb 2020

Australian Institute of Health and Welfare (2016) Dementia. https://www.aihw.gov.au/reports-data/health-conditions-disability-deaths/dementia/overview. Accessed 10 Feb 2020

Australian Law Reform Commission (2013) Equality, capacity and disability in commonwealth laws, Issues Paper 44

Australian Law Reform Commission (2014) Equality, capacity and disability in commonwealth laws. Summary Report 124

Australian Law Reform Commission (2017) Elder abuse: a national legal response. Report No 131

Berg JW, Appelbaum PS, Grisso T (1996) Constructing competence: formulating standards of legal competence to make medical decisions. Rutgers Law Rev 48:345–371

Blanck P, Martinis JG (2015) "The right to make choice": The national resource center for supported decision-making. Inclusion 3(1):24–33

British Medical Association and the Law Society (2015) Assessment of mental capacity: guidance for doctors and lawyers, 4th edn. Law Society Publishing, London

Browning M, Bigby C, Douglas J (2014) Supported decision making: understanding how its conceptual link to legal capacity is influencing the development of practice. Res Pract Intellect Develop Disabil 1(1):34–45

Cairns R et al (2005) Reliability of mental capacity assessments in psychiatric in-patients. Br J Psychiatry 187(4):372–378

Carney T, Keyzer P (2007) Planning for the future, arrangements for the assistance of people planning for the future of people with impaired competency. Queensland Univ Technol Law Justice J 7(2):255–278

Carney T et al (2019) Realising 'will, preferences and rights': reconciling differences on best practice support for decision-making? Griffith Law Rev. https://doi.org/10.1080/10383441.2019.1690741

Cockerill J, Collier B, Maxwell K (2005) Legal requirements and current practices. In: Collier B, Coyne C, Sullivan K (eds) Mental capacity, powers of attorney and advance health directives. Federation Press, Annandale

Collier B, Coyne C, Sullivan K (eds) (2005) Mental capacity, powers of attorney and advance health directives. Federation Press, Annandale

Creyke R (1995) Who can decide? Legal decision-making for others. Australian Government Publishing Service

Croucher RF (2009) An interventionist, paternalistic jurisdiction'? The place of statutory wills in australian succession law. Univ NSW Law J 32(3):674–698

Dārziņš P, Molloy DW, Strang D (eds) (2000) who can decide? The six step capacity assessment process. Memory Australia Press, Adelaide

Dunn LB et al (2006) Assessing decisional capacity for clinical research or treatment: a review of instruments. Am J Psychiatry 163(8):1323–1334

Earnst KS et al (2001) Loss of financial capacity in Alzheimer's disease: the role of working memory. Aging Neuropsychol Cogn 8(2):109–119

Falk E, Hoffman N (2014) The role of capacity assessments in elder abuse investigations and guardianships. Clin Geriatr Med 30(4):851–868

Frost M, Lawson S, Jacoby R (2015) Testamentary capacity law, practice, and medicine. Oxford University Press, Oxford

Ganzini L et al (2003) Pitfalls in assessment of decision-making capacity. Psychosomatics 44(3):237–243

Gooding P (2013) Supported decision-making: a rights-based disability concept and its implications for mental health law. Psychiatry, Psychol Law 20(3):431–451

Grisso T (2003) Evaluating competencies: forensic assessments and instruments. perspectives in law and psychology, 2nd edn. Kluwer Academic/Plenum Publishers, New York

Gurrera RJ et al (2006) Cognitive performance predicts treatment decisional abilities in mild to moderate dementia. Neurology 66(9):1367–1372

Hall MI (2012) Mental capacity in the (civil) law: capacity autonomy and vulnerability. Gerontologist 52(1):61–94

Harada CN, Natelson Love MC, Triebel KL (2013) Normal cognitive ageing. Clin Geriatr Med 29(4):737–752

Horswill MS et al (2013) How realistic are older drivers' ratings of their driving ability? Accid Anal Prev 50:130–137

Jacoby R, Steer P (2007) How to assess capacity to make a will. BMJ 335:155–157

Jourdan JB, Glickman L (1991) Reasons for requests for evaluation of competency in a municipal general hospital. Psychosomatics 32(4):413–416

Kapp MB (2015) Evaluating decision making capacity in older individuals: does the law give a clue? Laws 4(2):164–172

Karlawish JHT et al (2005) The ability of persons with Alzheimer disease (AD) to make a decision about taking an AD treatment. Neurology 64(9):1514–1519

Kerridge I, Lowe M, Stewart C (2009) Ethics and law for the health professions, 3rd edn. Federation Press, Annandale

Kershaw MM, Webber LS (2004) Dimensions of financial competence. Psychiatry Psychol Law 11(2):338–349

Kershaw MM, Webber LS (2008) Assessment of financial competence. Psychiatry Psychol Law 15(1):40–55

Kitamura T, Takahashi N (2007) Ethical and conceptual aspects of capacity assessments in psychiatry. Curr Opin Psychiatry 20(6):578–581

Lai JM, Karlawish J (2007) Assessing the capacity to make everyday decisions: a guide for clinicians and an agenda for future research. Am J Geriatr Psychiatry 15(2):101–111

Law Reform Committee, Parliament of Victoria (2010) Inquiry into powers of attorney final report of the victorian law reform committee https://www.parliament.vic.gov.au/images/stories/committees/lawrefrom/powers_of_attorney/Report_24-08-2010.pdf. Accessed 10 Feb 2020

Law Society of New South Wales (2016) When a client's mental capacity is in doubt: a practical guide for solicitors (2016) https://www.lawsociety.com.au/sites/default/files/2018-03/Clients%2520mental%2520capacity.pdf. Accessed 10 Feb 2020
Lin SY, Lewis FM (2015) Dementia friendly, dementia capable, and dementia positive: concepts to prepare for the future. The Gerontologist 55(2):237–244
Liptzin B et al (2010) Testamentary capacity and delirium. Int Psychogeriatr 22(6):950–956
Lock SL (2016) Age-friendly banking: how we can help get it right before things go wrong. Public Policy Aging Rep 26(1):18–22
Lonie J, Purser K (2017) Assessing testamentary capacity from the medical perspective. Aust Bar Rev 44(3):297–317
MacArthur Research Network on Mental Health and the Law (2004) The MacArthur treatment competence study executive. https://www.macarthur.virginia.edu/treatment.html. Accessed 10 Feb 2020
Marson DC et al (2000) Assessing financial capacity in patients with Alzheimer disease a conceptual model and prototype instrument. Arch Neurol 57(6):877–884
Marson DC, Huthwaite JS, Hebert KR (2004) Testamentary capacity and undue influence in the elderly: a jurisprudent therapy perspective. Law Psychol Rev 28:71
Marson DC, Savage R, Phillips J (2006) Financial capacity in persons with schizophrenia and serious mental illness: clinical and research ethics aspects. Schizophr Bull 32(1):81–91
Marson DC, Hebert KR (2007) Financial capacity. In: Cutler BL (ed) Encyclopedia of psychology and law. Sage, Los Angeles 1:313–316
Marson DC, Sabatino CP (2012) Financial capacity in an aging society. J Am Soc Aging 36(2):6–11
Marson DC (2013) Clinical and ethical aspects of financial capacity in dementia: a commentary. Am J Geriatr Psychiatry 21(4):382–390
Moye J, Marson DC (2007) Assessment of decision-making capacity in older adults: an emerging area of practice and research. J Gerontol B Psychol Sci Soc Sci 62(1):3–11
Moye J, Marson DC, Edelstein B (2013) Assessment of capacity in an aging society. Am Psychol 68(3):158–171
Myers P (2014) An increase in mental capacity claims and challenges to wills. Br J Neurosci Nurs 10(1):44–45
O'Connor D (2010) Personhood and dementia: toward a relational framework and assessing decisional capacity. J Mental Health Training, Educ Prac 5(3):22–30
Okonkwo OC et al (2008) Awareness of deficits in financial abilities in patients with mild cognitive impairment: going beyond self-informant discrepancy. Am J Geriatr Psychiatry 16(8):650–659
O'Neill N, Peisah C (2011) Capacity and the law. Sydney University Press, Sydney
Papageorgiou SG et al (2018) Testamentary capacity assessment tool (TCAT): a brief instrument for patients with dementia. J Alzheimer's Dis 61:985–994
Peisah C (2005) Reflections on changes in defining testamentary capacity. Int Psychogeriatr 17(4):709–712
Peisah C, Brodaty H (1994) Dementia and the will-making process: the role of the medical practitioner. Med J Aust 161(6):381–384
Perlin ML (2013) Striking for the guardians and protectors of the mind: the convention on the rights of persons with mental disabilities and the future of guardianship law. Penn State Law Rev 117(4):1159–1190
Pinsker DM et al (2010) Financial capacity in older adults: a review of clinical assessment approaches and considerations. Clin Gerontologist 33(4):332–346
Population Reference Bureau (2015) Population bulletin 70(2) December 2015. https://assets.prb.org/pdf16/aging-us-population-bulletin.pdf. Accessed 14 Feb 2020
Posener HD, Jacoby R (2008) Testamentary capacity. In: Jacoby R et al (eds) Oxford textbook of old age psychiatry. Oxford University Press, Oxford
Purser K, Magner E, Madison J (2009) Competency and capacity: the legal and medical interface. J Law Med 16(5):789–802

Purser K (2015) Assessing testamentary capacity in the 21st century: is banks v goodfellow still relevant? Univ NSW Law J 38(3):854
Purser K (2017) Capacity assessment and the law: problems and solutions. Springer International, Switzerland
Purser K et al (2018) Alleged financial abuse of those under an enduring power of attorney: an exploratory study. Br J Soc Work 48(4):887–905
Purser K, Rosenfeld T (2014) Evaluation of legal capacity by doctors and lawyers: the need for collaborative assessment. Med J Aust 201(8):483–485
Purser K, Rosenfeld T (2016) Too ill to will? Deathbed wills: assessing testamentary capacity near the end of life. Age Ageing 45(3):334–336
Purser K, Lonie J (2019) Mapping dementia and cognitive decline in testamentary capacity. Int J Law Psychiatry 66:101450
Purser K, Sullivan K (2019) Capacity assessment and estate planning—the therapeutic importance of the individual. Int J Law Psychiatry 64:88–98
Qin W, Xiang X, Taylor H (2019) Driving cessation and social isolation in older adults. J Aging Health. https://doi.org/10.1177/0898264319870400
Queensland Law Society, Allens Linklaters and Queensland Advocacy Incorporated (2014) Queensland handbook for practitioners on legal capacity https://www.qls.com.au/Knowledge_centre/Ethics/Resources/Client_instructions_and_capacity/Queensland_Handbook_for_Practitioners_on_Legal_Capacity. Accessed 10 Feb 2020
Queensland Law Reform Commission (2008) Shaping Queensland's guardianship legislation: principles and capacity discussion paper, Working Paper No 64
Queensland Law Reform Commission (2010) A review of Queensland's guardianship laws. Report No 67, vol 1
Schopp RF (1996) Therapeutic jurisprudence and conflicts among values in mental health law. In Wexler DB, Winick BJ (eds) Law in a therapeutic key: developments in therapeutic jurisprudence. Carolina Academic Press, Durham
Setterlund D, Tilse C, Wilson J (2002) Older people and substitute decision making legislation: limits to informed choice. Australas J Ageing 21(3):128
Shulman KI et al (2009) Contemporaneous assessment of testamentary capacity. Int Psychogeriatr 21(3):433–439
Shulman KI et al (2015) Cognitive fluctuations and the lucid interval in dementia: implications for testamentary capacity. J Am Acad Psychiatry Law 43(3):287–292
Sprehe DJ, Kerr AL (1996) Use of legal terms in will contests: implications for psychiatrists. Bull Am Acad Psychiatry Law 24(2):255–265
Standing Committee on Social Issues, Parliament of New South Wales (2010) Substitute decision-making for people lacking capacity
Sturman ED (2005) The capacity to consent to treatment and research: a review of standardized assessment tools. Clin Psychol Rev 25(7):954–974
Sullivan K (2004) Neuropsychological assessment of mental capacity. Neuropsychol Rev 14(3):131–142
Suto WMI et al (2005) Capacity to make financial decisions among people with mild intellectual disabilities. J Intellect Disabil Res 49(3):199–209
Then S et al (2018) Supporting decision-making of adults with cognitive disabilities: the role of law reform agencies—recommendations, rationales and influence. Int J Law Psychiatry 61:64–75
UK Government (2015) Policy paper: prime minister's challenge on dementia 2020. https://www.gov.uk/government/publications/prime-ministers-challenge-on-dementia-2020/prime-ministers-challenge-on-dementia-2020. Accessed 10 Feb 2020
Victorian Equal Opportunity and Human Rights Commission (2012) Rights on the road: the experiences of older drivers. https://www.humanrightscommission.vic.gov.au/home/our-projects-a-initiatives/past-projects/item/485-rights-on-the-road-the-experiences-of-older-victorian-drivers. Accessed 17 Feb 2020

Wadley VG, Harrell LE, Marson DC (2003) Self- and informant report of financial abilities in patients with Alzheimer's disease: reliable and valid? J Am Geriatr Soc 51(11):1621–1626
Webber LS et al (2002) Assessing financial competence. Psychiatry, Psychol Law 9(2):248–265
Wong IY, Smith SS, Sullivan KA (2016) Psychosocial factors significantly predict driving self-regulation in Australian older adults. Australasian J Ageing 35(2):133–138
Wong IY, Smith SS, Sullivan KA (2018) Validating an older adult driving behaviour model with structural equation modelling and confirmatory factor analysis. Transp Res Part F: Traffic Psychol Behav 59(Part B):495–504
World Health Organization (2015) World report on ageing and health. https://apps.who.int/iris/bitstream/handle/10665/186463/9789240694811_eng.pdf;jsessionid=156CD1FA4F3C1063E492164A9B6D7CE3?sequence=1. Accessed 19 Dec 2019

International Instruments

Convention on the Rights of Persons with Disabilities (2006) Opened for signature 13 December 2006, entered into force 3 May 2008. 2515 UNTS 3
International Covenant on Civil and Political Rights (1966) Opened for signature 16 December 1966, entered into force 23 March 1976. 999 UNTS 171
International Covenant on Economic, Social and Cultural Rights (1966) Opened for signature 16 December 1966, entered into force 3 January 1976. 993 UNTS 3
Universal Declaration of Human Rights (1948) Adopted by United Nations General Assembly Resolution 217A (III). UN Doc A/810 (10 December 1948)

Cases

Bailey v Bailey (1924) 34 CLR 558
Banks v Goodfellow (1870) LR 5 QB 549
Re Beaney deceased [1978] 2 All ER 595
Boughton v Knight (1873) LR 3 P & D 64
Bull v Fulton (1942) 66 CLR 295
Re Caldwell [1999] QSC 182
Challen v Pitt [2004] QSC 365
GAU v GAV [2014] QCA 308
Gibbons v Wright (1954) 91 CLR 423
Hall v Hall (1868) LR 1 P & D 481
Hoff v Atherton [2005] WTLR 99
In the estate of Park deceased [1954] P 89
Johnson v Buttress (1936) 56 CLR 113
Re K (Enduring Power of Attorney) [1988] Ch 310
Lloyds Bank v Bundy [1974] 3 All ER 757
Re Loxston, Abbott v Richardson [2006] WTLR 156
Masterman-Lister v Brutton [2003] WTLR 259 CA
Re Matsis [2012] QSC 349
Newman v Smith (1918) 77 Fla. 633
Nicholson v Knaggs [2009] VSC 64
Parker v Felgate (1883) 8 PD 171
Read v Carmody [1998] NSWCA 182 (23 July 1998)

Sargent & Anor v Brangwin [2013] QSC 306
Shaw v Crichton [1995] NSWCA BC 9505228 (23 August 1995)
Timbury v Coffee (1941) 66 CLR 277
Tippet v Moore (1911) 13 CLR 248
Union Bank v Whitelaw [1906] VLR 711
Wingrove v Wingrove (1885) 11 PD 81

Legislation

Guardianship and Administration Act 2000 (Qld)
Mental Capacity Act 2005 (UK)
Powers of Attorney Act 1998 (Qld)

Chapter 7
Elder Abuse

Elder abuse is a global problem. However, notwithstanding its suspected prevalence, it remains relatively hidden, frequently occurring in the familial environment. Abuse against older persons violates a number of human rights including, for instance, the rights to security of the person; freedom from cruel, inhuman or degrading treatment; freedom of movement; an adequate standard of living; and the highest attainable standard of health. It is also irreconcilable with the core human rights values of dignity, autonomy and liberty. The law has an essential role to play in promoting respect for these values and protecting the human rights of older persons, particularly in relation to promoting pathways to access both formal and informal mechanisms for justice. A multitude of stakeholders, however, will be needed to successfully address elder abuse and the repercussions this can have on older persons, including promoting the participation of older persons themselves in developing strategies and interventions. Authentic engagement from institutional stakeholders is also essential, notably banking and financial institutions, as well as aged care facilities. This chapter builds upon previous work to apply the foundational human rights framework developed in Chap. 3 to the elder abuse context. It argues that a human rights framework has the potential to achieve the socio-cultural paradigm shift necessary to effect real change in successfully preventing elder abuse and addressing the ageist attitudes that enable such abuse.

7.1 Introduction

Elder abuse presents one of the most significant threats to the human rights of older people and can encompass a violation of a wide range of those rights. It is a global problem prevalent in both the developed and the developing world.[1] It is also a significant public health problem as victims can experience severe health consequences

[1] World Health Organization (2002a); Yon et al. (2017); 147.

© Springer Nature Singapore Pte Ltd. 2020

B. Lewis et al., *The Human Rights of Older Persons*,
https://doi.org/10.1007/978-981-15-6735-3_7

including, for instance, an increased risk of illness, depression, mortality and admission into formal health and aged care settings.[2] Health and aged care settings can further facilitate abuse, particularly the use of physical and chemical restraints in aged care (as will be discussed in Chap. 10). Notwithstanding its suspected prevalence and significance, elder abuse nevertheless remains a relatively hidden phenomenon with abuse often occurring in the familial environment.

Abuse against older persons represents a number of specific human rights violations. Where older persons experience physical abuse, it will be a clear violation of the rights to security of the person and freedom from cruel, inhuman or degrading treatment.[3] The latter can also be breached by other forms of cruel, degrading or humiliating treatment perpetrated against older persons, even where non-physical in nature.[4] The right to freedom of movement, which includes the freedom to choose one's place of residence, is clearly affected where an older person is denied the opportunity to unreservedly make decisions, absent incapacity, about their living arrangements or where their freedom to travel is curtailed.[5] The right to an adequate standard of living is also relevant to issues of housing and financial security, both areas in which abuse can be perpetrated.[6] Furthermore, as discussed in Chap. 8, financial independence is instrumental to the enjoyment of a wide range of human rights. Given that human rights are interdependent and indivisible, financial abuse or exploitation can therefore have a significant and detrimental effect on the full range of human rights, and on an older person's quality of life generally. Similarly, the right to the highest attainable standard of health ('the right to health') can be infringed where a person is denied accessible, appropriate and safe health and aged care, which takes on particular significance in light of the Royal Commission into Aged Care and Quality in Australia (discussed in Chap. 10).[7] Fundamentally, abuse perpetrated against older persons is clearly incompatible with the core values of human rights—respect for dignity, autonomy and liberty. Promoting respect for these values within law and policy is thus an essential part of effectively responding to elder abuse.

The World Health Organization (WHO) called upon states to address elder abuse as early as 2002.[8] However, although attention and public awareness is increasing, there still remains a dearth of knowledge about abuse perpetrated against older people—both the frequency and impact on the victims but also why the abuse was perpetrated and the circumstances in which it occurred. Significantly, there is also a lack of a theoretical and legal framework for identifying, preventing and responding to elder abuse, although a human rights framework is a prime candidate. Discussions are therefore continuing around whether abuse against older people should be criminalised with discrete offences, or whether general legal and equitable principles are

---

[2] Yon et al. (2017), 147. See also: Elder (2007); Dong and Simon (2013); Lachs et al. (1998).

[3] *International Covenant on Civil and Political Rights* (1966), arts 7, 9 ('*ICCPR*').

[4] Ibid., arts 7, 9; Human Rights Committee (1992, 2014), para 3.

[5] *ICCPR*, art 12; Human Rights Committee (1999).

[6] *International Covenant on Economic, Social and Cultural Rights* (1966), art 11 ('*ICESCR*').

[7] Ibid., art 12; Royal Commission into Aged Care Quality and Safety (2019a).

[8] World Health Organization (2002a).

adequate given the vast array of circumstances in which abuse can be perpetrated.[9] This then leads to the question of whether older people who have experienced, or who are experiencing, abuse have adequate and affordable avenues through which to access justice, both informal and formal, including the remedial relief available.

In considering how to best safeguard against elder abuse, the genuine involvement of a multitude of stakeholders is required. Abuse against older persons truly is everyone's problem.[10] Principally amongst those who should have a voice are older people themselves. Their participation is essential on a number of levels. Not only does it demonstrate respect for their autonomy and dignity by prioritising their inclusion, it also respects the value of the lived experiences of older people, which is essential to developing effective responses. Other stakeholders who should be involved range from family members and friends through to organisations such as financial institutions and those charged with the care of older people. Health and aged care workers in particular have a significant role to play as they regularly come into contact with cases of abuse against older people.[11] However, they, along with other institutions such as banks, often fail to recognise abuse—calling into question the current responses of a number of important institutional stakeholders.[12]

Education and dissemination of information are fundamental in preventing and responding to abuse against older people—both through professional education (such as, for instance, continuing legal and medical education) as well as through popular media, which can help educate families and older persons themselves about the risks and signs of elder abuse, as well as the best ways to respond to it. Such education campaigns could also help to address the ageist stereotypes that pervade society (discussed in Chap. 5), and which are thought to underpin much of the elder abuse that occurs.[13] Drawing upon a number of different perspectives is fundamental, however, to effectively target education campaigns. Such perspectives include, for instance, an appreciation of cultural and linguistic differences, the needs of those experiencing cognitive decline and those people in the 'old, old' category who may also be at an increased risk of experiencing abuse. It is particularly important to consider the gendered dimensions of elder abuse, as older women are at greater risk of experiencing abuse and other human rights violations.[14]

Successfully combating elder abuse requires a cultural shift to a society in which intergenerational harmony is cultivated, and abuse of older people is genuinely understood to be unacceptable and stigmatised accordingly.[15] Countries need to achieve appropriate recognition of the human rights of older persons whilst developing systemic approaches to address the abuse of older individuals. Such services need to be able to identify and appropriately respond to elder abuse with a view to, ultimately,

---

[9] Ibid.; Lacey (2014).

[10] Australian Human Rights Commission (2018).

[11] See, for example: Cooper et al. (2009); Australian Law Reform Commission (2017), 105.

[12] World Health Organization (2002a); Purser et al. (2020).

[13] World Health Organization (2002a).

[14] Ibid.

[15] Ibid.

preventing it. They will necessarily include multidisciplinary implementation across a number of wide-ranging professions including, for instance, legal, health, financial, social work and policing.[16]

The Madrid International Plan of Action on Ageing clearly acknowledges the significance of elder abuse and recognises the importance of human rights in informing our responses to abuse and other issues affecting older persons.[17] This book develops that approach further by proposing a foundational human rights framework for all law and policy relating to older persons and the ageing population generally. A human rights paradigm is especially suited to preventing elder abuse because it addresses a broad range of causes and effects of abuse. It also has potential to achieve the sort of cultural paradigm shift necessary to effect real change in successfully preventing elder abuse and the ageist attitudes that enable it. The approach suggested in this book goes beyond merely recognising the importance of human rights to demonstrate ways in which those rights can be practically and meaningfully translated into law and policy reform (discussed in Chap. 11).

This chapter therefore expands on how a human rights-based approach ought to be applied to elder abuse. First, some comments will be made in relation to the definitions adopted here, including the different forms of elder abuse and the definitional problems that exist. These definitional problems can affect understanding of the prevalence of elder abuse, which will also be briefly discussed. The risk factors for elder abuse will then be examined, including the increased risk to older women, before considering access to justice issues. The ability to access justice is a key principle within a human rights-based approach and an essential component in addressing elder abuse effectively. Challenges relating to accessing justice, including not only the available avenues for redress but also the potential remedial outcomes available, will therefore be identified. This will include the feasibility of elder mediation as a dispute resolution method in cases of abuse against older persons. Some comments will then be made in relation to safeguarding against elder abuse with a view to not only protecting, but positively promoting, the human rights of older people to live free from abuse.

7.2 Definitions

As a preliminary issue, it should be noted that the terms 'elder abuse' and 'victim' can be problematic. Despite the growing body of literature, there remains no one universal definition of what constitutes 'elder abuse'. As noted in Chap. 1, there is also debate in the literature about whether 'elder' or 'older' should be the preferred term when discussing older persons. 'Elder' can have particular cultural connotations, while defining the class of 'older' persons can be arbitrary. Some have suggested that the use of 'elder abuse' is inappropriate and anachronistic with the phrase 'abuse of (older)

[16]Ibid.

[17]*Madrid International Plan of Action on Ageing* (2002).

people' being one possible alternative.[18] This terminological 'unease' is attracting attention in, for example, Australia with the term 'abuse of older people' gaining traction.[19] Similarly, the use of the word 'victim' can have negative implications which deny the agency and empowerment of individuals and potentially reinforce their abuse. That being said, the term 'elder abuse' has become accepted terminology to describe the recognised phenomenon of abuse or exploitation of older persons, and it is used here for this reason. Similarly, the term 'victim' is used for convenience, but neither term is used with the intention of obscuring important considerations about the power of language, or the need to avoid defining older persons purely by their experiences of abuse.

With this in mind, a working definition of elder abuse is required to be able to discuss both the abuse perpetrated against older persons as well as its human rights implications. The WHO definition is the most widely accepted. That is, elder abuse is defined as 'a single, or repeated act, or lack of appropriate action, occurring within any relationship where there is an expectation of trust which causes harm or distress to an older person.'[20] Notably, the WHO highlights the importance of the existence of an '*expectation* of trust', whereas other definitions have instead noted that there should be an '*implication* of trust'—arguably a more general requirement.[21] This is because the need to establish an 'expectation of trust' debatably excludes abuse that occurs outside of a trusting relationship, thus potentially ignoring abuse perpetrated by unknown third parties through, for instance, internet scams.[22] This is not to deny that elder abuse frequently occurs in circumstances in which the abuser is well known to the victim, such as in familial and care situations, or to ignore the particular harm that flows from a breach of trust but, rather, to emphasise the wide gamut of settings in which elder abuse is perpetrated.[23]

The narrow definition based on trust does help, however, to highlight some of the common relationships in which abuse can occur, as well as some of the responses which are available. For example, when abuse occurs in relation to substitute decision-making through the misuse of an enduring power of attorney (EPA), the element of trust is central to the fiduciary relationship existing between the principal (the donor of the power) and attorney(s) (the donee).[24] It is the nature of the relationship that gives rise both to specific duties on the part of the attorney and to remedies which the older person, as principal, or someone on their behalf if they lack legal capacity, can pursue such as, for instance, claims for breach of fiduciary obligation, undue influence or unconscionable conduct.[25]

---

[18] Kaspiew et al. (2019), 8.

[19] Ibid.

[20] Ibid.

[21] Kaspiew et al. (2016), 2.

[22] Alzheimer's Australia (2016), 3; Hervey Bay Seniors Legal and Support Service (2016), 1.

[23] World Health Organization (2002a).

[24] *Smith v Smith* [2017] NSWSC 408.

[25] Fischer (2012), 85.

Drawing on the WHO definition and other literature, five main components therefore exist in examining the abuse of older people.[26] First, is the person who experiences the abuse. Secondly, the actual act(s) or omission(s) need to be considered, that is, the type of abuse and at what stage abuse or harm occurs. Moreover, taking into account the type and stage of abuse, are there identifiable thresholds for abuse which could help in establishing definitional consistency when undertaking prevalence studies?[27] Thirdly, focus also needs to be had on the perpetrator(s), including any risk factors for them engaging in abusive behaviour. Fourthly, the circumstances of the abuse are a key concern, both the environment of the perpetrator(s) as well as that of the person who is experiencing the abuse. This will also include investigating the relationship between the perpetrator and the victim as well as, for example, any care arrangements. This can particularly become an issue in assets for care situations which can foster financial abuse, especially where the presumption of advancement applies (discussed in Chap. 8). Briefly, the presumption states that any transfer from a parent to a child is deemed to be a gift out of natural love and affection.[28] In the abuse context, consider, for instance, an older person who has transferred ownership of her house to her daughter in return for care, but her daughter fails to fulfil her end of the agreement, instead taking the financial benefit without providing the agreed for care, perpetrating elder financial abuse. The presumption of advancement is significant because not only does the mother need to successfully run the equitable claim (absent a contract) to seek remedial relief, but she must first overcome the presumption. Although it is a rebuttable presumption, it nevertheless adds an extra layer of complexity, and cost. Finally, the consequences of the abuse need to be examined. This is a twofold question scrutinising not only what the outcome of the abuse is but also what consequences should occur. The consequences then need to be investigated from a number of different viewpoints. These include not only those of the perpetrator but also the person experiencing the abuse. Furthermore, the nature of any impact, such as the legal and remedial outcomes, emotional and financial repercussions, as well as any impact on health, care arrangements, relationships and social interaction should be assessed.

There is also some debate about whether elder abuse should be classified as a subset of domestic or family violence.[29] The latter discussion has predominantly arisen as a result of the idea that if seen as a 'category' of domestic or family violence, elder abuse will then attract more attention and, critically, funding. While it is crucial to recognise the problem of inter-personal abuse of older people, and the particular components and subtleties this entails, elder abuse occurs in a wider range of environments extending beyond the familial dynamic. Furthermore, elder abuse perpetrators do not come solely from within familial units. Therefore, while it might be practically convenient to utilise the public awareness about family violence, there

---

[26] See, for example: Kaspiew et al. (2019), 10–5.

[27] On the issue of thresholds and the international approach to developing them see, for example: Kaspiew et al. (2019), 18–21.

[28] *Nelson v Nelson* [1995] 184 CLR 538; *Johnson v Johnson* [2009] NSWSC 503.

[29] Wydall et al. (2018), 963–5; Australian Law Reform Commission (2016), 33; Fischer (2012), 85.

is a justified concern that adopting this approach would overlook and, consequently, further marginalise victims of abuse perpetrated against older people. This is particularly the case for older persons suffering abuse in other contexts beyond the family setting, for example, where it is a singular incident or episodic, involving strangers, or linked more to institutionalised ageism, such as abuse in aged care. Consequently, elder abuse, if characterised as a subset of family violence, may lose its identity and therefore risk not being recognised as the significant issue that it is.

In addition to the definitional challenges, elder abuse also exists in many forms. The categories of abuse can generally be stated to include: physical abuse, financial abuse, psychological or emotional abuse, sexual abuse, chemical abuse and neglect.[30] The discrete categories can, however, vary, as can the elements necessary to meet each type of abuse.[31] Abandonment, for example, is sometimes categorised as a separate form of abuse but has been included here within the neglect category.[32] Chemical abuse is also not commonly listed as a distinct form of elder abuse as it tends to be referenced in terms of chemical restraints used in the aged care setting, and is dealt with as such here (see Chap. 10).[33] Nevertheless, it is another form of abuse that is increasingly being recognised that involves using prescription and non-prescription medications to sedate an older person when the medications do not have any other justifiable therapeutic purpose.[34]

The following section will outline the general forms of elder abuse before making some comments on prevalence. It is important to identify the definitional problems in this area, particularly as establishing clarity around what constitutes each type of abuse is necessary in order to begin to collect rigorous prevalence data.[35] A human rights-based approach can assist with this definitional clarity, as specific human rights can be used to help identify instances of mistreatment, abuse and neglect, even where that behaviour might blur the lines between specific categories of abuse or prima facie appear acceptable. The categories of abuse below all represent violations of human rights, particularly rights to physical security and freedom from cruel, inhuman and degrading treatment, the right to health and freedom of movement, and the general principles of autonomy and dignity. As attention is increasingly being focused on the need for definitional precision, definitional ambiguity and the resultant problems (most notable in the prevalence data) will ideally become less of an issue.[36]

---

[30] Australian Law Reform Commission (2017), 19.

[31] Kaspiew et al. (2019), 18–9.

[32] See, for example: The American Psychological Association, Elder Abuse.

[33] On this, see further: Royal Commission into Aged Care Quality and Safety (2019b).

[34] Australian Law Reform Commission (2016), 18.

[35] Kaspiew et al. (2019), 18–9.

[36] Pillemer et al. (2016), S195.

7.2.1 Physical Abuse

Physical abuse is generally defined to include acts performed with the intention of inflicting physical pain and/or harm.[37] Such abuse can include, for example, hitting, punching, pushing or rough handling of any nature. Along with chemical abuse, physical abuse can often be caught up in the use of restrictive practices in aged care facilities (discussed in Chap. 10).

7.2.2 Financial Abuse

Elder financial abuse is defined by the WHO as, 'the illegal or improper exploitation or use of funds or resources of the older person.'[38] It encompasses any and all facets of an individual's financial affairs and can assume numerous forms including, for example, mismanaging assets (money and/or property), establishing loans where the older person acts as a guarantor without independent legal advice, and/or stealing.[39] It can also include failing to carry through on obligations arising under an assets for care arrangement (discussed in Chap. 8) and unduly influencing, coercing and/or forcing an older person to execute, revoke and/or amend testamentary and enduring documents, particularly wills and EPAs. This last example of abuse seems to be growing in frequency being informed by (incorrect) notions of entitlement to older people's assets, known as 'inheritance expectation'.[40] The misuse of EPAs is also believed to be one of the main contributing factors to the prevalence of elder financial abuse.[41]

There is no minimum monetary amount required for elder financial abuse to be categorised as such.[42] It is also reasonably common for financial abuse to occur in combination with the other types of abuse.[43] In fact, even despite the suspected underreporting and prevalence issues, financial abuse, as well as emotional abuse, are estimated to be two of the most frequently perpetrated forms of abuse against older people.[44]

---

[37] Ibid.

[38] World Health Organization (2013).

[39] Australian Law Reform Commission (2017), 42–3; Australian Law Reform Commission (2016), 33.

[40] Australian Law Reform Commission (2017), 42–3; See also: Setterlund et al. (1999); Setterlund et al., Cockburn and Hamilton (2011); Gilhooly et al. (2013); World Health Organization (2017b).

[41] Joosten et al. (2015), 19.

[42] Setterlund et al. (2007), 599.

[43] Ibid.

[44] World Health Organization (2017b).

The perpetrators of elder financial abuse are often family members as well as carers.[45] Significantly, abuse can be perpetrated irrespective of the abuser's intentions, whether malevolent or benevolent.[46] This point is important because perpetrators may not even realise that their actions constitute elder abuse. For example, in the Australian state of Queensland, strict conflict rules exist governing the exercise of power by attorneys under EPAs.[47] These legislative rules state that any transaction between the principal and the attorney, or a family member, close associate and/or friend of the attorney will automatically attract the statutory presumption of undue influence.[48] That is, the attorney is presumed to have breached their strict obligations and have, in effect, committed an act of abuse. This statutory presumption can be rebutted through, for example, evidence of the principal's fully informed consent based on the full and free disclosure of the attorney or, if this is not possible given that the principal has lost capacity, then an application can be made to an appropriate court or tribunal to seek permission for the transaction.[49] Nevertheless, if attorneys are unaware of their strict obligations and provisions such as these, then they may not know to take appropriate measures and thus may inadvertently perpetrate abuse. Consequently, particularly in light of the onerous legal and equitable obligations placed upon people who are managing someone else's financial affairs, financial abuse can be perpetrated against older people by both well-intentioned individuals as well as those deliberately setting out to financially exploit an older person.

7.2.3 Psychological or Emotional Abuse

Psychological or emotional abuse is generally defined to include intentional acts performed to inflict psychological or emotional pain and/or damage.[50] Psychological or emotional abuse can involve verbal abuse, harassment, bullying and/or calling an older person by insulting names. Infantilising an older person is another example of inflicting psychological or emotional abuse, and one which can arise as a result of ageist attitudes, that is, an older person is child-like and therefore needs to be treated as such. Other examples of psychological or emotional abuse also include constantly saying that the older person has dementia and/or is experiencing cognitive impairment and threatening them with placement in a 'nursing home' or aged care facility. Threatening to withdraw affection is yet another illustration of this type of abuse, as is preventing an older person from seeing family and friends (which may also be classed as social abuse).[51]

---

[45] Australian Law Reform Commission (2017), para 2.61.

[46] For abuse of enduring powers of attorney see, for example: Purser et al. (2018, 2020).

[47] *Powers of Attorney Act 1998* (Qld) s 73.

[48] Ibid., s 87; *Smith v Glegg* [2004] QSC 443; *Pinter v Pinter* [2016] QSC 314.

[49] *Powers of Attorney Act 1998* (Qld) s 118(2). See also: Purser et al. (2018, 2020).

[50] Pillemer et al. (2016), S195.

[51] Australian Law Reform Commission (2017), 19.

7.2.4 Sexual Abuse

Sexual abuse includes the criminal offences of rape and sexual assault, and any other unwelcome sexual contact or advances. Indecent assault, inappropriate sexual handling and touching, as well as sexually offensive language can also constitute sexual abuse.[52] Sexual harassment, for instance inappropriate comments about appearance and/or behaviour, and any behaviour on the part of the perpetrator that makes an older person feel uncomfortable about either their body and/or their gender also constitutes sexual abuse.[53]

7.2.5 Neglect

Neglect occurs when an individual (often a carer) fails 'to provide someone with such things as food, shelter or medical care.'[54] It involves deliberately or unintentionally failing to provide the 'necessities of life'.[55] Unintentional neglect occurs where the individual charged with the care of an older person lacks the requisite knowledge, skill and/or ability to be able to capably take care of another person or persons who are dependent upon them.[56] Intentional neglect occurs where the person caring for the older person abandons them. This includes failing to competently provide adequate sustenance, clothing, shelter and healthcare.[57]

In addition to neglect perpetrated by others against an older person, neglect can also be self-inflicted.[58] Self-neglect presents as a failure, inability or refusal to provide oneself with appropriate water, food, shelter, clothing, hygiene and/or medication.[59] It can also include failure to take adequate safety precautions where necessary, and can result in damaging consequences for the older individual in question such as physical and psychological well-being as well as, possibly, death.[60] Self-neglect can arise for a number of reasons, most notably physical and/or cognitive impairments.[61] This again also serves to heighten the significance of any undetected mentally disabling conditions, such as depression (discussed in Chap. 6).

[52]Ibid.

[53]Elder Abuse Prevention Unit. What is Elder Abuse? https://www.eapu.com.au/elder-abuse. Accessed 17 February 2020.

[54]Australian Law Reform Commission (2017), 6.

[55]Ibid.; Elder Abuse Prevention Unit (2019).

[56]Elder Abuse Prevention Unit (2019).

[57]Ibid.

[58]Dong (2017), 954.

[59]Ibid.

[60]Ibid.

[61]Ibid.

Given the lack of systematic research in this area, the risk factors for self-neglect are difficult to determine with any certainty.[62] What is clear, however, is that it is a significant form of abuse. It has been estimated, for example, to be the primary type of abuse experienced by older people reported to the Adult Protective Services in the United States of America (41.9%).[63] Although, again, definitions vary between jurisdictions and prevalence rates remain difficult to ascertain despite the increasing research being conducted in this area.[64]

7.3 Prevalence

Turning then to the question of prevalence of abuse perpetrated against older persons. Despite the gravity of the human rights violations and wider effects of elder abuse, significant gaps remain in understanding the frequency with which it occurs—both generally as well as when considering the specific forms of abuse.[65] Further, there is also a lack of understanding around the prevalence of the risks of elder abuse and how they present. The subsequent ability of victims to then seek redress through accessing both formal and informal justice mechanisms also remains relatively unquantifiable, including what barriers and enablers to accessing justice exist, and in what frequency as well as to what effect.

One of the perhaps most notable ramifications of a lack of definitional clarity and consistency is the impact this has on the ability to determine prevalence rates. If an issue lacks a clear, universally accepted definition, then it becomes very difficult to determine how common that particular issue is. This is also compounded by the problems presented by differing definitions of what it is to be 'older' and when someone attains this label (discussed in Chap. 1). For instance, national annual estimates of elder abuse have varied from 2.6% in the United Kingdom[66] to 29.3% in Spain.[67] Is the abuse perpetrated against Spanish older people so much more frequent than it is in the United Kingdom, or do definitional anomalies have a role to play in explaining the discrepancy? The lack of definitional consistency could also be an influential factor in the lack of quantitative syntheses that have occurred globally.[68] This is where data is even available to conduct such analysis.[69]

---

[62] Ibid.

[63] Ibid.

[64] Ibid.

[65] Yon et al. (2017), 147.

[66] Ibid.; Biggs et al. (2009).

[67] Garre-Olmo et al. (2009); Yon et al. (2017), 147.

[68] Yon et al. (2017), 147. On reviews of prevalence in this area, see, for example: Cooper et al. (2008); Dong (2015).

[69] For example, in Switzerland it has been stated that although data was collected, particularly in relation to aged care facilities, it was not reported prior to 2016, see: Lacher et al. (2016).

Although statistical certainty remains elusive, it is believed that the rates of abuse perpetrated against older people in middle to high income countries ranges generally from 2.2 to 14% globally.[70] Interestingly, particularly given the discussion in Chap. 6 about the importance of capacity as people age, and the link between diminished or lost capacity and the increased risk of elder abuse, this approximation does not include older people experiencing cognitive decline or mentally disabling conditions more broadly. In fact, there appear to be few studies internationally devoted to identifying the prevalence of elder abuse in people with dementia.[71] Notwithstanding this, the rate of psychological abuse of older people with dementia is estimated to range from about 28–62%, with physical abuse occurring in approximately 3.5–23% of cases.[72] These estimates do not take into account financial abuse, for which cognitive impairment is a significant risk factor, especially when considering the role of substitute and supported decision-making (also discussed in Chap. 6). Given the growing incidents of mentally disabling conditions, particularly the dementing illnesses, and the heightened risk of elder abuse where cognitive impairment exists, it is therefore important to be able to identify the prevalence of the various risk factors in planning, developing and implementing an effective response to elder abuse.

In a recent review of the relatively few prevalence studies in existence, the combined prevalence rate estimate for elder abuse (after examining studies published between 2002 and 2015 internationally) was approximately one in six older people, or 15.7%.[73] It is estimated that this, when considered in conjunction with the population figures, amounts to roughly 141 million older people who are victims of abuse every year.[74] Further, given these rates, and provided they remain constant, it is estimated that there will be 330 million older victims of abuse by 2050.[75] The prevalence estimate rate also varies significantly between and within regions.[76] For example: in Asia, it is anticipated that the rate is approximately 14% in India[77] ranging to 36.2% in China;[78] and in Europe the range recorded was as low as 2.2% in Ireland[79] to 61.1% in Croatia.[80] The rate was somewhere between 10 and 11.7% in the United States of America,[81] rising to 79.7% in Peru.[82]

---

[70] World Health Organization (2015), 74.

[71] Seven studies were identified in the quantitative review conducted by Yon et al. (2017), 150.

[72] World Health Organization (2015), 75.

[73] Yon et al. (2017), 150. See pp. 151–52 for a useful summary of all prevalence studies reviewed. This is similar to the estimates in Pillemer et al. (2016).

[74] Yon et al. (2017), 152.

[75] Ibid.

[76] Ibid.; Dong (2015).

[77] Chokkanathan and Lee (2005); Yon et al. (2017), 153.

[78] Wu et al. (2012); Yon et al. (2017), 153.

[79] Naughton et al. (2012); Yon et al. (2017), 153.

[80] Ajdukovic et al. (2009); Yon et al. (2017), 153.

[81] Acierno et al. (2010); Yon et al. (2017), 153.

[82] Silva-Fhon et al. (2015); Yon et al. (2017), 153.

Looking at the particular forms of abuse, the prevalence estimate for physical abuse was 2.6%, 0.9% for sexual abuse, and 4.2% for neglect.[83] As stated, psychological and emotional abuse (11.6%), and financial abuse (6.8%) are two of the most common forms of abuse.[84] In relation to financial abuse, it is estimated that approximately $14 million has been inappropriately obtained from older people in Australia, although this figure is more likely to be around $1.8 billion when accounting for the suspected under-reporting.[85] Further, it is estimated that somewhere between 0.5 and 5% of people aged over sixty-five years have experienced elder financial abuse on an annual basis.[86]

Staying with Australia as an example, a significant amount of data is collected by the Elder Abuse Prevention Unit (EAPU) which is situated in the state of Queensland. The EAPU operates an elder abuse hotline to provide assistance to older victims of abuse. The majority of callers, however, are not victims and are instead calling to report or obtain information about elder abuse. This highlights some of the problems with determining prevalence rates, that is, in addition to the suspected under-reporting, many people who do report elder abuse are witnesses rather than victims. Consequently, there is a necessary reliance upon these third-party accounts which are informed by the understanding, perceptions, interpretation and judgement about the abuse made by the witness.[87] Witness accounts can therefore be deliberately or unintentionally unreliable. What is clear, however, is that elder abuse is occurring on a grand scale globally and is therefore an international public health issue demanding immediate action. It is suggested here that a human rights-based framework may be able to assist with formulating and implementing such necessary action.[88]

7.4 Risk Factors

In order to develop and implement effective intervention and redress strategies, however, it is necessary to first understand the risk factors for abuse of older people—both for the victim as well as the perpetrator. It is therefore important to consider the relationship between the victim and the perpetrator, as well as the community and social circumstances of the victim.[89] It is also necessary to appreciate that the level of risk involved will depend upon individual circumstances, whether there are multiple risk factors in play, and the type of abuse. Pillemer et al., for example, have categorised risk factors as strong, potential or contested depending on the level

[83] Yon et al. (2017), 152.

[84] Ibid.

[85] Elder Abuse Prevention Unit (2015a).

[86] Chesterman (2017).

[87] Elder Abuse Prevention Unit (2015b), 6.

[88] Pillemer et al., 194.

[89] Ibid., 194, 198, 200.

and standard of evidence available to support them.[90] That is, strong risk factors are evidenced by a substantial evidence base, potential risk factors have a mixed evidence base, and there is a lack of evidence in relation to contested risk factors.[91] Risk factors can also be understood in terms of human rights, as factors which contribute to the likelihood of abuse frequently coincide with human rights-limiting circumstances. Conversely, strong protections of human rights, including meaningful respect for dignity and autonomy, and effective access to justice, can be valuable in helping to minimise the risk of elder abuse. Areas where human rights can contribute to our understanding of risk factors are identified in the discussion below.

Functional dependence, cognitive impairment, dementing illnesses and/or disability are all common risk factors for the individual.[92] That is, any loss of capacity can result in a heightened risk for the older person. This again emphasises the importance of satisfactory assessment processes as discussed in Chap. 6. Dependence, disability and cognitive impairment are strong risk factors for all types of elder abuse, but particularly for physical and financial abuse. Poor physical health and low socio-economic status are two other significant risk factors which can also result in all types of abuse, as is poor mental health, particularly depression.[93] In fact, the significance of depression, particularly when untreated, cannot be underestimated (discussed in Chap. 6).[94]

These risk factors are all closely related to the right to health. The Committee on Economic, Social and Cultural Rights explains that this right includes both physical and mental health and, importantly, acknowledges that the right goes beyond healthcare to include the determinants of health, including being safe from abuse (discussed in Chap. 10).[95] While some risk factors relating to cognitive decline or other mentally disabling conditions may be unavoidable, the right to health requires that treatment which can minimise the effect of these conditions ought to be made available. Further, the right to health demands that health services be accessible, affordable, appropriate and of sufficient quality. Appropriate public health and educational policies should also be implemented to try to promote the best levels of physical and mental health amongst older persons. Following these human rights, guidelines can therefore assist in minimising the health-related risk factors for elder abuse.

Other risk factors, which Pillemer et al. suggest fall within the 'potential' classification, include gender, age, financial dependence, race/ethnicity.[96] The inclusion of gender in the potential rather than the strong category is somewhat less surprising when considering the findings of not only the Pillemer et al. research, but also that of

[90] Ibid., 194, 198.

[91] Ibid.

[92] Cripps et al. (2002), Zannettino et al. (2015); World Health Organization (2017a); Pillemer et al. (2016); 198.

[93] Pillemer et al. (2016), 198.

[94] Ibid., 194, 199. On the effects of depression on the capacity assessment process, see, for example: Purser and Sullivan (2019); Purser et al. (2017).

[95] *ICESCR*, art 12; Committee on Economic, Social and Cultural Rights (2000).

[96] Pillemer et al. (2016), 199.

Yon et al., wherein the elder abuse perpetrated against older women was not demonstrated to have been significantly more frequent than that experienced by older men (which is discussed further below).[97] However, it is unexpected when considering that older women are traditionally more likely to relinquish financial control to others.[98] Consequently, gender may be a significant risk factor—this is another example of the need for rigorous prevalence data to provide useful and reliable information in order to inform adequate responses. A human rights-based approach would demand that laws and policies which address elder abuse aim to support all persons equally and without discrimination, pointing again to the need for accurate and current data on the prevalence of abuse against different cohorts so that effective and targeted strategies can be developed. An awareness of the interdependence of human rights would also help to improve our understanding of the way that different factors (including not just gender but also, for instance, race, religion and sexuality) might intersect to create particular vulnerabilities for women or other marginalised groups. The specific impact of elder abuse on older women will be considered in more detail below.

The mention of age amongst Pillemer et al's list of risk factors is interesting. On the one hand, there is an increased risk of cognitive decline as people age, particularly for those in the 'old, old' cohort and thus the risks associated with diminished or lost capacity increase.[99] However, on the other hand, assuming that there is an increased risk of abuse merely because someone has reached a specific chronological age is to actually perpetuate ageist assumptions reinforcing a protectionist rather than a rights-based approach to ageing. Financial dependence would also, it is suggested, be a significant factor in financial abuse—again highlighting the individual nature of the risk factors in relation to each form of abuse.

Other risk factors for abuse perpetrated against older people include visual and/or auditory impairment; social, cultural and linguistic factors; prior history of traumatic life events; family conflicts; and relational support as well as dependence (either physical and/or mental), including shared housing with the abuser.[100] In Australia, for example, dependence, particularly on the perpetrator, is a significant risk factor for abuse with statistics demonstrating that reliance on a family member for care was identified as a major risk factor in 81% of reported cases of elder financial abuse.[101] Shared housing, especially when there is an assets for care arrangement in place is a growing risk factor for abuse.[102]

Ensuring meaningful respect and protection of human rights, including the right to adequate housing and the right to privacy, along with respect for autonomy generally,

---

[97] Yon et al. (2017).

[98] Peri et al. (2009).

[99] Moye and Marson (2007), 3; Harada et al. (2013), 737–8; Moye et al. (2013), 162.

[100] For a discussion see: Cripps et al. (2002); Zannettino et al. (2015); World Health Organization (2017a).

[101] Bagshaw et al. (2013).

[102] Australian Law Reform Commission (2017), 19.

will help to reduce the potential of these risk factors to lead to elder abuse. At the same time, it must be recognised that some of these arrangements are entered into by older persons by choice, and respect for individual dignity and autonomy means that such choices should not be overridden on the basis that they might create a risk of abuse. Older persons should therefore be supported to make these decisions with the benefit of appropriate information and effective processes should be established to help them in the event that abuse occurs. Rather than minimise risk by limiting older individuals' choices, the risk of abuse in these situations of dependence needs to be reduced through addressing ageist attitudes and other cultural as well as behavioural factors which render these situations risky in the first place. Again, this is something that a human rights-based approach can assist with by promoting respect for older persons and positioning them as active rights holders, rather than dependent and passive recipients of care.

What is often not well understood is that in addition to the risk factors for individuals, there are also risk factors for perpetrators committing acts of abuse. However, knowledge about the risk factors for perpetrators remains a significant gap in the literature.[103] This is likewise a gap that is unlikely to be addressed given the methodological and ethical difficulties involved in collecting such data, not to mention the challenge of locating perpetrators of elder abuse who are then willing to acknowledge both what they have done and the reasons for their actions.[104] Having said this, risk factors for perpetrators generally include alcohol and/or drug abuse,[105] as well as whether the perpetrator themselves is experiencing financial difficulties. Significantly, the impact of dependence is also a risk factor for the perpetrator, particularly when the perpetrator is in a relationship with the victim characterised by financial, emotional and/or relational reliance.[106] Mental illness is also potentially a contributing factor, particularly depression which appears to be relatively common amongst perpetrators of abuse against older persons.[107]

In examining who the perpetrators are, the relationship between victim and abuser varies according to a number of factors such as the type of abuse as well as cultural background. For instance, in the United States of America, Europe and Israel, where data is available, the perpetrators were most commonly the spouse of the victim.[108] In Asia, however, the perpetrators are generally the children or the spouse of the children.[109] The literature is mixed on whether being married or single places an older person at risk (or a greater risk) of abuse.[110] Again, a wider commitment to promoting human rights may help reduce some of these risk factors for perpetrators, particularly where abuse is the result of the perpetrator's own vulnerability.

---

[103] Pillemer et al. (2016), 199.

[104] Ibid.

[105] von Heydrich et al. (2012), 75; O'Keeffe et al. (2007).

[106] World Health Organization (2015); Sethi et al. (2011).

[107] Pillemer et al. (2016), 199; Wiglesworth (2010); Vandeweerd et al. (2006).

[108] Burnes et al. (2015); Lowenstein et al. (2009); Pillemer et al. (2016), 200.

[109] Chokkanathan and Lee (2005); Pillemer et al. (2016), 200.

[110] Pillemer et al. (2016), 200.

Community and social considerations can also have a significant effect on the risk of elder abuse. Social isolation of an older person within the community as well as loneliness can be significant factors.[111] They can also be heightened by geographic remoteness. However, some literature suggests that it is instead older people in metropolitan areas that are at an increased risk of elder abuse.[112] The country in which a person lives can also increase the risk of abuse.[113] The normalisation of violence has been suggested as a further risk factor,[114] as have ageism and discrimination (discussed in Chap. 5).[115] These risk factors point again to the importance of promoting social inclusion and participation by older persons (Chap. 4). This will not only help to shift ageist attitudes but also to ensure that abuse does not go unnoticed and that older individuals have supportive communities around them, thus ideally reducing dependence and, importantly, providing avenues for assistance.

The risk factors for elder abuse are therefore varied. They depend upon a number of both disparate and interconnected factors taking into account both the victim and the perpetrator, as well as the relationship between the abused and the abuser. Community and societal dynamics can also significantly influence the risk of elder abuse.

7.5 Violence Against Older Women

Gender is a one such risk factor and a recurring theme throughout this examination of the application of a human rights framework to elder law. Violence against older women constitutes a significant human rights infringement.[116] Although the Sustainable Development Goals (SDGs) of the United Nations (UN) call for the end of violence against all women and a decrease in violence in general, a significant criticism is that they do not specifically address violence against older women, who may have a heightened vulnerability to abuse.[117] The fact that women are living longer than men and there are more women than men in the UN member states,[118] combined with the fact that women face considerable discrimination and injustice based on their gender which, in turn, increases their risk of poverty, poor health and violence suggests that older women may also be at an increased risk of abuse,

[111] For a discussion see Cripps et al. (2002); Zannettino et al. (2015); World Health Organization (2017a).

[112] Brozowski and Hall (2004; 2010); Pillemer et al. (2016).

[113] Pillemer et al. (2016), 200.

[114] Ibid.; Penhale et al. (2000).

[115] Sethi et al. (2011); Pillemer et al. (2016), 200.

[116] World Health Organization (2013); Yon et al. (2019), 245.

[117] United Nations General Assembly (2015), Targets 5.2 and 16.2 ('*UNSDGs*'); Yon et al. (2019), 245.

[118] United Nations Department of Economic and Social Affairs, Population Division (2015).

particularly given the rates of intimate partner violence (despite the prevalence rates discussed above).[119]

It is strongly suspected therefore that violence against older women is more frequent than that against older men. Surprisingly, as discussed above, research has discovered that in the reported prevalence studies there is no significant gender variance in the frequency of abuse perpetrated against older people.[120] However, some other studies have found there to be disparate rates of abuse between the genders.[121] While certain studies determined that abuse decreases as people age, other research has suggested that violence in fact continues at the same rates as people age.[122] Further, intimate partner violence against older women was traditionally ignored in research, although there is an increasing focus on this area that acknowledges the different experiences of older women from younger women.[123] It is therefore apparent that abuse against older women is yet another area in which prevalence data is severely lacking and desperately needed.[124] Notwithstanding these issues, however, it is estimated that abuse against older women ranges from roughly 0.8% in the United States of America to 3.8% in the United Kingdom through to 69.5% in China, with few multinational studies existing.[125]

In considering women, ageing and abuse, women face specific challenges as they age. For example, they can experience a lack of financial security (discussed in Chap. 8) and issues arising as a result of sexual assault, such as problems with reporting the assault and/or lack of access to justice—even if the assault is believed and then able to be reported.[126] This can, in turn, increase the risk to older women of abuse. There is a further danger that the issues older women experience may also disappear in the face of the focus on intimate partner violence, which tends to focus on younger women, and elder abuse more broadly. Consequently, it is clear that further research is needed in this area. Such research needs to address not only the prevalence of violence against older women, but also adequate safeguards as well as intervention programs in response to such violence, abuse and neglect bearing in mind the unique challenges facing older women.

---

[119] Lundy and Grossman (2009); Yon et al. (2019), 245.

[120] Yon et al. (2017), 153.

[121] Ibid.; Yon et al. (2014).

[122] Yon et al. (2019), 245; Wilke and Vinton (2005); Zink et al. (2003).

[123] See, for example: Band-Winterstein (2015).

[124] Ten studies of subpopulations (older women and minorities) were included in a recent quantitative review of prevalence studies, see: Yon et al. (2017), 150.

[125] Yon et al. (2019), 245. See also: Fritsch et al. (2005); Biggs et al. (2009); Su et al. (2011); De Donder et al. (2011).

[126] Fileborn (2017).

7.6 Access to Justice

As introduced previously (both above and in Chap. 1) access to justice is a serious issue facing older people generally, but particularly when they have experienced abuse. A person's ability to access justice has been identified as 'their knowledge, skills and readiness to act in response to legal problems.'[127] A person's legal capacity to be able to participate is also relevant, especially when considering that people with diminished or lost capacity can also be at a risk of increased vulnerability. This is because their rights to make legally enforceable decisions may be reduced or even removed, sometimes erroneously if experiencing unsatisfactory assessment processes.[128] Access to justice is one of the key principles underpinning a human rights-based approach, flowing from the recognised entitlement that all persons have to receive an effective remedy when their human rights have been infringed.[129] It is also a key component of the rule of law, which demands that all persons be equally entitled to the protection of the law and that states should provide transparent, fair, effective, non-discriminatory and accountable services for all.[130]

An older person may also be at an increased risk of ongoing or escalating abuse if they are unable to access legal and justice systems in a timely and effective way when experiencing exploitative or abusive behaviour. In fact, there are a number of interrelated challenges which can make effectively accessing justice difficult in the context of elder abuse. First, there is the invisibility of older persons in society generally, and the legal system more specifically, often resulting from ageist attitudes and assumptions.[131] Older members of the LGBTIQA community are arguably even more invisible, particularly fearing discrimination in aged care.[132] Members of the LGBTIQA community are more likely to live alone, less likely to be partnered or to have children, and have increased rates of isolation and loneliness, all of which can contribute to the risk of elder abuse.[133] There are also unique challenges faced by people who are members of multiple minority groups, that is, older LGBTIQA people from culturally and linguistically diverse backgrounds.[134] Secondly, there can be a lack of awareness on behalf of the older person about their rights and/or an unwillingness and/or inability to enforce them. This can arise for a number of reasons including, for example, a lack of information about how to act or feelings of not wanting to be a burden on the system, which are reinforced by ageist assumptions and practices. Geographic isolation can also serve as a barrier to accessing justice given that it may be difficult to practically access affordable and quality advice if there are no locally based services or avenues through which to access such services. Victims

[127] Law Council of Australia (2017), 8.

[128] See, for example: Purser (2017).

[129] *ICCPR*, art 2(3).

[130] United Nations, Access to Justice.

[131] Lacey (2014), 109.

[132] See, for example: Gutman et al. (2017); Johnson et al. (2005).

[133] Ibid.

[134] For more on this see, for example: Kum (2017).

are often vulnerable and may be isolated from support systems, making speaking up more difficult. The financial cost involved can also be prohibitive when attempting to access justice. In addition to financial cost, there can also be an emotional cost. This can result from, for example, feelings of shame, fear and/or embarrassment on behalf of the person who has experienced the abuse, particularly in relation to financial and sexual abuse, or potentially through the anxiety that may arise in anticipation of the loss of the relationship with the abuser. Thirdly, there can often be a lack of accessible and effective mechanisms of dispute resolution ranging in formality, which can again be compounded by geographic isolation. Finally, fear of repercussions from reporting the abuse, either to themselves or to the perpetrator, can also serve as a barrier to accessing justice.

There are several international instruments that are relevant when considering access to justice for older people from a human rights perspective. The *International Covenant on Civil and Political Rights* (ICCPR) guarantees that all people should be able to exercise their rights, which includes holding perpetrators and decision-makers accountable and being treated equally before the law.[135] The *Convention on the Rights of Persons with Disabilities* (CRPD) explicitly declares that states should enforce effective and equal access to justice for persons with disabilities, including through the implementation of appropriate procedural and age-related provisions.[136] Under article 12 of the CRPD, states parties are obliged to ensure that persons with disabilities have equal recognition before the law and equal access to legal support to ensure that their rights can be upheld.[137] These obligations would obviously apply to any older person who lives with a disability, but are equally applicable to all older persons under the general right to equality before the law and the right to a remedy recognised in the ICCPR.

Regional human rights regimes also emphasise the central importance of access to justice for ensuring older persons' rights, including in situations of elder abuse, and have elaborated on the obligations of states in delivering access to justice. The *Inter-American Convention on Protecting the Rights of Older Persons* ('the Inter-American Convention') lists access to justice among the general duties of states found in article 4. This is elaborated on in article 31, which explicitly requires that states parties ensure access to justice for older persons on an equal basis with others, including by providing appropriate procedural accommodations. The *African Protocol on the Rights of Older Persons* includes similar provisions to ensure both equal treatment of older persons and appropriate assistance to enable access to justice. Both the African and Inter-American regimes recognise that access to justice also requires effective training of judicial, administrative and law enforcement personnel to ensure that

[135] *ICCPR*, arts 2, 26, 16.

[136] *Convention on the Rights of Persons with Disabilities* (2006), art 13.

[137] Committee on the Rights of Persons with Disabilities (2014).

those working within the justice system perform their work with a view to upholding the rights of older persons.[138]

The Inter-American Convention further requires steps to ensure 'differentiated and preferential treatment for older persons in all areas.' Article 31 states that 'judicial action must be particularly expedited in instances where the health or life of the older person may be at risk.'[139] The European Court of Human Rights has similarly determined that administrative and judicial authorities are obliged to act within a 'reasonable time'. That is, where there is an applicant of advanced age there should be 'exceptional diligence' applied to determining the matter when considering both the age, as well as the health of the applicant. This is deemed to be a requirement under the right to a fair trial, found in the *European Convention on Human Rights*.[140] A similar interpretation would likely apply to the right to a fair trial articulated in article 14 of the ICCPR, under which the interests of justice are generally considered to require that legal proceedings be pursued without undue delay.[141]

The lack of domestic enforceability of human rights instruments (discussed in Chaps. 1, 2 and 3) is also problematic when promoting effective access to justice. This is because a lack of affordable and/or accessible enforcement mechanisms fundamentally restricts the vehicles through which older people can attempt to enforce their rights and thus effectively access justice. While international and regional human rights regimes do provide enforcement mechanisms, these are frequently unavailable until domestic measures have been exhausted and are not designed to substitute for effective domestic processes. To comply with their human rights obligations, states therefore need to ensure that laws and procedures are available at the domestic level to guarantee older persons' equal access to justice.

Given the previous discussion focusing on older women and abuse, it is also important to note that older women may face unique barriers when attempting to access justice. Women are more likely to experience gender-based discrimination (see Chap. 5), and poverty and/or financial insecurity (see Chap. 8) which can make them more likely to lack the necessary resources to be able to fund quality legal assistance if required. Such barriers can then reinforce not only the hidden nature of elder abuse but also the perpetuation of the cycle of abuse. This is because the experiences of women are often dismissed or downplayed not only by those who have been sought out for help, but also by those who are witnesses to the abuse.[142]

One method of promoting accessibility and affordability of dispute resolution methods is through elder mediation. Without effective management, conflicts, especially those arising as a result of dysfunctional family dynamics, can escalate and lead

---

[138] *Inter-American Convention on Protecting the Rights of Older Persons* (2015), art 31(b) ('*IACROP*'); *Protocol to the African Charter on Human and Peoples' Rights on the Rights of Older Persons* (2016), art 4(3).

[139] IACROP, art 31.

[140] *European Convention for the Protection of Human Rights and Fundamental Freedoms* (1950), art 6(1).

[141] Human Rights Committee (2007).

[142] Fileborn (2017).

to abuse. Elder mediation is an alternative dispute mechanism designed to resolve such conflicts without recourse to costly and time consuming (more) formal legal proceedings.[143] It is a dispute resolution method designed specifically for older people and is intended to be complementary to other forms of conflict resolution intervention such as advocacy and legal determinations. It should also be viewed in conjunction with available health, psychiatric, and social work assistance, where necessary. Elder mediation is frequently conducted by specialist mediators, often with a social gerontology background. As a process, it is intended to disrupt the passivity and dependency that are often pre-conditions of abuse, neglect and self-neglect.[144] Adopting a less formal dispute resolution pathway should, ideally, foster constructive family discussions resulting in outcomes promoting the rights and needs of the older person while nurturing positive familial relationships. Elder mediation is therefore meant to be a preventative and positive process focusing on promoting an older person's autonomy, self-determination, independence, dignity and participation in the process of making decisions that will impact them. As with any form of mediation, however, it is important to bear in mind power imbalances which can negatively affect the process.[145]

Irrespective of the formality of the dispute resolution method involved, and the vehicle through which access to justice is sought, the central question of cost remains. Funding is essential to any serious discussion around improving effective access to justice, including the affordability and quality of such mechanisms. In any context appropriate funding raises difficult questions to which there are no easy answers. What is clear is the need for some form of legal centre based within communities to ensure ease of access for older people to, at the very least, useful information. Such centres must be adequately staffed to ensure that the legal advice being accessed is 'quality' advice (in and of itself a significant issue). This is especially the case for those older people in regional, rural and remote areas who may experience not only geographical isolation but also poor quality of, for example, the internet. It is thus important to remember that the solution is not always technology based.[146] Furthermore, heavy reliance on technology can actually serve to erode the trust that is necessary between older people and their legal advisers (as well as in legal systems), thus jeopardising effective access to justice if the older person is not comfortable or mistrusts the legal actors and/or systems. Certainty and continuity over funding for various dispute resolution methods, ranging from informal through to formal, would therefore assist in the promotion of a human rights framework to accessing justice. Inadequate funding and/or (threatened or actual) funding cuts, as well as poor quality services, can result in increased workloads for courts, police, hospital and other community-based organisations who may already be overloaded, not to mention the potential significant infringement on an older person's human rights.

---

[143] Kornfeld-Matte (2016), para 9.

[144] Ammerman and Herson (1991), cited in Craig (1994), 86–7.

[145] National Alternative Dispute Resolution Advisory Council (1997); Williams (2013); Crampton (2013).

[146] See, for example: Sharkey (2012); Bennett et al. (2017).

Such outcomes only serve to foster frustration, anxiety, suspicion and/or distress amongst the older people (who may be vulnerable) attempting to access the services, thus endangering access to justice.

To effectively promote access to affordable and quality mechanisms through which justice can be sought, it is therefore vital to critically review relevant laws, policies, practices, systems, as well as the role of legal actors, that is both lawyers and the judiciary.[147] Legal actors in particular have an important role to play in making justice accessible—including in understanding how to effectively relate to vulnerable older clients in order to maximise the individual's ability to access justice. While technology can sometimes serve as a barrier to creating the human relationships that are critical in this area, it can also be used to positive effect, for example, in fostering access to quality legal advice for older people in geographically remote locations—when handled sensitively and remembering the limitations of technology.[148] It is also important to remember that the 'law' can be imposing, intimidating and expensive to access—particularly to both locate and then obtain 'quality' advice. Accordingly, governments, policy-makers, legal practitioners and the judiciary all have important roles to play in overcoming this and in encouraging authentic avenues through which to access justice, which, as discussed, is a fundamental human right and a basic principle in the rule of law.

7.7 Safeguarding Against Abuse

Despite all the work currently being undertaken, still more is needed to effectively safeguard against the abuse being perpetrated against older people. Such work is necessary to not only protect human rights but to also address the growing global public health problems presented by the increasing rates of abuse which have not attracted the same degree of attention as intimate partner violence or public health concerns more generally.[149] For example, as mentioned above, despite the SDGs addressing violence against women, there is no mention of elder abuse specifically.[150] There are also few evidence-based interventions addressing elder abuse given that the lack of rigorous data makes it difficult to identify appropriate interventions, as does the absence of international comparisons assessing the effectiveness of intervention programs.[151] Compounding this problem even further is the lack of information about cost, a key component in the success of any measure that is adopted.[152] Consequently, both developing and implementing effective interventions to address elder abuse must

---

147 Law Council of Australia (2017), 6.

148 See, for example, Sharkey (2012); Bennett et al. (2017); World Health Organization (2017c); Garçon et al. (2016); Cocco (2011).

149 Yon et al. (2017), 154–5.

150 Ibid., 153; UNSDGs.

151 Pillemer et al. (2016), 200.

152 Ibid.

become a priority for global communities.[153] One of the ways in which to further this work is by clearly demarcating elder abuse as an infringement of human rights and establishing a human rights-based legal and policy framework in which to begin to establish enforceable domestic mechanisms to protect those rights. Developing a more detailed understanding of the relationship between human rights, ageism and elder abuse is a crucial first step, followed by meaningful progress on improving protections for all human rights of older persons in order to minimise risk factors, ensure access to justice and ultimately engender the sort of cultural shift that will help to eliminate abusive behaviours. This section outlines a number of recommendations which could assist in the implementation of a human rights-based approach to prevent elder abuse.

What does seem to be emerging from the literature is the fact that increased social connectivity and support lessens an older person's risk for abuse.[154] Similarly, the accommodation (see Chap. 9) and care arrangements (Chap. 10) of the older person play a significant role in the risk of abuse, and thus considering ways in which to effectively address this source of potential risk must be a priority.[155] Interventions therefore should consider not only accommodation for older people, but also interventions for health and aged care-givers.[156]

Education programs and public information campaigns could also prove to be invaluable across all forms of abuse, but particularly in relation to the provision of health and aged care, and financial management.[157] Such programs could, and should, be aimed at both the older person as well as family members and care providers. Consider, for example, elder financial abuse. As mentioned above, one of the major vehicles for this type of abuse to occur is through the misuse of an EPA. Often the abuse is unintentional and occurs because the attorney does not understand their obligations.[158] Effective education programs could assist. The key word is, however, *effective*—and much evidence-based work is needed to determine how to successfully achieve this. Helplines, such as the EAPU discussed above, are also thought to be useful interventions, mainly because of their ease of use and accessibility—particularly when staffed by multidisciplinary teams with relevant expertise.[159]

Many suggestions for protective mechanisms to safeguard against elder abuse have been made. For example, it is necessary to not only raise awareness about the issue of elder abuse within wider society, but to also sensitise the community and society as to its effects, as well as how to identify and protect against it. The public information campaigns and education programs mentioned above, for example, could be useful in achieving this. Adopting a law reform and legislative agenda designed

[153] Yon et al. (2017), 155.

[154] Pillemer et al. (2016), 200; Dow and Johnson (2012).

[155] Pillemer et al. (2016), 200; Peterson et al. (2014).

[156] Pillemer et al. (2016), 201.

[157] Ibid.

[158] Purser et al. (2018).

[159] Pillemer et al. (2016), 201.

to protect older people is also necessary—such an agenda will necessarily require powers of investigation, redress and enforcement while respecting the independence and autonomous right of individuals to participate in their own decision-making where possible.[160] However, in order for these safeguards to be effective, there is a significant issue in relation to not only domestic enforcement but also funding that will first need to be addressed.

Facilitating access to appropriate legal, health and social care services is also fundamental to effectively intervene in abuse. This then depends on the provision of better access to information about the available supports and services.[161] Increasing available emergency accommodation is one such service given that one of the significant risks for older people who have experienced abuse is loss of accommodation and the security that accompanies that. Providers of services dealing with older victims of abuse should also be given appropriate training because dedicated programs have yielded positive outcomes in helping to provide support for at-risk older people.[162] In fact, training more broadly is required, including, for instance, for legal actors and the judiciary, as well as healthcare and social workers.[163] Moreover, those involved with law enforcement need training in order to be able to effectively respond to allegations of elder abuse.[164] Transportation is another important consideration for older people attempting to leave abusive situations or access justice, particularly for those older people with physical and cognitive impairments that prohibit driving.[165]

In relation to violence against women, such emergency accommodation and transportation options are vital, particularly shelters that are responsive to age-specific issues. Further, it is imperative that programs are developed that can support older women who have been the victims of abuse—distinct from the need to develop programs in response to elder abuse more generally.[166] Promoting and protecting human rights provides the framework within which to encourage empowerment—which is an important aspect that can be incorporated into programs being designed and developed in response to elder abuse generally, but also in relation to abuse against older women in particular.[167]

On the national level, action is slowly occurring, although there is a question as to the practical utility of much of the dialogue. For example, a key recommendation of the recent Australian Law Reform Commission (ALRC) Inquiry into Elder Abuse was the adoption of a national plan as the key mechanism through which to combat elder abuse in Australia, and that this plan be developed within a human rights framework.[168] The plan was then to incorporate reforms to laws relating to aged care, EPAs,

---

160 United Nations Department of Economic and Social Affairs (2019).

161 Ibid.

162 Yon et al. (2019), 245; Heck and Gillespie (2013); Pillemer et al. 2016

163 United Nations Department of Economic and Social Affairs (2019).

164 Ibid.

165 Yon et al. (2019), 245; Straka and Montminy (2006).

166 Yon et al. (2019), 245; Straka and Montminy (2006); McDonald and Collins (2000).

167 Yon et al. (2019), 245; Straka and Montminy (2006).

168 Australian Law Reform Commission (2017).

superannuation or accrued retirement benefits, banking, and wills and guardianship, which currently vary across the different Australian states and territories.[169] However, as noted by the ALRC, the actual implementation of such a plan falls within the purview of others, many others, who sometimes have competing objectives, limited power to impact legislation and/or policy, and/or limited funding.[170]

This is not the only problem faced by the adoption of an elder abuse plan and human rights framework. In Australia, for example, the ALRC did not go far enough in its recommendations in that the human rights approach was not woven through the recommendations. Rather, the proposed framework sits separately from the recommendations and thus is easily discounted—a problem facing not just Australia, but many jurisdictions without strong domestic human rights protections or integrated legal frameworks. Further, if adopted into practice, such national plans and recommendations are at risk of being even further removed from the human rights and underlying principles they were intended to protect. To ensure a meaningful and durable human rights-based approach it is thus essential not only that human rights are adequately protected within domestic legal frameworks, but also that those protections are translated into authentic implementation within elder abuse law and policy.

Fundamental also to the success of any policies, programs and/or systems developed to safeguard against the abuse of older people is the need to address the pervasive issues of not only ageism, but also the gender and power dynamics arising from social norms.[171] Such norms have traditionally been instrumental in informing normative opinions, beliefs and actions. Consequently, effecting any genuine and long-standing change will require the modification of such behaviours and, at its core, the constitution of socially acceptable interactions and behaviours.[172] Such a paradigm shift facilitating a truly human rights-focused framework will help to build greater respect for the dignity of older persons and human rights generally. A key component to be able to achieve this is the collection of relevant data in relation to elder abuse broadly, including prevalence data. Such data should also represent the voices of older people, particularly those who have been victims of abuse.[173] Their participation is fundamental in engendering a meaningful human rights-based paradigm. It should also involve multidisciplinary teams of researchers who can draw on the skills of different fields, notably legal, medical and health professionals but also social workers. Front-line services and agencies should be encouraged to participate in rigorous research as their experiences in assessing the practical utility of the interventions being offered will be invaluable.[174] Any research will also need to address the core concern of funding.

---

[169] Ibid.

[170] Ibid., 15.

[171] Yon et al. (2019), 245; Straka and Montminy (2006).

[172] Yon et al. (2019), 245; Straka and Montminy (2006).

[173] See, for example: World Health Organization (2002b); Wydall et al. (2018), 964; Purser and Sullivan (2019).

[174] Baker et al. (2017), 346.

7.8 The Role of Legal Professionals in Addressing Elder Abuse

As well as improving national and international human rights protections, adequately addressing elder abuse has to occur from the ground up.[175] Legal professionals therefore have a significant role to play in not only recognising and responding to elder abuse, but also in its prevention.[176] Legal professionals have strict ethical, fiduciary, tort and contractual obligations. However, failure to meet any of these standards can result in enabling the abuse to continue to occur. The question is, to what extent, and how, should legal professionals be involved in responding to elder abuse within a human rights framework bearing in mind their relevant legal and ethical responsibilities? This question is a fruitful area for future research, particularly considering potential obligations which legal or other professionals might bear in response to identifying and reporting elder abuse experienced by their clients.

It is clear that increasing the education of legal professionals in relation to identifying and effectively responding to elder abuse will help support strategies designed to respond and prevent the abuse from occurring.[177] Developing and then implementing best practice paradigms is one avenue through which to attempt to make a positive impact. Such best practice guidelines would not only be in relation to engaging with older clients but should also take into account capacity assessments when considering the role of impaired or lost capacity in increasing the vulnerability of older people to abuse.[178] Access to information, in this case information about abuse as well as the legal redress and remedial relief available, is a significant component of effectively facilitating access to justice. This then subsequently serves as a vehicle for the promotion of autonomy, respect for individual dignity and participation. It has been suggested that best practice would involve the following: private face-to-face meetings with the client asking open-ended questions; carefully documenting the instructions as well as the scope of the retainer; refusing to act for more than one client, noting the relevant legal, professional and ethical rules; suggesting independent legal advice where appropriate; considering issues in relation to diminished or lost capacity, including the necessity of an assessment and who should be involved (for example, health professionals) and referring to relevant context-specific best practice guidelines; and considering speaking to health as well as medical professionals where the client is in aged care or a hospital setting.[179]

In addition to best practice, it has been suggested that screening measures could be adapted from other disciplines, such as for health professionals, in order to facilitate

[175] Lacey (2014), 105.

[176] See, for example: Purser et al. (2020); Ries (2018).

[177] Ries (2018).

[178] On capacity assessment and best practice guidelines, see, for example: Purser et al. (2017).

[179] Cockburn and Hamilton (2009); Hamilton and Cockburn (2008); Purser et al. (2020); Lonie and Purser (2017); Purser and Rosenfeld (2014).

the effective identification and prevention of abuse.[180] These measures include being aware of signs of elder abuse, understanding the situation as a potential form of abuse, determining whether action is within the scope of the legal professional's responsibility, knowing strategies for action and then deciding to act.[181] It would seem that the suggested best practice paradigms and the screening measures therefore cover similar theoretical territory. Consequently, the real issue then seems to concern how to effectively implement such measures in practice to facilitate respect for individual autonomy, dignity and participation. Fundamental to this is the need to promote efficient and cost-effective access to justice through the identification of appropriate avenues of legal redress and remedial relief, be they the less formal mechanisms offered through, for example, elder mediation, or the formal judicial determinations made by courts.

7.9 Conclusion

Elder abuse occurs in many forms, ranging from neglect of care and emotional abuse to financial and physical abuse. All forms represent a violation of human rights, if not directly, for example as a breach of the right to liberty and security of the person, then at least indirectly, for instance if financial exploitation undermines the older person's ability to enjoy their other human rights. These rights are all guaranteed under international human rights law, but currently there is no law which recognises the specific impacts of elder abuse, its origins in ageist community attitudes, or the particular challenges which exist in achieving access to justice for victims. Elder abuse therefore demonstrates very clearly the need for a dedicated convention on the rights of older persons (CROP—discussed in Chap. 2).

Without a dedicated convention, a human rights-based approach to elder abuse relies on the various rights found principally in the ICCPR and ICESCR and, importantly, their implementation within national legal frameworks. Currently this implementation is lacking in many respects, and stronger protection as well as enforcement mechanisms are needed to guarantee all older persons' rights to be safe from abuse. Notwithstanding this, a human rights-based approach is still a strong foundation on which to anchor the standards and systems necessary to protect vulnerable older people from abuse. This is because such a framework transcends national jurisdictional boundaries and instead seeks to establish clear principles that can then be adapted within domestic borders. Human rights provide a set of standards of treatment which can be used to help identify neglect, abuse and exploitation, as well as a language for articulating the nature of harms which emphasises that every older person is entitled to be safe, to be free from exploitation and intimidation, and to have their dignity and autonomy respected.

[180] Ries (2018), 29. On screening measures for health professionals see, for example: Ries and Mansfield (2018).

[181] Ries (2018), 29.

Work is needed, however, to enhance the benefits of a human rights-approach to elder abuse. Rigorous research is required into prevalence rates of all forms of abuse against older people, including rates of self-neglect, in order to ensure design and implementation of effective responses. Such research is predicated upon establishing definitional clarity around the types of abuse and whether there is a threshold for what is considered abuse. A comprehensive understanding of what it means for older people to effectively access justice is also required, so that barriers to justice can be identified and addressed, and pathways for remedial relief established and supported. Further consideration also needs to be given to specific groups of older people, such as women, who may be at an even higher risk for abuse.

The complex and important task of addressing the causes and effects of elder abuse would be significantly bolstered by the adoption of enforceable obligations in the form of a CROP which would offer an objective, international yardstick for countries to use in harmonising their respective laws in response to elder abuse. The CROP would provide an opportunity to articulate the particular human rights harms embodied in elder abuse and recognise its foundations in ageism and disregard for fundamental dignity. By signing up to a new convention, governments would signal their commitment to preventing elder abuse as well as to protecting and promoting the human rights of older people, and would accept an international legal obligation to do so within their respective laws. The widespread prevalence of elder abuse across jurisdictions and the very real harm that it causes justifies the need for an international response which can lead governments to more meaningful protection and promotion of older persons' human rights.

References

Acierno R, Hernandez MA, Amstadter AB et al (2010) Prevalence and correlates of emotional, physical, sexual, and financial abuse and potential neglect in the United States: the national elder mistreatment study. Am J Public Health 100:292–297

Ajdukovic M, Ogresta J, Rusac S (2009) Family violence and health among elderly in Croatia. J Aggr Maltreat Trauma 18:261–279

Alzheimer's Australia (2016) Submission No 80 to the Australian law reform commission. Elder Abuse. https://www.alrc.gov.au/inquiry/elder-abuse-2/elder-abuse-submissions/. Accessed 11 Feb 2020

American Psychological Association. Elder abuse. https://www.apa.org/pi/prevent-violence/resources/elder-abuse. Accessed 10 Dec 2019

Ammerman R, Herson M (eds) (1991) Case studies in family violence. Springer, New York

Australian Law Reform Commission (2016) Elder abuse. Discussion Paper 83. https://www.alrc.gov.au/wp-content/uploads/2019/08/dp83.pdf. Accessed 10 Dec 2019

Australian Law Reform Commission (2017) Elder abuse—a national legal response. Final report No 131. https://www.alrc.gov.au/wp-content/uploads/2019/08/elder_abuse_131_final_report_31_may_2017.pdf. Accessed 10 Dec 2019

Australian Human Rights Commission (2018) Elder abuse is everybody's business. https://www.humanrights.gov.au/about/news/speeches/elder-abuse-everybodys-business. Accessed 11 Feb 2020

Bagshaw D et al (2013) Financial abuse of older people by family members: views and experiences of older Australians and their family members. Aust Soc Work 66(1):86–103
Baker PR et al (2017) Interventions for preventing elder abuse: applying findings of a new Cochrane review. Age Ageing 46(3):346–348
Band-Winterstein T (2015) Aging in the shadow of violence: a phenomenological conceptual framework for understanding elderly women who experienced lifelong IPV. J Elder Abuse Negl 27(4–5):303–327
Bennett B et al (2017) Assistive technologies for people with dementia: ethical considerations. Bull World Health Organ 95(11):749–755
Biggs S et al (2009) Mistreatment of older people in the United Kingdom: findings from the first national prevalence study. J Elder Abuse Negl 21(1):1–14
Brozowski K, Hall DR (2004) Growing old in a risk society: elder abuse in Canada. J Elder Abuse Negl 6(3):65–81
Brozowski K, Hall DR (2010) Aging and risk: physical and sexual abuse of elders in Canada. J Interpers Violence 25(7):1183–1199
Burnes D et al (2015) Prevalence of and risk factors for elder abuse and neglect in the community: a population-based study. J Am Geriatr Soc 63(9):1906–1912
Chesterman J (2017) Taking control: putting older people at the centre of elder abuse response strategies. Aust Soc Work 69(1):115–124
Chokkanathan S, Lee AE (2005) Elder mistreatment in urban India: a community based study. J Elder Abuse Negl 17(2):45–61
Cocco J (2011) Smart home technology for the elderly and the need for regulation. Pittsburgh J Environ Public Health Law 6(1):85–108
Cockburn T, Hamilton B (2009) Assessment of capacity: disciplinary issues and potential liability. Proctor (March):15
Cockburn T, Hamilton B (2011) Equitable remedies for elder financial abuse in inter vivos transactions. Queensland Lawyer 31:123–138
Cooper C, Selwood A, Livingston G (2008) The prevalence of elder abuse and neglect: a systematic review. Age Ageing 37:151–160
Cooper C, Selwood A, Livingston G (2009) Knowledge, detection, and reporting of abuse by health and social care professionals: a systematic review. Am J Geriatr Psychiatry 17(10):826–838
Craig Y (1994) Elder mediation: can it contribute to the prevention of elder abuse and the protection of the rights of elders and their carriers? J Abuse Negl 6:83
Crampton A (2013) Elder mediation in theory and practice: study results from a national caregiver mediation demonstration project. J Gerontol Soc Work 56(5):423–437
Cripps D et al (2002) Abuse of older people: issues for lawyers. Elder Law Rev 1:14
De Donder L et al (2011) Perpetrators of abuse against older women: a multi-national study in Europe. J Adult Prot 13:302–314
Dong XQ (2017) Elder self-neglect: research and practice. Clin Interv Aging 12:949–954
Dong X (2015) Elder abuse: systematic review and implications for practice. J Am Geriatr Soc 63:1214–1238
Dong X, Simon MA (2013) Elder abuse as a risk factor for hospitalization in older persons. JAMA Internal Med 173:153–162
Dow B, Johnson M (2012) Understanding elder abuse: a social rights perspective. Int Psychogeriatr 24(6):853–855
Elder Abuse Prevention Unit. What is elder abuse? https://www.eapu.com.au/elder-abuse. Accessed 17 Feb 2020
Elder Abuse Prevention Unit. Neglect. https://www.eapu.com.au/elder-abuse/neglect. Accessed 10 Dec 2019
Elder Abuse Prevention Unit (2015a) The cost of elder abuse in Queensland: who pays and how much? https://www.eapu.com.au/publications/research-resources. Accessed 10 Dec 2019
Elder Abuse Prevention Unit (2015b) The EAUP helpline: results of an investigation of five years of call data. https://www.eapu.com.au/publications/research-resources. Accessed 10 Dec 2019

Elder BM (2007) Mistreatment: risk, vulnerability, and early mortality. J Am Psychiatr Nurses Ass 12:313–321

Fileborn B (2017) Sexual assault and justice for older women: a critical review of the literature. Trauma, Violence, & Abuse 18(5):496–507

Fischer AR (2012) Elder abuse: a private problem that requires private solutions. J Health Biomed Law 8:81

Fritsch TA et al (2005) Intimate partner violence against older women in Kentucky. J Ky Med Assoc 103:461–463

Garçon L et al (2016) Medical and assistive health technology: meeting the needs of aging populations. The Gerontologist 56(S2):S293–S302

Garre-Olmo J et al (2009) Prevalence and risk factors of suspected elder abuse subtypes in people aged 75 and older. J Am Geriatr Soc 57:815–922

Gilhooly ML et al (2013) Framing the detection of financial elder abuse as bystander intervention: decision cues, pathways to detection and barriers to action. J Adult Prot 15(2):54–68

Gutman GM et al (2017) Elder abuse in the LGBT community: a hidden problem. Innov Aging 1:753

Hamilton B, Cockburn T (2008) Capacity to make a will and enduring power of attorney: issues new and old. Queensland Law Soc J (December):14

Harada CN, Natelson Love MC, Triebel KL (2013) Normal cognitive ageing. Clin Geriatr Med 29(4):737–752

Heck L, Gillespie G (2013) Interprofessional program to provide emergency sheltering to abused elders. Adv Emerg Nurs J 35:170–181

Hervey Bay Seniors Legal and Support Service (2016) Submission No. 75 to the Australian Law Reform Commission. Elder abuse. https://www.alrc.gov.au/inquiry/elder-abuse-2/elder-abuse-submissions/. Accessed 11 Feb 2020

Johnson MJ et al (2005) Gay and Lesbian perceptions of discrimination in retirement care facilities. J Homosex 49(2):83–102

Joosten M, Dow B, Blakey J (2015) Profile of elder abuse in Victoria: analysis of data about people seeking help from seniors rights Victoria. National Ageing Research Institute and Seniors Rights Australia. https://seniorsrights.org.au/wp-content/uploads/2014/03/Summary-Report_Profile-of-Elder-Abuse-in-Victoria_Final.pdf. Accessed 11 Feb 2020

Kaspiew R, Carson R, Rhoades H (2016) Elder abuse understanding issues, frameworks and response. Research Report No. 35. Australian Institute of Family Studies

Kaspiew R et al (2019) Elder abuse national research—strengthening the evidence base: research definition background paper. Australian Institute of Family Studies, Melbourne

Kornfeld-Matte R (2016) Report of the independent expert on the enjoyment of all human rights by older persons. UN Doc A/HRC/3344

Kum S (2017) Gay, gray, black, and blue: an examination of some of the challenges faced by older LGBTQ people of color. J Gay Lesbian Mental Health 21(3):228–239

Lacey W (2014) Neglectful to the point of cruelty? Elder abuse and the rights of older persons in Australia. Sydney Law Rev 36(1):99–130

Lacher S et al (2016) Types of abuse and risk factors associated with elder abuse. Swiss Med Weekly 146(w14273):1

Lachs MS et al (1998) The mortality of elder mistreatment. JAMA 280:428–432

Law Council of Australia (2017) The justice project introduction and consultation questions. Consultation Paper. https://www.lawcouncil.asn.au/files/web-pdf/Justice%2520Project/Consultation%2520Papers/Introduction%2520and%2520Consulation%2520Questions.pdf. Accessed 11 Feb 2020

Lonie J, Purser K (2017) Assessing testamentary capacity from the medical perspective. Aust Bar Rev 44(3):297–317

Lowenstein A et al (2009) Is elder abuse and neglect a social phenomenon? Data from the first national prevalence survey in Israel. J Elder Abuse Negl 21:253–277

Lundy M, Grossman SF (2009) Domestic violence service users: a comparison of older and younger women victims. J Fam Violence 24:297–309
McDonald L, Collins A (2000) Abuse and neglect of older adults: a discussion paper (No. H72-21/162-1998E). Family Violence Prevention Unit, National Clearinghouse on Family Violence, Ottawa, Canada
Moye J, Marson DC (2007) Assessment of decision-making capacity in older adults: an emerging area of practice and research. J Gerontol B Psychol Sci Soc Sci 62(1):3–11
Moye J, Marson DC, Edelstein B (2013) Assessment of capacity in an aging society. Am Psychol 68(3):158–171
National Alternative Dispute Resolution Advisory Council (1997) Issues of fairness and justice in alternative dispute resolution. Discussion paper. https://www.ag.gov.au/LegalSystem/AlternateDisputeResolution/Documents/NADRAC%2520Publications/Issues%2520of%2520Fairness%2520and%2520Justice%2520in%2520Alternative%2520Dispute%2520Resolution.PDF. Accessed 10 Dec 2019
Naughton C et al (2012) Elder abuse and neglect in Ireland: results from a national prevalence survey. Age Ageing 41(1):98–103
O'Keeffe M et al (2007) UK study of abuse and neglect of older people: prevalence survey report. Department of Health, London
Penhale B, Parker J, Kingston P (2000) Elder abuse: approaches to working with violence. BASW/Venture Press, Birmingham
Peri K, Fanslow J, Hand J (2009) Keeping older people safe by preventing elder abuse and neglect. Soc Policy J New Zealand 35:159–172
Peterson JC et al (2014) Financial exploitation of older adults: a population-based prevalence study. J Gen Intern Med 29:1615–1623
Pillemer K et al (2016) Elder abuse: global situation, risk factors, and prevention strategies. The Gerontologist 56(S2):194–205
Purser K, Rosenfeld T (2014) Evaluation of legal capacity by doctors and lawyers: the need for collaborative assessment. Med J Aust 201(8):483
Purser K (2017) Capacity assessment and the law: problems and solutions. Springer International Publishing, Switzerland
Purser K et al (2018) Alleged financial abuse of those under an enduring power of attorney: an exploratory study. Brit J Soc Work 48(4):887–905
Purser K, Sullivan K (2019) Capacity assessment and estate planning—the therapeutic importance of the individual. Int J Law Psychiatry 64:88–98
Purser K, Cockburn T, Ulrick E (2020) Examining access to formal justice mechanisms for vulnerable older people in the context of enduring powers of attorney. Elder Law Rev 12(2)
Ries NM (2018) Elder abuse and lawyers' ethical responsibilities: incorporating screening into practice. Legal Ethics 21(1):23–45
Ries NM, Mansfield E (2018) Elder abuse: the role of general practitioners in community-based screening and multidisciplinary action. Aust J Gen Pract 47(4):235–238
Royal Commission into Aged Care Quality and Safety (2019a) Interim report: neglect, vols 1–3. https://agedcare.royalcommission.gov.au/publications/Documents/interim-report/interim-report-volume-1.pdf. Accessed 10 Dec 2019
Royal Commission into Aged Care Quality and Safety (2019b) Restrictive practices in residential aged care in Australia. Background Paper No. 4. https://agedcare.royalcommission.gov.au/publications/Documents/background-paper-4.pdf. Accessed 10 Dec 2019
Sethi D et al (2011) European report on preventing elder maltreatment. World Health Organization, Geneva, Switzerland
Setterlund D, Tilse C, Wilson J (1999) Substituted decision-making and older people. Trends Issues Crim Justice 139. https://aic.gov.au/publications/tandi/tandi139. Accessed 11 Feb 2020
Setterlund D et al (2007) Understanding financial elder abuse in families: the potential of routine activities theory. Ageing Soc 27(4):599–614

Sharkey A (2012) Granny and the robots: ethical issues in robot care for the elderly. Ethics Inf Technol 14:27

Silva-Fhon JR et al (2015) Domestic violence in older people living in the district of Breña, Peru. Rev Fac Med 63:367–375

Straka S, Montminy L (2006) Responding to the needs of older women experiencing domestic violence. Violence Against Women 12:251–260

Su P et al (2011) The prevalence and influencing factors of abuse and negligence against elderly in rural areas of Anhui Province. Zhonghua Liu Xing Bing Xue Za Zhi 32(2):110–115

United Nations, Department of Economic and Social Affairs, Population Division (2015) World population prospects: the 2015 revision. https://www.un.org/en/development/desa/publications/world-population-prospects-2015-revision.html. Accessed 11 Feb 2020

United Nations Department of Economic and Social Affairs (2019) We are UN DESA towards sustainable development for all. https://www.un.org/development/desa/family/wp-content/uploads/sites/23/2019/06/Corrected_Amal_WEAAD2019Presentation.pdf. Accessed 11 Feb 2020

United Nations, Access to Justice. https://www.un.org/ruleoflaw/thematic-areas/access-to-justice-and-rule-of-law-institutions/access-to-justice/. Accessed 6 Feb 2020

Vandeweerd C, Paveza GJ, Fulmer T (2006) Abuse and neglect in older adults with Alzheimer's disease. Nurs Clin North Am 41:43–55

von Heydrich L, Schiamberg LB, Chee G (2012) Social-relational risk factors for predicting elder physical abuse: an ecological bi-focal model. Int J Aging Hum Dev 75(1):71–94

Wiglesworth A et al (2010) Screening for abuse and neglect of people with Dementia. J Am Geriatr Soc 58:493–500

Wilke DJ, Vinton L (2005) The nature and impact of domestic violence across age cohorts. Affilia 20:316–328

Williams K (2013) Elder mediation in Australia. Elder Law Rev 7:6

World Health Organization (2002a) The Toronto declaration on the global prevention of elder abuse. https://www.who.int/ageing/publications/toronto_declaration/en/. Accessed 6 Feb 2020

World Health Organization (2002b) Missing voices: views of older persons on elder abuse. https://www.who.int/ageing/projects/elder_abuse/missing_voices/en/. Accessed 10 Dec 2019

World Health Organization (2013) Prevalence and health effects of intimate partner violence and non-partner sexual violence. Switzerland, Geneva

World Health Organization (2015) World report on ageing and health. https://apps.who.int/iris/bitstream/handle/10665/186463/9789240694811_eng.pdf;jsessionid=156CD1FA4F3C1063E492164A9B6D7CE3?sequence=1. Accessed 19 Dec 2019

World Health Organization (2017a) Elder abuse fact sheet. https://www.who.int/mediacentre/factsheets/fs357/en/. Accessed 6 Dec 2019

World Health Organization (2017b) Abuse of older people on the rise—1 in 6 affected. https://www.who.int/mediacentre/news/releases/2017/abuse-older-people/en/. Accessed 11 Feb 2020

World Health Organization (2017c) Global priority research agenda for improving access to high-quality affordable assistive technology. https://apps.who.int/iris/bitstream/10665/254660/1/WHO-EMP-IAU2017.02-eng.pdf. Accessed 11 Feb 2020

Wu L et al (2012) Prevalence and associated factors of elder mistreatment in a rural community in People's Republic of China: a cross-sectional study. PLoS ONE 7:7e33857

Wydall S et al (2018) Domestic abuse and elder abuse in wales: a tale of two initiatives. Brit J Soc Work 48:962–981

Yon Y et al (2014) A comparison of spousal abuse in mid-and-old aged: is elder abuse simply a case of spousal abuse grown old? J Elder Abuse Negl 26:80–105

Yon Y et al (2017) Elder abuse prevalence in community settings: a systematic review and meta-analysis. Lancet Glob Health 5:e147–e156

Yon Y et al (2019) The prevalence of self-reported elder abuse among older women in community settings: a systematic review and meta-analysis. Trauma, Violence, & Abuse 20(2):245–259

Zannettino L et al (2015) The role of emotional vulnerability and abuse in the financial exploitation of older people from culturally and linguistically diverse communities in Australia. J Elder Abuse Negl 27(1):74–89

Zink T et al (2003) Cohort, period, and aging effects: a qualitative study of older women's reasons for remaining in abusive relationships. Violence Against Women 9:1429–1441

International Instruments

Committee on Economic, Social and Cultural Rights (2000) *General Comment No. 14: The right to the highest attainable standard of health (Art 12)* UN Doc E/C.12/2000/4 (11 August 2000)

Committee on the Rights of Persons with Disability (2014) *General Comment No. 1: Equality Before the Law (Art 12)* UN Doc CRPD/C/GC/1 (19 May 2014)

Convention on the Rights of Persons with Disabilities (2006) Opened for signature 13 December 2006, entered into force 3 May 2008. 2515 UNTS 3

European Convention for the Protection of Human Rights and Fundamental Freedoms (1950) Opened for signature 4 November 1950, entered into force 3 September 1953. 213 UNTS 221

Human Rights Committee (1992) *General Comment No. 20: Prohibition of Torture, or Other Cruel, Inhuman or Degrading Treatment or Punishment (Art 7)* (10 March 1992)

Human Rights Committee (1999) *General Comment No. 27: Freedom of Movement (Art 12)* UN Doc CCPR/C/21/Rev.1/Add.9 (2 November 1999) (23 August 2007)

Human Rights Committee (2007) *General Comment No. 32: The Right to Equality Before Courts and Tribunals and to a Fair Trial (Art 14)* UN Doc CCPR/C/GC/32 (23 August 2007)

Human Rights Committee (2014) *General Comment No. 35: Liberty and Security of Person (Art 9)* UN Doc CCPR/C/GC/35 (23 October 2014)

Inter-American Convention on Protecting the Rights of Older Persons (2015) Opened for signature 15 June 2015, entered into force 11 January 2017. 55 ILM 985

International Covenant on Civil and Political Rights (1966) Opened for signature 16 December 1966, entered into force 23 March 1976. 999 UNTS 171

International Covenant on Economic, Social and Cultural Rights (1966) Opened for signature 16 December 1966, entered into force 3 January 1976. 993 UNTS 3

Madrid International Plan of Action on Ageing (2002) Adopted by Second World Conference on Ageing, Madrid, 8–12 April 2002

Protocol to the African Charter on Human and Peoples' Rights on the Rights of Older Persons (2016) Opened for signature 30 January 2016 (not yet in force)

United Nations General Assembly (2015) *Transforming Our World: the 2030 Agenda for Sustainable Development*. Resolution 70/1. UN Doc A/RES/70/1 (21 October 2015)

Legislation

Powers of Attorney Act 1998 (Qld)

Cases

Johnson v Johnson [2009] NSWSC 503

Nelson v Nelson [1995] 184 CLR 538

Pinter v Pinter [2016] QSC 314

Smith v Glegg [2004] QSC 443

Smith v Smith [2017] NSWSC 408

Chapter 8
Financial Security and Ageing

Poverty is one of the greatest threats to older people's human rights and is common, even amongst wealthy states, despite the right to social security. Some older people are at a heightened risk of financial insecurity, depending on a number of factors including, for example, socio-economic status, location, gender, as well as cultural or linguistic background. The level of risk is also influenced by ageist attitudes, and inadequate or non-existent laws, regulations and policies which are often poorly enforced or lack remedial clout. Financial security, an umbrella term including both income and wealth security, is fundamental to being able to access associated rights such as safe housing and healthcare. Pension schemes, both non-contributory and contributory, are one of the main avenues through which governments around the world attempt to fulfil the right to social security. However, the threat of poverty and homelessness remains, particularly for older women and those from culturally and ethnically diverse backgrounds. Although significant progress has been made in developing and progressively extending the legal pension coverage for all older people, this chapter argues that a human rights framework has the ability to be utilised to great effect to ensure better financial security by providing international obligations and global scrutiny. Calls are also made for further research, particularly for rigorous prevalence data and genuinely understanding the voice of older people. Appreciating their lived experiences of financial *in*security is essential to promoting income and wealth security.

8.1 Introduction

The legal and regulatory framework which promotes or hinders financial security in 'retirement' is complex, interactive and hugely influential on a person's life in old age. Importantly, these laws and regulations start having an effect early in the life course. Therefore, it is a landscape in which embedding a human rights-based approach is crucial, but one that requires nuanced and well-thought out responses

© Springer Nature Singapore Pte Ltd. 2020

B. Lewis et al., *The Human Rights of Older Persons*,
https://doi.org/10.1007/978-981-15-6735-3_8

tailored to each jurisdiction's particular circumstances. The wide variety of legal responses across different legal systems reflects the many different ways the law can impact on older people and serves to reinforce how wide-ranging the eclectic nomenclature of 'elder law' is. It also again raises the issue of whether 'elder law' is based on the idea of 'the law *of* older people' or whether it is about the 'law *and* older people'.[1]

The issue of financial security for older people was first introduced in Chap. 1 in relation to some of the key concerns associated with the globally ageing population. To briefly recap, given the current rates of ageing, as well as the increasingly rapid rate of technological advancement, people are living well beyond 'retirement' age. However, this is often without the financial security necessary to be able to access, for example, adequate health and long-term aged care (discussed further in Chap. 10), accommodation (Chap. 9), food and water. Financial insecurity also interferes with access to justice in order to remedy abuses of basic human rights. Financial security therefore serves as a precondition for the realisation of numerous human rights, including the rights to an adequate standard of living (which incorporates the rights to adequate food, water and housing);[2] the highest attainable standard of health;[3] freedom of movement;[4] liberty and security (which acknowledges the heightened risk of abuse for those older people who find themselves financially dependent upon others);[5] privacy;[6] work and education;[7] as well as social and cultural participation.[8] In recognition of this interrelationship between financial security and wider human rights protections, international human rights law guarantees the right to social security as a means of ensuring that all individuals are protected from the negative human rights impacts of poverty.[9] It is also recognised as being essential for human dignity and the enjoyment of all other human rights.[10]

This chapter will therefore focus on human rights issues arising in relation to financial security for older people post 'retirement'. The concept of 'financial security' is complex and has been explored here in two broad categories, that of income security and asset or wealth security (that is, the accumulated assets of an individual throughout their life course). Income security is largely predicated upon paid

---

[1]Williams (2003), 107.

[2]*International Covenant on Economic, Social and Cultural Rights* (1966), art 11 ('*ICESCR*').

[3]Ibid., art 12.

[4]Ibid.

[5]*International Covenant on Civil and Political Rights* (1966), arts 7 and 9 ('*ICCPR*').

[6]Ibid., art 17.

[7]ICESCR, arts 6 and 7.

[8]Ibid., art 15, and ICCPR, art 25, Committee on Economic, Social and Cultural Rights (2008); Sepúlveda Carmona (2010); Alston (2014).

[9]*Universal Declaration of Human Rights* (1948); art 22 ('*UDHR*') and ICESCR art 9, Kornfeld-Matte (2016) para 10; Sepúlveda Carmona (2010).

[10]UDHR, arts 22 and 25, ICESCR, art 9, Committee on Economic, Social and Cultural Rights (2008); paras 1–3, *López Rodríguez v. Spain* E/C.12/57/D/1/2013, Committee on Economic, Social and Cultural Rights (4 March 2016), paras 10.1 and 10.2; *Marcia Cecilia Trujillo Calero v. Ecuador* E/C.12/63/D/10/2015, Committee on Economic, Social and Cultural Rights (14 November 2018).

employment, hence ‘retirement’ being a milestone event. Whilst acknowledging the difficulties with defining exactly what ‘retirement’ is and when it occurs given the many stages it can incorporate, the definition adopted here is the cessation of paid employment eliminating that source of income security for older persons.[11]

Ideally, retirement should be a time during which older people can focus on their own wellbeing—whatever that may consist of, and however that looks—and a time which is considered to be a natural and appealing part of ageing.[12] The negative attitudes historically displayed towards ‘retirement’ are slowly starting to change as they are often interconnected with the ageist attitudes and practices prevalent throughout society more generally.[13] However, as discussed in Chap. 5, identifying, acknowledging, addressing and, ultimately, altering those attitudes will take multi-generational perseverance. A central argument of this book is that a move towards greater recognition and implementation of human rights can help achieve this attitudinal shift, thus creating a culture that emphasises the autonomy, dignity and worth of older persons. This includes acknowledging and combating ageist and discriminatory attitudes that tie a person’s ‘value’ to their ability to continue making economic contributions to society through paid work, as discussed in Chap. 4.

In addition to ageist attitudes, gender, location, cultural background, education and socio-economic status can all influence financial security as people age because each factor can affect an individual’s ability to obtain income security and, consequently, to accumulate wealth (security) throughout their life.[14] As introduced in Chap. 1, the questions raised by the notion of financial security for older people are diverse, complex and challenging, ranging across a wide variety of financial areas.[15] Not only does planning for financial security after retirement include issues around what a person wants to do during their working life, it also deals with when a person wants to cease work, or indeed what happens if they are forced to stop work because, for example, they are physically unable to continue. Questions also arise as to what happens if an older person is not able to provide for themselves once they do retire, particularly when considering the increased health issues that can occur as people age, especially in the ‘old, old’ cohort.[16] Significantly, a comparatively low number of people will accumulate adequate assets to be financially secure, or self-funded, when they retire.[17] Thus, there can often be increased pressure on older people to continue working past a point where they would otherwise have chosen to retire because of the risk of financial insecurity. This extension of an older person’s working life may also be detrimental to their health and other human rights.

Ensuring financial security is not easy. It requires forethought and planning. It also generally requires access to paid formal employment, that is, employment with some

[11]On the difficulties defining retirement see, for example: Hershenson (2016).

[12]*Vienna International Plan of Action on Ageing* (1982).

[13]Williams (2003), 103.

[14]United Nations Department of Economic and Social Affairs Programme on Ageing (2016), 1, 3.

[15]Ibid., 1.

[16]Ibid.

[17]Ibid.

sort of contract or agreement rather than cash-in-hand arrangements, which are more precarious. The resultant income security throughout the life course is necessary to have the freedom to plan for an (ideally self-funded) future beyond paid work (wealth security). Quality pension schemes (both contributory and non-contributory) are also essential for ensuring financial security, especially for people who have limited or no income security, but also for people whose wealth security is inadequate to ensure their needs are met post-paid employment. Illustrative of this point is the fact that, for example, one-third of the 'baby boomer' generation in the United States of America in 2014 had no savings in retirement plans and little time to commence building the necessary funds to ensure financial security beyond retirement.[18]

There are two main categories of income sources when examining financial security for older people—persons who are self-funded and those in receipt of government assistance, for example, pensions. Depending on the national scheme, older people may be self-funded through part of their retirement and receive a government funded pension for the remainder—that is, until the end of their life. Where available, pension schemes can be non-contributory or contributory, or a mixture of both.[19] Non-contributory schemes require no contributions from the individual in question, and the pension is funded entirely by the government. Under a contributory pension model, the individual contributes to their own pension throughout their working life. This then determines the entitlement benefits payable and whether the person will be entirely self-funded in retirement or whether they will require government assistance in the form of a pension or, more likely, a mixture of the two.[20] The right to social security and the principle of non-discrimination also do not prevent a state from having dedicated pensions for specific groups, such as war veterans, although authentically adopting a human rights approach requires that the core content of the right still needs to be met. That is, any pension that is offered cannot discriminate on the grounds of age, sex, gender or ethnicity. In addition to being individually financially secure and/or being able to access government support, older people may also rely on familial support provided, of course, that such support is accessible.[21]

The consequences of not being financially secure can be substantial including, as mentioned above, experiencing a heightened risk of poverty and homelessness, as well as the increased threat of abuse of human rights more broadly. This is particularly significant for older people when considering that they may also be physically and/or mentally impaired, or otherwise in poor health. Consequently, promoting financial security, and the basic human right to social security, is essential, as they play a crucial role in protecting other human rights. This is necessary not only for the older person specifically but also for society generally, especially when considering the availability of adequate resources and systemic supports required for older people who are not able to be self-funded.

---

[18] Stanford Center on Longevity (2018).

[19] See, for example, Australian Government (2019).

[20] Ibid.

[21] United Nations Department of Economic and Social Affairs Programme on Ageing (2016), 1–2.

This chapter will therefore consider a range of conceptual issues relating to poverty and ageing including the notions of income and wealth security, estate planning, as well as non-contributory and contributory pension schemes as pathways for financial security for older people. It should be noted, however, that given the incredible complexity of the law in each of these areas, this chapter is not intended to be, nor can it be, a detailed examination of all of these issues. The following discussion will therefore deliberately be restricted to a theoretical exploration of select themes arising in the application of a human rights framework to financial security in ageing generally rather than an examination of the intricacies of particular pension schemes. One such theme discussed here is the heightened risk to older women of experiencing financial insecurity and the increased threat of human rights abuses of older women more broadly. Given that one of the main criticisms of a human rights framework is the lack of practicability of some of the principles, this chapter also highlights the applied relevance of adopting a human rights framework to promote the financial security of older people.

8.2 Poverty and Ageing

Although there are different ways to measure poverty, no matter the measurement tool adopted, it is one of the greatest threats to both older people and society more generally. This is because socio-economic inequality weakens social cohesion and can drive instability.[22] This section will detail the prevalence rates of poverty for older persons before examining the risk factors as well as the fundamental role of 'financial security' in successfully adopting a human rights framework to elder law.

8.2.1 Prevalence

Unfortunately, the same problems with obtaining prevalence data in relation to older people more generally also exist in relation to ascertaining the extent of poverty. For instance, the definitional issues around who is 'older' and at what chronological age this occurs, as well as what constitutes concepts such as 'retirement' and 'poverty' make it challenging to obtain globally comparable data. A lack of harmonised data collection also means that it is difficult to discuss prevalence with any sense of certainty.[23] Consequently, the data that exist in relation to poverty rates for older people tend to be nationally or regionally based, with little information being collected in lower socio-economic countries.

[22]United Nations General Assembly (2015) ('*UNSDGs*'), Goal 1: End poverty in all its forms everywhere ('*SDGs*').

[23]United Nations Department of Economic and Social Affairs Programme on Ageing (n.d.), 1.

Notwithstanding the difficulties in data collection, it is estimated that 14.7% of people aged over seventy-five years throughout OECD countries live in poverty, up by 3.5% from those in the sixty-six to seventy-five age bracket.[24] Data also show that the rates of poverty for the cohort aged sixty years and over ranged from an estimated 2% in the Netherlands in 2013, for instance, to a high of approximately 80% of people in Zambia in 2005.[25] Australia, as another example, had a poverty rate of about 34% in 2012,[26] with one in four Australians aged sixty-five years and over (the commencement of data collection at sixty-five rather than sixty again evidencing the difficulty with obtaining comparable data) estimated to live in poverty.[27] In fact, it is believed that approximately 7% of those experiencing homelessness in Australia are older people, with a majority of those people experiencing homelessness for the *first* time after having had an 'orthodox' housing record.[28] In many cases this occurred because of unexpected events such as needing to care for grandchildren; the loss of paid employment which, when combined with age discrimination, can make it difficult to locate other employment; experiencing an accident resulting in large medical bills; difficult family situations, including the breakdown of a relationship; and/or needing to care for others.[29]

Older people in lower socio-economic countries have an even greater risk of experiencing poverty and homelessness given that the social protection (pension) coverage in less developed countries is often inadequate or non-existent.[30] This is because the pension coverage is relative to the strength of the national economy.[31] In fact, two-thirds of older people live in developing nations where employment largely occurs on an informal basis—thus negatively impacting any chance to access state-funded age-based pension schemes which can be connected to participation in formal employment.[32]

8.2.2 Risk Factors

Despite difficulties obtaining consistent and comparable data, there is no doubt that the rates of older persons experiencing poverty and homelessness present a considerable threat to human rights. The next question arises as to the risk factors driving this. Those who have experienced a lifetime of poverty and older people who are

[24] OECD (2015).

[25] Ibid.

[26] See Footnote 23.

[27] Australian Human Rights Commission (2015).

[28] Ibid., Petersen and Jones (2011); Homelessness Australia (2016), 31.

[29] Petersen and Jones (2011); Homelessness Australia (2016), 31.

[30] United Nations Department of Economic and Social Affairs Programme on Ageing (n.d.), 2.

[31] Ibid.

[32] Ibid., 3.

vulnerable for other reasons may be at particular risk.[33] In fact, older people face an increased danger of either becoming or remaining poor.[34] Consider, for example, a married couple of thirty years who have two children aged twenty-seven and twenty-three. The husband is aged sixty and has been in paid employment his entire life, now holding a senior management position after having progressed in his career. The wife, who is also sixty, has been out of paid work since the birth of their eldest child instead deciding, jointly with her husband, that she would care for their children full time. The husband decides to leave his wife. There is a property settlement pursuant to which all assets are divided. This necessitates the sale of the matrimonial home. Although most jurisdictions seek to ensure a just outcome for the parties after the property settlement this, in reality, is not always the case. Whereas the husband has paid employment guaranteeing his income security, the wife does not and, when coupled with age discrimination, may find it difficult to re-enter the workforce. She thus lacks not only the income security that the husband possesses, but she also lacks the ability to acquire it. Some jurisdictions, such as Australia have attempted to address this inequality through family law legislation.[35] However, despite the legislation, it is difficult to adequately or accurately compensate a spouse who has forgone their career to be a full-time homemaker and parent. As a result, there have been calls to redress the hidden nature of homelessness for older women. This example helps to illustrate why financial insecurity of older women warrants particular attention, as discussed in more detail below.[36]

These circumstances demonstrate the often precarious financial security of older people—even for those who have not previously appeared to be at risk, such as the wife in the above example. Downwards trends have been noted in the overall financial security of older people, particularly in relation to reduced or lost income security which then negatively impacts an older person's ability to establish wealth security, including obtaining secure and affordable housing.[37] This is especially relevant when considering that as people age, they either need, for example owing to poor health or reduced mobility, or want to reduce their working hours, or to cease work entirely.[38] Negative and erroneous rhetoric promulgated in relation to the causes of poverty and homelessness—'it was their own fault' or they 'did it to themselves'—must be addressed and removed from the collective thoughts of society at large.[39] Instead, it is necessary to recognise that poverty can result from numerous causes including unemployment, underemployment, social exclusion, vulnerability and abuse—and that 'fault' has no role to play in protecting the rights of vulnerable older persons.[40]

---

[33]Ibid., 1; *Madrid International Plan of Action on Ageing* (2002) para 45; UNSDGs Goal 1; Cox CB (2015), 41.

[34]See Footnote 23.

[35]Family Law Act 1975 (Cth), s 75(2).

[36]Australian Human Rights Commission (2019).

[37]Stanford Center on Longevity (2018), 3.

[38]See Footnote 23.

[39]UNSDGs Goal 1; Brünner and Andersen (2018), 114.

[40]UNSDGs Goal 1, Brünner and Andersen (2018), 114; Virick (2011); Slack and Jensen (2008).

The effects of poverty on older people are not uniform, however, nor are the risk factors. Nevertheless, as people grow older, there generally comes a tipping point in the ageing process at which an older person's ability to earn money reduces or stops altogether. This inability to earn money needs to be appreciated in the context of, often, growing health and personal care issues. Such issues may require costly treatment, often at a time when older people may have already expended or be close to exhausting their financial resources.[41] This can particularly be an issue for those in the 'old, old' category which, as discussed in Chap. 1, is the fastest growing cohort of older persons, with the number of people in this category expected to more than triple by 2050 to 425 million people.[42] Significantly, two out of three people in the 'old, old' category will live in developing and lower socio-economic regions,[43] thus further increasing their risk of poverty and endangering their human rights, including the right to social security.[44] In fact, older members of ethnic and racial minorities in general experience an increased risk of financial insecurity.[45] It is also important to understand that financial security tends to be approached from a western viewpoint and thus authentic attempts to embed the right to social security worldwide will have to be done so on a specific jurisdictional basis respecting the local cultural and social norms.

There are other, often unexpected, issues that can also arise as people age that function as a financial drain. Such concerns can compound the risks for older people of experiencing poverty and/or homelessness, particularly if there is a lack of established income and/or wealth security. Take, for example, the role of grandparents in providing care for their grandchildren.[46] Grandparents are increasingly becoming involved as the long-term care providers for grandchildren where parents are unable to continue in that role for various reasons.[47] This can be an economic burden on the grandparents who may no longer be in paid employment and who may lack wealth security. As discussed in Chap. 5, age discrimination can also affect the ability of an older person to re-enter the workforce in order to re-establish lost income security or to rebuild dissipated wealth security resulting from unforeseen circumstances. Therefore, there are a number of different factors that can combine to heighten the threat of poverty (and homelessness) for older people.

41 United Nations Department of Economic and Social Affairs, Population Division (2015).

42 United Nations (n.d.); United Nations Department of Economic and Social Affairs, Population Division (2017), 1.

43 United Nations Department of Economic and Social Affairs, Population Division (2017), 1.

44 UDHR, art 22 and ICCPR, art 9.

45 Cox (2015), 46.

46 See, for example: Hayslip et al. (2019).

47 Ibid.

To understand fully the role of financial security (in the form of both income and wealth security) in combating the risk of poverty we need to appreciate the methods available to acquire that security—either through self-funded means or the social protection mechanisms that are provided by governments, such as pension schemes.[48] When considering the interplay between self- and government-funded methods, it is significant to note that in many countries, if an older person does not have sufficient savings to self-fund their retirement there may be no government funded social security schemes to fill that gap.[49] Consequently, given the difficulty in acquiring the wealth security necessary to self-fund post-paid employment, a lack of government supported systems providing income security in old age means that many older people will lack the overall financial security necessary to protect their basic human rights.[50] What is also clear is that there are limited avenues for older people to escape this cycle when considering that often their ability to earn through formal employment decreases over time, especially when taking into consideration the impact of age discrimination, and the increased possibility of physical and mental impairments.[51]

8.3 Understanding Financial Security

As discussed, as people age, they are either self-funded or supported by external systems—be they familial or government-based.[52] The problem arises when none of these three sources exist or when they are inadequate because they are not accessible or because the benefit entitlement level, that is, the level of support an older person is entitled to, fails to meet their needs. As noted above, financial security in older age depends on two separate, though closely related, aspects. The first is income security, which, as the name suggests, relates to the funds which the individual receives either through paid employment or social security benefits. The second is wealth or asset security, which relates to the accumulation of financial resources in various forms, such as property, shares or savings, which a person acquires throughout their life.[53] Some comments will now be made in relation to income and wealth security in light of the application of a human rights framework.

48 See Footnote 23.

49 Ibid.

50 Ibid.

51 Ibid.

52 Ibid., 2.

53 Ibid., 6–7.

8.3.1 Income Security

As has been seen above, ensuring a source of secure income for older people is a crucial obligation on societies worldwide, especially as governments and policy-makers seek to protect the right to social security.[54] Income security is also an important factor when protecting the other human rights of older people, for instance in relation to healthcare, long term aged care, and safe housing. This is because if older people lack income security, it can heighten their vulnerability to the risk of poverty and/or homelessness which is exacerbated if these other services, particularly healthcare, residential and long term aged care and housing, are not accessible and adequate.[55]

8.3.2 Wealth Security

As noted above, wealth security refers to the assets a person has accumulated throughout their life. The existence of wealth security means that, if necessary, the assets can be used as income producing sources or sold or otherwise utilised, for instance as security for a loan, to provide a larger cash injection. Take, for example, an investment property. It can either be used as an income producing asset (rent) or it can be sold to reap a more significant return. It is, however, important to note that home ownership is decreasing in some countries, such as the United States of America where the rate of home ownership fell by 50% in the years leading up to 2016.[56]

Where a person does own their home, it can also be used as collateral to secure loans, for example through the controversial ‘reverse mortgage’ scheme adopted in Australia. A reverse mortgage, as the name suggests, is a form of loan that can help older people access the equity in their homes. It allows the older person to use their home as security for a loan that can be taken in a lump sum, as a regular source of income, or a combination of the two.[57] As with other loans, compound interest is charged. Different from other loans, however, is the fact that repayments do not need to be made while the older person remains living in the home.[58] The total loan (including fees and interest) becomes repayable when the older person sells the house, moves out of the house or dies.[59] A common repayment trigger point is when the older person moves into aged care.[60]

[54] International Labour Organization (2017), 76.

[55] Ibid.

[56] Stanford Center on Longevity (2018), 4.

[57] Australian Securities and Investments Commission (n.d.).

[58] Ibid.

[59] Ibid.

[60] Ibid.

The reverse mortgage scheme is controversial because of the often-unforeseen financial difficulties it can place older people in. Take, for example, the loan repayment trigger of moving into aged care. In Australia, fees are generally required to be paid in order to gain entry into aged care facilities (discussed in more detail in Chap. 10). If an older person needs to sell their home to pay these fees, but they have previously taken out a reverse mortgage over their home, the older person may not realise the full cost involved in paying out the loan. That is, in an attempt to gain early access to the equity in their homes—often because they lack income security—they may have opened themselves up to a debt which, once satisfied, leaves them without the resources necessary to provide for themselves after the sale of the home—that is, they lose their wealth security in the process. Other problems with reverse mortgages include: higher ongoing fees and interest rates than for normal mortgages; the use of compound interest which can result in significant debt accumulating quickly; houses are subject to market fluctuations which can, in turn, negatively affect the amount of equity in the home; and, finally, the costs to break the agreement can be substantial.

The nature of home ownership can also be problematic. For instance, what happens in a situation where a husband and wife own the property jointly and have previously agreed to take out a reverse mortgage, however the husband needs to enter aged care and cannot meet the intake costs without the sale of the home? If there is a sale, the loan becomes repayable. The wife may then not have anywhere to live if the remaining equity in the home (if any) after the reverse mortgage is paid out is inadequate to cover her future housing costs. Further if a person is the sole owner but another person lives with them, that other person will have to leave the property if the older person vacates the premises.[61] Additional research is therefore required into the potential economic efficacy of reverse mortgages and the associated risks attached to them in relation to protecting the human rights of older people.[62]

As people age the existence of 'wealth security' assumes increasing significance. This is because, as discussed, people either do not want to continue to work or may be unable to do so—bearing in mind, too, the ageist assumptions and practices leading to discriminatory hiring practices that contribute to the job and income insecurity of older people (discussed in Chap. 4). Wealth security is therefore especially important when faced with the growing jobless trend and resultant income insecurity.[63] Disparities in wealth security are actually believed to be higher than income disparities.[64] Yet wealth security almost always depends on obtaining the initial income in order to accumulate assets—without income security, financial insecurity in older age is almost a self-fulfilling prophecy.

Further, even where there is wealth security, the assets may not be easily realisable and thus may almost amount to a false economy. Take, for example, the existence of a family discretionary trust. A discretionary trust places the ownership of

---

[61] Ibid.

[62] See, for example: Ong (2008); Hancock (1998). For more on reverse mortgages generally, see: Steen and Taggart (2018).

[63] United Nations Department of Economic and Social Affairs Programme on Ageing (n.d.), 6–7.

[64] Ibid.

identified assets in the names of trustees to hold those assets for the benefit of the beneficiaries named in the trust deed. This is the archetypal example of a fiduciary relationship wherein the trustee(s) must act in the best interests of the beneficiaries and cannot engage in any activities in which their personal interests conflict with those of the beneficiaries. These types of trusts are incredibly common estate planning mechanisms. The problem with discretionary trusts, however, is that they are also complicated legal vehicles through which to achieve certain aims, such as asset protection and tax minimisation (depending on the jurisdiction). Take a husband and wife who establish a discretionary trust and name themselves as trustees, and themselves, along with their four children as beneficiaries. For all intents and purposes the husband and wife will use the assets in the trust as their own in order to best facilitate the effective management of that wealth to benefit their family—that is one of the significant factors driving the establishment of the trusts in the first place.

The difficulty arises, however, in relation to individual understanding of the trust deed and the ownership of trust property. Property contained in a discretionary trust is property of the trust, not the trustees, who have a fiduciary duty to act in the best interests of *all* the beneficiaries when managing the trust.[65] This is often misunderstood, resulting in, for instance, the husband and wife from the above example, believing the assets to be theirs to do with as they please—which is not the case. To act in such a manner is contrary to the fiduciary responsibility they bear as trustees, as well as statutory obligations imposed upon them in relation to their trustee obligations.[66] Trusts, therefore, serve as a useful illustration of situations where, although an older person may experience wealth security (for example as trustee and beneficiary of a discretionary trust), it is a question of how that wealth has been structured and whether, as well as how, they can then access it. This demonstrates the importance of estate planning (discussed below) but also of ensuring that legal mechanisms created for the purposes of managing wealth are adequately understood by the people relying on those assets. This enables them to make informed decisions and avoids potential risks to their human rights which would flow from wealth and overarching financial insecurity.

8.4 Planning for the Future

Preparations for retirement are informed by individual personalities, work-life history, health, people's own perceptions of what 'retirement' is, as well as what they are allowed to do with their income and wealth security, and how long it will last. This all sits within an overarching and jurisdictionally specific legal and policy

[65]See, for example: Keech v. Sandford (1726) Sel Cas T King 61; 25 ER 223.

[66]See, for example: Heydon and Leeming (2016); Keech v. Sandford (1726) Sel Cas T King 61; 25 ER 223; and *Trusts Act 1973* (Qld).

framework within which a person lives and earns an income and/or owns assets.[67] It also includes the policy agenda discussed above which may incentivise working longer or otherwise reduce reliance on state-funded pensions—an approach which needs to be balanced correctly in order to ensure that no negative impacts on human rights occur. There is therefore no one model for how people do, or should, approach retirement, particularly given that the world has never before seen a generation like the 'baby boomers'.[68]

'Baby boomers', for example, have expressed a desire to control their time and to find purpose in the activities they undertake—whether in paid or unpaid positions and including post- 'retirement'.[69] Being able to decide for oneself when to retire and how that time will be spent is an important exercise of autonomy, and supports the enjoyment of a full range of human rights—having those options curtailed due to financial factors can undermine the enjoyment of rights and have negative impacts on self-esteem, as well as on mental and physical health. Therefore, an older person's experience of retirement depends very much on whether the income and wealth security exist to support their vision of what retirement should be.

Financial security is thus a fundamental consideration in post-paid employment life. As discussed above, however, life occurrences, cultural norms, gender, socio-economic status, and location can all have an impact on a person's financial security in retirement. What is constant is that the law has a role to play. It is an authoritative vehicle for facilitating and protecting an older person's ability to obtain financial security, including their right to social security as well as other associated human rights. It is also an effective agent through which to implement social change.[70]

In addition to pension schemes (discussed below), one of the ways in which the law is significant when planning for financial security in retirement is through the implementation of estate or future planning. These terms are often used interchangeably, and both have been used here. As discussed in Chap. 1, the estate planning undertaken by legal practitioners differs from the financial planning undertaken by financial planners and accountants who have their own set of legal, ethical and professional rules and regulations to abide by.[71]

Estate planning is work undertaken by legal professionals to plan for the effective growth, management, protection and transmission of assets. It is concerned with ensuring financial security. There is a misconception that estate planning is only for older people and that the planning necessary varies depending upon the age

[67] In Australia, for instance, the rules of conflict of laws govern situations where a deceased person owned assets in a different jurisdiction to one in which they lived. In this situation, immovable property is governed by the laws of the jurisdiction in which the immovable property is situtated (*lex situs*) whereas movable property is governed by the laws of the deceased person's domicile at the time of their death (*lex domicilli*). For a recent case see: *Public Trustee of Queensland as Administrator of the Estate of KH, Deceased* [2017] QSC 48.

[68] Kojola and Moen (2016).

[69] Ibid.

[70] Prindiville (2015).

[71] Royal Commission into Misconduct in the Banking, Superannuation and Financial Services Industry (2019); Recommendation 2.10; Fyrdenberg and Hume (2019).

of the person, whereas the underlying considerations are actually largely the same irrespective of age. Another misconception is that estate planning is predominantly concerned with the execution of wills.[72] It is not. A will may not even be the main document through which the transmission of assets will occur, depending on the planning implemented.[73] Estate planning instead utilises, in addition to testamentary documents, a myriad of other legal structures such as enduring powers of attorney, advance health directives, corporate entities and trusts.[74] Removing assets from the estate is an often-deliberate attempt to avoid potential will contestations through, for example, family provision applications. That is, if there are no assets in the estate then it makes contesting the will a less attractive option.[75] While such planning is generally undertaken in response to anticipated death and/or loss of capacity, it can also include *inter vivos* transactions such as the establishment of an *inter vivos* discretionary (or family) trust or transfers of land. The transmission of assets is often intergenerational in nature which has given rise to the increasingly common idea that older people *must* transfer their assets to younger generations, also known as inheritance expectation (that is, younger generations are 'entitled' to the assets). This, in turn, can potentially result in elder abuse in situations where a person's response is abusive if their feelings of 'entitlement' are not met (discussed in Chap. 7).[76]

Interestingly in relation to transfers of property from a parent to a child, the presumption of advancement applies, and it will be assumed that the transfer was a gift. This is because the presumption represents the idea that parents want to further their children in life and thus any transaction will be out of natural love and affection, not one requiring payment—fair market value or otherwise.[77] This can be problematic in situations where the transaction was not intended to be a gift. The presumption is able to be rebutted but doing so can incur additional legal fees, thus potentially raising access to justice as well as evidential issues.

Assets for care arrangements provide useful examples of the problems that can arise as a result of the presumption. Take, for instance, a mother aged eighty-eight years who is in need of care. She owns her own home. Her son has just married, and he and his partner are unable to afford to buy a property. The mother and the son reach a verbal agreement that in exchange for the son caring for his mother she will transfer her property into his name. The transfer is executed and registered in the son's name. He refuses, however, to care for his mother and tells her to move into an aged care facility. The mother commences legal proceedings to recover her

---

[72] James (2015), 27–28.

[73] Ibid.

[74] For a discussion of the frequency of the usage of trusts in the United States of America, see: James (2015), 28–29.

[75] However, some jurisdictions, such as New South Wales in Australia, have notional estate provisions. These are provisions which, in effect, 'claw back' into the estate any property which was the subject of a transaction designed to defeat the will contestation provisions (known as family provision legislation in Australia). See, for example: *Succession* Act 2006 (NSW) part 3.3.

[76] See on this, for example: Joosten et al. (2015), 19; Purser et al. (2018, 2020).

[77] See, for example: Johnson v Johnson [2009] NSWSC 503; Nelson v. Nelson (1995) 184 CLR 538.

property but will first have to rebut the presumption of advancement that she gave her property to her son through an act of natural love and affection. It then becomes necessary to consider the evidence required to negate this presumption and the cost of obtaining such evidence.

Given that estate planning can be complex and costly, its value is predicated upon having the income security to pay for it and the wealth security to make it worthwhile. Effective estate planning can nevertheless help to strengthen income and wealth security through implementing measures designed to maximise wealth management and protection. The security provided can then, in turn, help to ensure protection of the rights to, for example, housing and healthcare. At a minimum, however, all people should have a will and enduring power of attorney in the event of a loss of capacity and/or death. These documents can help to ensure that a person's wishes are respected in the event that they can no longer express them for themselves.

8.5 The Right to Social Security

The right to social security generally requires countries to provide financial support sufficient for (at least) essential healthcare, basic housing and education, water, sanitation and food.[78] As discussed in Chap. 2, this right extends to 'old age'. Protection of this right ensures that older people are able to live both securely and with dignity, two key components of the United Nations' (UN) development agenda.[79]

The *International Covenant on Economic, Social and Cultural Rights* (ICESCR) establishes that states are required to take steps towards the progressive realisation of the right to social security. While this standard of progressive realisation allows states to tailor their activities to suit their available economic resources, the Committee on Economic, Social and Cultural Rights has made it clear that states must not delay taking the first steps, that they must provide a minimum essential level of security and, importantly, they must not be discriminatory in the way that rights are protected and fulfilled.[80] For older persons, this means that social security must not be structured in a way that negatively disadvantages people on the basis of age. In fact, states are required to identify particularly marginalised or vulnerable groups, which could include older persons, and provide specific protections suited to their needs.[81] The ultimate goal is the complete implementation of the right to social security for all people, and states should develop a national strategy to achieve this outcome, including through the allocation of appropriate resources and services. As

---

[78]Sepúlveda Carmona (2010), para 87; *Madrid Plan of Action*, paras 52–53.

[79]Cox (2015), 51; Hokenstad and Roberts (2013), 77.

[80]Committee on Economic, Social and Cultural Rights (2008), para 41; *López Rodríguez v. Spain*, para10.3; *Trujillo v Ecuador*, para 11.3; See also Sepúlveda Carmona (2010), para 87.

[81]Committee on Economic, Social and Cultural Rights (2008), para 31; *Trujillo v Ecuador,* para 13.1.

stated by the Committee on Economic, Social and Cultural Rights, article 9 of the ICESCR:

> implicitly recognizes the right to old-age benefits. States are obligated to pay particular attention to promoting and protecting the economic, social and cultural rights of older persons, to which end they must take appropriate measures to establish general regimes of compulsory old-age insurance.[82]

Whereas few countries had social protection systems in place a century ago, virtually all countries now have such systems, with extension of coverage a continual aim.[83] Significantly, the ICESCR prohibits states from relying exclusively on contributory pensions for older persons given the problems explained below, particularly because of the risk that these will be inadequate for maintaining an adequate standard of living.[84] Instead, non-contributory schemes must be made available for all people who lack the means to make sufficient contributions for their own protection.[85] The requirement for progressive realisation also demands that states not take retrogressive action. Consequently, states are encouraged to provide non-contributory pensions subject to available resources and, where pension schemes have been established, they should not be wound back.[86]

In addition to the ICESCR, several treaties specifically refer to the protection of older persons through enforcement of the right to social security.[87] Regional provisions also exist recognising the geographically unique characteristics and vulnerability of older persons in relation to the right to social security.[88] Further, the International Labour Organization (ILO) has adopted international standards which establish a normative framework to both protect as well as promote the right to social security.[89] There are 16 standards designed to guide the national implementation of social

[82] *Trujillo v Ecuador*, para 11.3; citing Committee on Economic, Social and Cultural Rights (1995).

[83] International Labour Organization (2017).

[84] HelpAge International (2010), 2.

[85] *Trujillo v Ecuador*, para 14.1; Office of the High Commissioner for Human Rights (2012), 16; Committee on Economic, Social and Cultural Rights (1995); Committee on Economic, Social and Cultural Rights (2008).

[86] See Footnote 84.

[87] In addition to articles 9, 10 and 11 of the ICESCR, see also UDHR, arts 22 and 25; article 5 (iv) of the *Convention on the Elimination of All Forms of Racial Discrimination* (1965); article 11 of the *Convention on the Elimination of All Forms of Discrimination Against Women* (1979) (*'CEDAW'*); article 27 of the *Convention on the Protection of the Rights of All Migrant Workers and Members of their Families* (1990); and article 28 of the *Convention on the Rights of Persons with Disabilities* (2006).

[88] See, for example, the *African Charter on Human and Peoples' Rights* (1981); The *European Social Charter* (1996); The *Additional Protocol to the American Convention on Human Rights in the Area of Economic, Social and Cultural Rights* (1988); and the International Labour Organization Convention C102 concerning Minimum Standards of Social Security (1952) and Convention C128 concerning Invalidity, Old-Age and Survivors' Benefits (1967). The ILO Convention 102 establishes minimum standards for all nine branches of social security including medical care, sickness benefit, old-age benefit, employment and injury benefit and invalidity benefit.

[89] International Labour Organization (2017), v.

protection policies.[90] The ILO Social Protection Floors Recommendation No. 202 of 2012 was the most recent standard which requires, at a bare minimum, that there be a basic level of universal social security available 'in the form of a nationally defined social protection floor' and that there should be a 'progressively wider scope and higher levels of protection'.[91] The Special Rapporteur on extreme poverty and human rights, Philip Alston, has recommended that social protection floors be implemented as a core objective of the broader human rights and development agenda, and these would have numerous benefits for older persons experiencing poverty.[92] Alston has emphasised the interdependence of human rights, explaining that social protection floors would not only help secure economic, social and cultural rights, but also improve the enjoyment of civil and political rights.[93]

The Sustainable Development Goals (SDGs), particularly number 1, to 'end poverty in all its forms everywhere', also demonstrate a commitment to developing and then progressively widening the supports available through social security policies and systems.[94] In relation to SDG number 1,[95] including the development of floors in social protection systems, the UN has stated that its aim,

> by 2030, [is to] ensure that all men and women, in particular the poor and the vulnerable, have equal rights to economic resources, as well as access to basic services, ownership and control over land and other forms of property, inheritance, natural resources, appropriate new technology and financial services, including microfinance.[96]

Although SDG 1 does not specifically target poverty experienced by older people, given the rates of ageing, particularly in the potentially more vulnerable 'old, old' category, addressing poverty for older people is fundamental to realising this goal. How to realise this though is becoming an increasingly pressing question, especially when considering the stated aim to achieve each and every one of the SDGs by 2030, only a decade away.[97]

If an older person cannot be self-supported in retirement and there is a lack of government social protection policies, the question then arises as to what familial or customary support exists. However, the support traditionally supplied by familial networks is increasingly being eroded given smaller family numbers and wider socio-economic changes, including the rate of the ageing population.[98] This issue is even

90 Ibid.

91 Ibid.

92 Alston (2014).

93 Ibid., para 6.

94 International Labour Organization (2017), v.

95 See also UNSDGs Goal 3.8, which calls for universal health coverage; Goal 5.4 in relation to gender equality; Goal 8.5 for decent work and economic growth, and Goal 10.4 regarding greater equality.

96 UNSDGs Goal 1, target 4.

97 See UNSDGs; United Nations Department of Economic and Social Affairs Programme on Ageing (n.d.), 2.

98 United Nations Secretary-General (2012); United Nations Department of Economic and Social Affairs Programme on Ageing (n.d.), 2.

more pronounced in low socio-economic countries where there is a greater concentration of older people, as well as poor government supported age-based social security. This, therefore, further highlights the need for 'real' solutions in order to achieve SDG number 1.

Obviously, there are significant costs involved with any policy involving systemic changes, particularly in relation to implementing non-contributory age-based pension schemes, and especially in low to middle socio-economic countries where the rates of older persons experiencing poverty and homelessness are greatest.[99] However, the age-old excuse of 'cost' cannot be allowed to see older, sometimes vulnerable people, living without access to basic human rights such as safe accommodation, food, water and healthcare. This is especially the case given the standard of obligation states bear under the ICESCR with regard to the progressive realisation of economic rights to the maximum of their available resources and in cooperation with other countries. Perhaps the solution, or part of the solution, lies in the wording of the targets of the SDGs. For example, the aim is to 'implement nationally appropriate social protection systems and measures for all, including floors, and by 2030 achieve substantial coverage of the poor and the vulnerable.'[100] This will necessarily involve re-evaluating the government support systems, such as pensions, and avenues for developing financial security and independence—particularly as people age. However, as stated, there is one notable exception in all of the targets, which is the lack of acknowledgment of the increased risk of poverty for older people. The unique needs of older people experiencing poverty and homelessness must therefore also be specifically addressed, including how their financial security is going to be recognised, promoted and, ultimately, achieved.

Securing the right to social security and the related right to an adequate standard of living, including housing, is therefore more than just a policy decision requiring the establishment or extension of social security systems. It is an international obligation to ensure the protection of the basic human rights of older persons, a cohort which, as has been seen, may be at increased risk of poverty and/or homelessness. It is through the adoption of a human rights framework, therefore, that the opportunity to begin to create and redevelop policy solutions to this issue can be realised—both at an international as well as a national level. A human rights-based approach is able to tap into the obligations that flow from membership of the international rules-based order, the normative and moral force of human rights language, and the supervision and enforcement mechanisms established in human rights treaties.

[99]United Nations Department of Economic and Social Affairs Programme on Ageing (n.d.), 2.
[100]UNSDGs, Goal 1.3.

8.6 Pension Schemes

It is unlikely that many older people will be in a position to be self-funded in retirement.[101] In a large number of developing countries, older people still rely heavily on family support arrangements.[102] However, decreased familial support of older people, when combined with the lack of individual income and wealth security necessary to self-fund retirement, means that the financial security of older people is therefore frequently dependent upon the availability of social security measures, such as pensions.[103] This section will briefly look at the availability of pensions throughout the world before making some comments about non-contributory and contributory schemes.

8.6.1 Availability and Viability

Pensions are a key element to meeting SDG 1 with 68% of eligible people globally estimated to be in receipt of a pension (both non-contributory and contributory).[104] Of the 192 countries for which there are data, 186 countries provide old-age pension schemes making periodic payments of cash benefits through at least one scheme or a combination of non-contributory and contributory schemes that are secured in national legislation.[105] The remaining six countries do not have periodic benefits, instead often offering a form of lump-sum payment or similar scheme.[106]

Out of the 186 countries, a number of countries have a mixture of contributory and non-contributory schemes. For example, fourteen countries (8%) have a contributory and non-contributory universal scheme, whereas sixty-four countries (34%) have a contributory and non-contributory means-tested scheme.[107] Seventy-two countries (39%) have contributory schemes only, which can be compared to two countries (1%) which have non-contributory means-tested schemes only and ten countries (5%) that have non-contributory universal schemes only.[108] Subject to the proper implementation and enforcement of relevant laws, it is anticipated that 67.6% of working age people internationally would be eligible for a pension. Interestingly, when considering the importance of both income and wealth security in retirement, 17.7% of

[101] On the situation in, for example, the United States of America see: Stanford Center on Longevity (2018), 19–27. See also, for example: Ranzin et al. (2004).

[102] International Labour Organization (2017), 75.

[103] United Nations Department of Economic and Social Affairs Programme on Ageing (n.d.), 2.

[104] See Footnote 102.

[105] Ibid., 76.

[106] Ibid., 77.

[107] Ibid., 77, Fig. 4.3.

[108] Ibid., 78. For earlier data see: International Labour Organization (2014).

working-age people globally have the option to make voluntary contributions to age-based pension schemes, yet few take advantage of this.[109] Further, despite the fact that pension schemes were expanded internationally between 2000 and 2012, there are still employment and healthcare patterns which result in older people in higher-income countries enjoying financial security in retirement far more than those in lower-income countries.[110] Moreover, less than 20% of eligible older people in most low-income countries receive a pension.[111]

In addition to the availability of pensions, questions also arise about eligibility and the level of entitlement benefits provided under the various schemes.[112] Notwithstanding recognition of the right to social security, and its importance for facilitating the enjoyment of other rights, it is still common for older people to receive inadequate or no pension entitlements.[113] However, this again is highly dependent upon both national and regional geographic location.[114]

The highest pension coverage appears to be in Europe and North America (approximately 90%).[115] In Australia, 80% of the older population (aged sixty-five years and over) are estimated to rely to some extent on government funded, age-based assistance.[116] Latin America and the Caribbean (56%), as well as Asian-Pacific countries (47%) are estimated to have moderate pension access,[117] while North Africa (37%), the Middle East (30%) and sub-Saharan Africa (17%) are estimated to have some of the lowest access globally to age-based, government funded pensions.[118]

What these statistics demonstrate is that, despite the 'universal' right to social security, this right looks substantially different depending upon location (including not only regional variances but also differences between metropolitan and regional, rural and remote areas), gender and the economic strength of the country in which the older person resides.[119] It also depends on the type of employment that a person has been engaged in, with those in informal employment (that is, unpaid work or 'cash-in-hand') situations (which are especially common in developing countries) at an even greater risk of poverty in old age. This is because systemic collection of income tax and/or other contributions to finance the public social security support systems are inadequate, ineffective and/or underdeveloped owing to a lack of ongoing

[109] See Footnote 102.

[110] International Labour Organization (2014); United Nations Department of Economic and Social Affairs Programme on Ageing (n.d.), 2.

[111] See Footnote 102.

[112] United Nations Department of Economic and Social Affairs Programme on Ageing (n.d.), 2.

[113] Ibid., International Labour Organization (2014).

[114] United Nations Department of Economic and Social Affairs Programme on Ageing (n.d.), 3.

[115] Ibid., International Labour Organization (2014).

[116] See Footnote 27.

[117] International Labour Organization (2014); United Nations Department of Economic and Social Affairs Programme on Ageing (n.d.), 3.

[118] See Footnote 117.

[119] United Nations Department of Economic and Social Affairs Programme on Ageing (2016), 1–2, 6; International Labour Organization (2017), 75.

financial contributions where there is high informal employment.[120] Consequently, older people in higher-income countries have better access to financial security and government assistance than those in lower socio-economic countries. Even in the stronger performing socio-economic countries, however, there remains a proportion of people engaged in informal employment thus also reducing their ability to access age-based social security when they become eligible, despite living in a country in which accessing age-based pensions might be easier.[121]

Countries have therefore begun to adopt strategies in an attempt to widen social security protections. This can occur, for example, through the extension of basic rate non-contributory schemes to a wider group of (generally) low-income workers such as cleaners and domestic workers.[122] However, such an approach is putting the fiscal viability of the schemes at risk given that, although the number of recipients is increasing, the contributions ensuring the sustainability of the schemes are not keeping pace with the outgoings.[123] States therefore need to exercise care when developing, reviewing and expanding pension schemes to ensure that they fulfil the mandate of protecting the right of older persons to social security.[124] Regular adjustments also need to occur so as to combat inadvertent (or deliberate) erosion of pension benefits over time as other goods and services increase in price.[125]

Significantly, however, in order to establish the long-term fiscal viability of age-based, government funded assistance schemes, governments in higher-income countries are frequently doing one of two things. One approach to ensure fiscal sustainability is to increase the statutory age at which a person can retire and become eligible for an age-based, state supported pension scheme—the justification being that people are living longer, so they can, in turn, work longer.[126] The second approach is to create incentives for people to undertake a longer working-life[127] through, for example, the promotion of flexible working hours and opportunities for older workers.[128] Investments are also being made in 'life-long' learning opportunities with a view to encouraging and assisting people to acquire skills to enable them to continue in paid employment for as long as possible.[129] Such schemes can include, for instance, campaigns in relation to anti-discrimination and 'ageing well' or 'healthy ageing'.[130]

---

[120] United Nations Department of Economic and Social Affairs Programme on Ageing (n.d.), 3.

[121] International Labour Organization (2017), 98–99.

[122] United Nations Department of Economic and Social Affairs Programme on Ageing (n.d.), 3.

[123] Ibid.

[124] See Footnote 121.

[125] Ibid.

[126] International Labour Organization (2014); United Nations Secretary-General (2012); United Nations Department of Economic and Social Affairs Programme on Ageing (n.d.), 3; Brown and Vickerstaff (2011).

[127] International Labour Organization (2014); United Nations Secretary-General (2012); United Nations Department of Economic and Social Affairs Programme on Ageing (n.d.), 3; Brown and Vickerstaff (2011).

[128] United Nations Department of Economic and Social Affairs, Population Division (2017), 2.

[129] Ibid.

[130] Ibid.

However, only creating a legislative and regulatory framework for social security is not sufficient to discharge the duty of states in relation to the right to social security. Much depends on how such a scheme operates in practice and whether individuals are meaningfully able to access adequate benefits to support their basic needs and other human rights.[131] This is particularly a problem where policy and internal guidelines have erroneously replaced the law, or the law is being incorrectly interpreted and/or applied when determining benefit entitlements. When this is coupled with inadequate or inaccessible opportunities for redress the human rights consequences can be serious.[132]

Further, policy settings and regulations aimed at reducing the economic cost of pension schemes must ensure that the benefit provided is at a minimum sufficient to allow older persons to afford the goods and services they require to meet 'the minimum core content of their economic, social and cultural rights'.[133] The former UN Special Rapporteur on the question of human rights and extreme poverty, Magdalena Sepúlveda Carmona, has explained that, 'while the cost of schemes can be reduced by increasing the age of eligibility or decreasing the size of the benefit, a balance should be struck between reaching all those in need and providing a benefit that would allow older people a minimum subsistence level.'[134]

Consequently, unless meaningful changes are made to state-funded social security pensions, many older people, particularly those in informal employment, will risk financial insecurity as they age. This will put them at a heightened risk of experiencing poverty and/or homelessness, bearing in mind the comments made above in relation to the often increasing healthcare costs as well as the physical and mental impairments that can impede or prevent older people from participating in paid employment, particularly in the 'old, old' category.[135] This is not just restricted to those older people in developing nations. People in countries with strong socio-economic development can also experience problems accessing government funded, age-based pensions. Elaborating on this general discussion, some comments in light of the application of a human rights framework will now be made with specific reference to both the non-contributory and contributory schemes.

---

[131] Prindiville (2015).

[132] Ibid.

[133] Sepúlveda Carmona (2010), para 87; see also *Madrid Plan of Action*, paras 52–53.

[134] Sepúlveda Carmona (2010), para 87.

[135] United Nations Department of Economic and Social Affairs Programme on Ageing (n.d.), 3; Jolly (2014), 51; United Nations Department of Economic and Social Affairs Programme on Ageing (2016), 2; Cox (2015), 41.

8.6.2 *Non-contributory Pension Schemes*

Non-contributory pension schemes normally do not require recipients to have made contributions throughout their lifetime in order to be able to access benefit entitlements.[136] They are often referred to as social pensions and are promising avenues through which to begin to address financial insecurity for older people, especially when considering that generally there are reviewable and enforceable legal provisions underpinning their establishment.[137] The very nature of non-contributory pensions makes them critical to ensuring fulfilment of the right to social security, as well as addressing the effects of extreme poverty. This is because they serve to support those most at-risk individuals who lack the means to contribute to their own future financial needs. Non-contributory pension schemes can be either universal or targeted. Under a universal scheme everyone above a specified age is eligible, whereas people entitled under a targeted scheme are eligible based on their income or poverty level (that is, they are means tested) in addition to their age.[138] Universal non-contributory pension schemes have been adopted in, for example, Samoa, Namibia, Brunei, Botswana, Nepal and Bolivia.[139] It appears, however, that only nine countries world-wide make access available to universal non-contributory pensions, with a further three countries providing benefit entitlements on a means-tested basis.[140]

Non-contributory pensions are generally more in line with human rights principles, especially the principles of equality, universality and non-discrimination, as they do not suffer from the innately discriminatory problems facing contributory schemes (discussed below).[141] Although this is true of both universal and targeted pensions, universal schemes are even better aligned with these human rights principles. This is because, although targeted schemes can be cheaper to implement than universal, 'targeting' specific income and/or poverty levels can result in exclusion errors given that the selection process can be a complex one resulting in controversial

---

[136] United Nations Department of Economic and Social Affairs Programme on Ageing (2016), 2.

[137] Ibid.

[138] Ibid.

[139] HelpAge International (2010), 3. The International Labour Organization states that 'universal pensions have been developed in Argentina, Belarus, the Plurinational State of Bolivia, Botswana, Cabo Verde, China, Georgia, Kyrgyzstan, Lesotho, Maldives, Mauritius, Mongolia, Namibia, Seychelles, South Africa, Swaziland, Timor-Leste, Trinidad and Tobago, Ukraine, Uruguay, Uzbekistan and Zanzibar (United Republic of Tanzania). Other developing countries, such as Azerbaijan, Armenia, Brazil, Chile, Kazakhstan and Thailand, are near universality'. It is not clear whether these are universal contributory or non-contributory schemes, however. International Labour Organization (2017), 75.

[140] United Nations Department of Economic and Social Affairs Programme on Ageing (2016), 2. See also the ILO which states that 'among the countries considered, in 12 cases pensions are provided exclusively through non-contributory schemes. Of these, the majority provide universal coverage.' On first glance it appears that the numbers are inconsistent but the ILO report states that it is a majority of the 12 countries for which data are available that provide universal non-contributory schemes. Accordingly, the specific number of 9 has been adopted. See International Labour Organization (2017), 77.

[141] HelpAge International (2010), 1.

inclusions and/or exclusions, or negative or even punitive consequences if an older person receives a benefit for which they are later deemed to have been ineligible.[142] Universal schemes also reduce any stigma that may be attached to 'needing' to receive financial assistance if everyone becomes eligible at a certain time (or age).[143] It is important to note as well, however, that to assume that every older person needs, or wants, assistance is to perpetuate ageist assumptions.

Non-contributory pension schemes therefore offer a vehicle through which countries can implement social protection systems and begin to address the significant threat to human rights that exists when older persons lack financial security. It is thus vital to recognise the need to develop effective policy implementation in relation to non-contributory pension schemes. Such systems will, in turn, help acknowledge and fulfil nation-states' obligations in responding to the rights of older persons to social security and an adequate standard of living.[144]

In order to achieve this, states must ensure that older people have access to adequate benefit entitlements, as well as associated goods and services, such as healthcare and access to justice. The positive effects of accessing social security can be negated if there is no way to enforce the right (that is, access to justice) or the burden of healthcare costs outstrips the benefits paid. In fact, the most successful non-contributory schemes are those founded in the legal and regulatory systems of countries which evidence both permanence and also enforceability.[145] To fulfil human rights obligations, such legal and regulatory systems need to clearly establish eligibility requirements, ensure transparency, define roles and responsibilities, and establish complaints mechanisms.[146] This last element should also provide information on redress mechanisms, depending upon the nature of the complaint, which will also assist in ensuring accountability and protection where the rights of an older person have been violated.[147] The human rights principles of equality and non-discrimination should also be at the fore of the development of any scheme in relation to access to social security, particularly in relation to (progressively) ensuring universal access, especially for disadvantaged and vulnerable older persons.[148]

This will assist in protecting not only the right to social security, but also the interconnected rights to an adequate standard of living and healthcare. However, when considering facilitating universal access, it is important to note that legal protections for equality of access and the enforcement of non-discrimination policies on

---

[142] Ibid. In an analogous example, a successful challenge was mounted to the automated 'Robodebt' program in Australia wherein the Federal Government abandoned its reliance on averaging incomes to determine debts owed by welfare recipients. Although the challenge involved a 34-year-old, this risk of punitive consequences nevertheless holds true for all welfare recipients in similar circumstances. See, for example: Karp (2019); *Deanna Amato v The Commonwealth of Australia* (27 Nov 2019) Federal Court of Australia No: VID611/2019.

[143] See Footnote 141.

[144] Ibid.

[145] Ibid.

[146] Ibid.

[147] Ibid.

[148] Ibid.

the basis of age (where they exist at all) remain reasonably scarce outside of the employment context. Older people may require, for instance, a loan and/or insurance, which can be more difficult to acquire as people age. Often age restrictions (or even prohibitions) on policies and/or additional fees can be imposed. Older people can therefore experience inequality of access when attempting to procure goods and services, particularly financial services.[149]

The lower numbers of non-contributory pension schemes may also be representative of the idea that such schemes present a fiscal drain on national economies. Conversely, however, particularly in light of the increasing rates of ageing, these social pensions may actually serve as 'economic stabilisers' addressing the effects of poverty and homelessness which, as discussed above, can have wider social, and economic, implications in relation to promoting social cohesion.[150] In fact, it has been demonstrated that universal schemes can be implemented in low-income countries,[151] and governments in developing countries have taken significant steps in improving the income security of their older people.[152] Policy-makers and legislators worldwide must therefore continue to challenge the assumption that non-contributory pensions are a burden to the economy. In doing so, it is also imperative to address the ageist attitude that older people are an economic drain preventing socio-economic growth. Instead, the role of older people in contributing to their families, wider communities and society more broadly should be acknowledged, respected and encouraged.[153] The promotion of non-contributory pensions can also inadvertently help address issues of sex-discrimination given the discriminatory effects of contributory pensions on women.[154]

This is an area that has, thus far, attracted little attention (from governments and researchers alike), particularly within a human rights framework, and one in which further work is needed.[155] For instance, the development of effective information and education programs about what government-funded vehicles and services are available is needed. Older people, particularly older women given issues of gender inequality, must also be participants in developing policy, systems and education programs for them to have maximum effectiveness.

---

[149] United Nations Department of Economic and Social Affairs Programme on Ageing (2016), 8.

[150] See Footnote 141.

[151] Ibid.; International Labour Organization (2017), 75.

[152] Barrientos (2014), 76.

[153] See Footnote 141.

[154] International Labour Organization (2014); United Nations Department of Economic and Social Affairs Programme on Ageing (2016), 6.

[155] United Nations Department of Economic and Social Affairs Programme on Ageing (2016), 8. See also United Nations Secretary-General Assembly (2012); Douglas et al. (2017).

8.6.3 Contributory Pension Schemes

Unlike non-contributory pension schemes, contributory pension schemes are predicated upon the premise that workers will make financial contributions throughout their lifetime to a fund. These contributions will then be used to determine the older person's entitlement in retirement.[156] Some countries have traditionally focused on promoting a policy of contributory pensions in order to establish financial security for people once they retire.[157] However, the policy behind the realisation of contributory pension schemes can differ markedly, depending upon the country in which an older person resides. Contributory schemes only are offered in 39% of countries with available data.[158] Of the countries for which data are available, seventy-seven of them have contributory schemes that only cover people who are engaged in formal work. Consequently, particularly in the low-income countries, many older people cannot access contributory schemes.[159] Contributory pension schemes are also estimated to have resulted in up to 80% of people worldwide being unable to access any form of pension.[160]

Where such schemes even exist, manifest inequalities and discrimination can occur hindering or preventing an individual from providing for their financial security into retirement. Such inequalities include: the older person's location; the socio-economic security of their country of residence; and the nature of work undertaken by the person throughout their life, that is formal or informal work. As discussed below in relation to older women and financial security, gender can also be a significant discriminatory factor in being able to make contributions given, for example, unequal pay and time taken out of work to have children and/or provide (unpaid) care to others. These inequalities also do not exist in isolation. For example, informal work is especially common in low-income countries which means that the workers are consequently less likely to be able to access contributory pension schemes.[161]

8.7 The Financial Security of Older Women

As explained, gender can be a significant factor in creating financial *in*security. Moreover, it has been an emergent theme of our analysis more generally, indicating that older women are at an increased risk of experiencing human rights abuse across all areas. The heightened risk of financial insecurity for older women exists despite the increase in governments attempting to introduce gender equity as well as fiscal

---

[156] United Nations Department of Economic and Social Affairs Programme on Ageing (2016), 2.

[157] See Footnote 141.

[158] International Labour Organization (2017), 77.

[159] United Nations Department of Economic and Social Affairs Programme on Ageing (2016), 2.

[160] See Footnote 141.

[161] United Nations Department of Economic and Social Affairs Programme on Ageing (2016), 2.

sustainability in their social security and pension schemes.[162] Older women disproportionately experience financial insecurity, in part, because their access to pensions, particularly contributory pension schemes, is often lower than it is for men.[163] This inequity is exacerbated by the fact that women generally live longer and therefore require greater financial security.[164] The Committee on Economic, Social and Cultural Rights has recognised the particular vulnerability of women in its recent decision in *Trujillo v Ecuador*, in which it stressed that states are obliged to establish non-contributory pension schemes and that these 'must also take account of the fact that women are more likely to live in poverty than men; that often they have sole responsibility for the care of children; and that it is more often they who have no contributory pensions'.[165]

Globally, the Arab States, as well as sub-Saharan and Northern Africa exhibit the greatest trends of gender inequity when it comes to access to pensions. Jordan and Egypt have the greatest discrepancies wherein men are seven to eight times more likely to be in receipt of a pension than women.[166] Although higher-income countries tend to have more equal coverage between men and women, the benefit levels can vary markedly. For instance, the benefit entitlement of pensions paid to men throughout the European Union is 40% higher than that paid to women.[167] This not only reflects the lower formal workforce participation of women (and higher participation in informal work) but also their over-representation amongst self-employed workers and/or unpaid carers.[168] Women carry out, for example, an estimated two and a half times more domestic and unpaid care than men.[169] Women's wages are also historically lower than those of their male counterparts earning, on average, approximately half as much as men throughout their lives.[170] This is despite women in all regions working longer hours than men when combining paid and unpaid work.[171] The inequities arising from the fact that women do not receive equal pay for equal work are therefore compounded for older women in retirement resulting in the reinforcement of life-long inequalities.[172] Women who are from culturally and linguistically diverse backgrounds or who have a disability can be even further

---

[162] Ibid.; Stanford Center on Longevity (2018), 3.

[163] United Nations Department of Economic and Social Affairs Programme on Ageing (2016), 1–5; Cox (2015), 44; Sepúlveda Carmona (2010), paras 19–21, 95–97; Sepúlveda Carmona (2013), paras 26, 50; *Madrid Plan of Action*, para 46; Centre for Workplace Excellence (2018).

[164] United Nations Department of Economic and Social Affairs Programme on Ageing (2016), 5.

[165] *Trujillo v Ecuador*, para 14.2.

[166] International Labour Organization (2017), 78.

[167] United Nations Department of Economic and Social Affairs Programme on Ageing (2016), 5; United Nations Entity for Gender Equality and the Empowerment of Women (2015).

[168] International Labour Organization (2017), 78; United Nations Department of Economic and Social Affairs Programme on Ageing (2016), 1, 5.

[169] UN Women (2015), 10–1.

[170] United Nations Department of Economic and Social Affairs Programme on Ageing (2016), 1, 5.

[171] See Footnote 169.

[172] Ibid.

disadvantaged and therefore experience a heightened risk of poverty and/or homelessness.[173] For example, in 2012 in the United States of America, single white older women had 9% less median wealth security than single white older men.[174] This figure increased substantially to evidence a 160% difference between single white older men and single older women of colour.[175]

Further compounding the challenges faced by older women, the benefit entitlements they receive are often much lower than those for men. This occurs largely because of a combination of work history and family circumstances, for instance, time taken away from paid employment to care for children and other family members.[176] Divorce or relationship breakdown can be another significant contributing factor to the increased risk of financial insecurity for older women.[177] It is important, however, to distinguish between discrete groups of unpartnered women, for instance, divorced, widowed and never married, as not all single women face the same risks for financial insecurity.[178] A lack of financial literacy also contributes to the increased threat of financial insecurity experienced by women. In the United States of America, for example, older women on average have a lower financial literacy than their male counterparts, as well as being less confident when handling finances.[179]

Discriminatory attitudes and practices in some countries also mean that not all women are guaranteed the right to own and control assets.[180] Inheritance and succession laws play a significant role in perpetuating these discriminatory practices. For example, in 35 out of 173 economies in the East and South Asian, Pacific, sub-Saharan and North African, and Middle East regions, female surviving spouses do not have equivalent inheritance rights to their male counterparts.[181] They may also feel pressure to hand over any inheritance to male family members.[182] These practices have obvious negative human rights implications, and in themselves represent a violation of the right to equality before the law. The *Convention on the Elimination of all forms of Discrimination Against Women* explicitly provides that states are to treat women and men equally in all legal matters, and that men and women are to be accorded the same legal rights, including in relation to ownership and administration of property.[183]

---

173 United Nations Department of Economic and Social Affairs Programme on Ageing (2016), 5, Westwood (2012) 510; *Madrid Plan of Action*, para 47.

174 United Nations Department of Economic and Social Affairs Programme on Ageing (2016), 8.

175 Ibid.

176 Ibid., 1, 5; Peeters and Wouter (2015), 1171.

177 Peeters and Wouter (2015), 1171–3.

178 Ibid., 1174.

179 Stanford Center on Longevity (2018), 47, 50.

180 United Nations Department of Economic and Social Affairs Programme on Ageing (2016), 8, see also United Nations Secretary-General (2015).

181 United Nations Department of Economic and Social Affairs Programme on Ageing (2016), 8, see also World Bank (2015).

182 See Footnote 181.

183 CEDAW, arts 15–16.

Given fewer opportunities to contribute, typically lower wages and a (generally) longer life span, the very nature of pension schemes therefore results in negative and unequal outcomes for older women. In recognition of this effect on older women's income and wealth security (and attendant negative human rights implications), significant work is being undertaken in order to promote gender equality in relation to inheritance and property rights, as well as overarching financial security worldwide.[184] Governments are increasingly recognising, for instance, paid maternity leave. It is also becoming more common for discriminatory aspects of old-age pensions to be challenged, such as the difference in retirement ages between men and women.[185]

However, despite increased attention in the literature and growing progress to address gender inequality in the workforce and social security systems, older women continue to experience financial insecurity and, in many jurisdictions, their concerns are still largely invisible within mainstream society.[186] Although general ideas exist as to why older women face increased risks of poverty and homelessness, there remains a need for detailed, up-to-date, research. Such research is necessary to explore the interconnected effects and prevalence of these contributing factors in order to adequately protect the rights of older women.[187]

8.8 Conclusion

The ageing population has significant consequences for all aspects of society and in all countries globally, including for labour and financial markets, as well as in relation to the supply and demand of goods and services, including financial services and social security.[188] Financial inequalities exist not only between older people but also between different age demographics.[189] These inequalities are compounded by socio-economic status, location, gender, ageist attitudes, and inadequate or non-existent laws, regulations and policies which are often poorly enforced.[190] They also negatively affect the ability of an older person to work, as well as their ability to accumulate resources to plan for a financially secure retirement. In addition to limiting economic stability, the inequalities can also reduce opportunities for participation

[184]United Nations Department of Economic and Social Affairs Programme on Ageing (2016), 8. See also, United Nations Economic and Social Council (2014).

[185]United Nations Department of Economic and Social Affairs Programme on Ageing (2016), 5–6, 9.

[186]United Nations Department of Economic and Social Affairs Programme on Ageing (n.d.); Jolly (2014), 51; United Nations Department of Economic and Social Affairs Programme on Ageing (2016), 2; Cox (2015), 41.

[187]Peeters and Wouter (2015), 1172, 1174.

[188]United Nations Department of Economic and Social Affairs, Population Division (2017).

[189]See Footnote 23.

[190]Ibid.

and social unity.[191] Furthermore, only a small minority of older people have accumulated sufficient financial assets to be able to ensure their own financial security as they age.[192]

Significant worldwide progress has been made, however, in developing and progressively extending the legal pension coverage for all older people. This is also true in low-economic countries where increased coverage is being seen, although significant deficits still exist. In higher economic countries, rather than extension of coverage, the focus tends to be on adequacy of benefit entitlements and the fiscal sustainability of such schemes.[193] However, the socio-economic security of a country does not necessarily guarantee access to adequate social security schemes and inequities are observed even in affluent states. Moreover, substantial regional and national differences occur in the implementation of the effective legal coverage of pensions. This arises for a multitude of reasons including, for instance: lack of economic capability—pension schemes, particularly universal non-contributory schemes, cost money; lack of political motivation to develop and realise such schemes; poor access to affordable quality goods and services for older people such as safe housing, healthcare, as well as long-term aged care; and high levels of informal employment, particularly in the lower to middle income countries.[194] There is a positive trend towards the increase of non-contributory schemes worldwide. However, instead of the non-discriminatory universal schemes, these often tend to be targeted ones which can, if directed too narrowly, mean that many people are not covered. An even bigger challenge is the need to also address socio-economic policies in an attempt to better recognise members of the informal workforce, particularly women and others who are shown to be at a greater risk of vulnerability as they age.

The SDGs of the UN, particularly Goal 1 focusing on the elimination of all forms of poverty, are significant when considering financial security, poverty and ageing. However, in order to ensure that 'no one is left behind' the characteristics of the people experiencing financial insecurity and poverty need to be examined. This includes the particular risk factors for older people in this situation. This is because the risk of poverty and/or homelessness is one of the greatest threats facing the growing numbers of older people worldwide. This is not just a problem for the less developed countries either, but is one confronting *all* countries. It is also one which violates the human rights of a cohort which may be at increased risk of vulnerability because of physical and/or mental impairment, particularly those in the 'old, old' category.

The human right to social security is therefore fundamental to helping to protect all rights of older people—not only to be able to experience income and wealth security, but also to acquire safe and suitable housing, as well as to access adequate health and/or aged care. Thus, financial security is one of the factors underpinning elder law generally and is one of the main mechanisms through which human rights can arguably be best protected given that it facilitates access to other rights such as

[191] Ibid.

[192] Ibid.

[193] See Footnote 121.

[194] Ibid.

accommodation, healthcare, aged care and justice. The need for financial security is therefore clear, and governments must act to ensure that it is available, adequate and appropriate for all older persons. Establishing financial security and addressing issues arising from old age and poverty will take considerable resources from all levels of government, both nationally and internationally. This is where a human rights framework has the ability to be utilised to great effect—by providing international obligations requiring that this be done, and global scrutiny if it is not.

There are also a number of areas of research that are still desperately needed. For instance, more information is needed in relation to the specifics of ageing and poverty—what are the risk factors, who is more likely to be at risk, what interventions actually assist and at what stage of life are they most effective? Underlying this is, again, the need for rigorous prevalence data comparable on an international basis but that is also valuable at a national level. Specific reference needs to be had to older people. It is imperative that this perspective be taken into consideration when both developing and implementing policies, laws and regulations in response to international obligations. Genuinely hearing the voice of older people is the most powerful way of understanding their lived experiences of financial *in*security in order to promote income and wealth security.

References

Alston P (2014) Report of the special rapporteur on extreme poverty and human rights. UN Doc A/69/297 (11 Aug 2014)

Australian Government (2019) retirement income review. https://treasury.gov.au/sites/default/files/2019-11/c2019-36292-v2.pdf. Accessed 17 Feb 2020

Australian Human Rights Commission (2015) Face the facts: older Australians. https://www.humanrights.gov.au/our-work/education/face-facts-older-australians. Accessed 6 Dec 2019

Australian Human Rights Commission (2019) Older women's risk of homelessness: background paper. https://www.humanrights.gov.au/sites/default/files/document/publication/ahrc_ow_homelessness2019.pdf. Accessed 13 Feb 2020

Australian Securities and Investments Commission (n.d.) Reverse mortgages. https://www.moneysmart.gov.au/superannuation-and-retirement/income-sources-in-retirement/home-equity-release/reverse-mortgages. Accessed 9 Dec 2019

Barrientos A (2014) Ensuring income security in old age. In: Age international (ed) Facing the facts: the truth about ageing and development, Age International, pp 74–77. https://www.ageinternational.org.uk/globalassets/documents/age-international-facing-the-facts-report.pdf. Accessed 13 Feb 2020

Brown P, Vickerstaff S (2011) Health subjectivities and labor market participation: pessimism and older workers' attitudes and narratives around retirement in the United Kingdom. Res Aging 33(5):529–550

Brünner N, Andersen S (2018) Making meaning of financial scarcity in old age. J Aging Stud 47:114–122

Centre for Workplace Excellence (2018) Work, care, health and retirement: ageing agenders project. https://www.wgea.gov.au/sites/default/files/documents/unisa_work-care-health--retirement-report_digital.pdf. Accessed 13 Feb 2020

Cox CB (2015) Social policy for an aging society: a human rights perspective. Springer, New York

Douglas E, Rutherford A, Bell D (2017) Pilot study protocol to inform a future longitudinal study of ageing using linked administrative data: healthy ageing in Scotland (HAGIS). BMJ Open 8(1):e018802

Frydenberg J, Hume J (2019) Taking action on the banking, superannuation and financial services industry royal commission—recommendation 2.10: New Financial Adviser Disciplinary System. Joint Media Statement of the Australian Treasurer and Assistant Minister for Superannuation, Financial Services and Financial Technology. https://ministers.treasury.gov.au/ministers/josh-frydenberg-2018/media-releases/taking-action-banking-superannuation-financial-1. Accessed 11 Dec 2019

Hancock R (1998) Housing wealth, income and financial wealth of older people in Britain. Ageing Soc 18(1):5–33

Hayslip B, Fruhauf CA, Dolbin-MacNab ML (2019) Grandparents raising grandchildren: what have we learned over the past decade? The Gerontologist 59(3):e152–e163

HelpAge International (2010) UN independent expert: social pensions are critical to human rights. https://www.helpage.org/silo/files/older-people-human-rights-and-socialprotection.pdf. Accessed 9 Dec 2019

Hershenson DB (2016) Reconceptualizing retirement: a status-based approach. J Aging Stud 38:1–5

Heydon JD, Leeming MJ (2016) Jacobs' Law of trusts in Australia, 8th edn. LexisNexis Butterworths, Australia

Hokenstad MC, Roberts AR (2013) The United Nations plans for a future free of ageism and elder invisibility. J Am Soc Aging 37(1):76–79

Homelessness Australia (2016). Homelessness and older people. https://www.homelessnessaustralia.org.au/sites/homelessnessaus/files/2017-07/Homelessness_and_Older_People.pdf. Accessed 19 Aug 2019

International Labour Organization (2014) World social protection report 2014/15: Building economic recovery, inclusive development and social justice. https://www.ilo.org/global/research/global-reports/world-social-security-report/2014/WCMS_245201/lang--en/index.htm. Accessed 6 Feb 2020

International Labour Organization (2017) World social protection report 2017–19: Universal social protection to achieve the sustainable development goals. https://www.ilo.org/wcmsp5/groups/public/-dgreports/-dcomm/-publ/documents/publication/wcms_604882.pdf. Accessed 9 Dec 2019

James RN (2015) The new statistics of estate planning: lifetime and post-mortem wills, trusts, and charitable planning. Estate Plan Community Property Law J 8(1):1–40

Jolly R (2014) Inequality and ageing. In: Age international (ed) Facing the facts: the truth about ageing and development, Age International, pp 48–50. https://www.ageinternational.org.uk/globalassets/documents/age-international-facing-the-facts-report.pdf. Accessed 13 Feb 2020

Joosten M, Dow B, Blakey J (2015) Profile of elder abuse in Victoria: analysis of data about people seeking help from seniors rights Victoria. National Ageing Research Institute and Seniors Rights, Australia. https://seniorsrights.org.au/wp-content/uploads/2014/03/Summary-Report_Profile-of-Elder-Abuse-in-Victoria_Final.pdf. Accessed 11 Feb 2020

Karp P (2019) Robodebt: the federal court ruling and what it means for targeted welfare recipients. The Guardian (28 Nov 2019) https://www.theguardian.com/australia-news/2019/nov/28/robodebt-the-federal-court-ruling-and-what-it-means-for-targeted-welfare-recipients. Accessed 10 Dec 2019

Kojola E, Moen P (2016) No more lock-step retirement: boomers' shifting meanings of work and retirement. J Aging Stud 26:59–70

OECD (2015) Pensions at a glance 2015: OECD and G20 indicators. https://www.oecd-ilibrary.org/social-issues-migration-health/pensions-at-a-glance-2015_pension_glance-2015-en. Accessed 14 Feb 2020

Office of the High Commissioner for Human Rights (2012) Normative standards in international human rights law in relation to older persons: analytical outcomes paper: https://social.un.org/ageing-working-group/documents/OHCHRAnalyticalOutcomePaperonOldePersonsAugust2012.doc. Accessed 9 Dec 2019

Ong R (2008) Unlocking housing equity through reverse mortgages: the case of elderly homeowners in Australia. Eur J Housing Policy 8(1):61–79

Peeters H, Wouter DT (2015) Lifecourses, pensions and poverty among elderly women in belgium: interactions between family history, work history and pension regulations. Ageing Soc 35(6):1171–1199

Petersen M, Jones A (2011) Homelessness and older australians: developing a research evidence base. Parity 24(9):31

Prindiville K (2015) The law—a crucial tool in fighting elder poverty. Aging Today 36(10):8–10

Purser K et al (2018) Alleged financial abuse of those under an enduring power of attorney: an exploratory study. Br J Soc Work 48(4):887–905

Purser K, Cockburn T, Ulrick E (2020) Examining access to formal justice mechanisms for vulnerable older people in the context of enduring powers of attorney. Elder Law Rev 12(2)

Ranzin R et al (2004) Independence and self-provision in old age: how realistic are these goals? Australas J Ageing 23(3):120–124

Royal Commission into Misconduct in the Banking, Superannuation and Financial Services Industry (2019) Final report. https://financialservices.royalcommission.gov.au/Pages/reports.aspx. Accessed 11 Dec 2019

Sepúlveda Carmona M (2010) Report of the independent expert on the question of human rights and extreme poverty on social protection and old age poverty. UN Doc A/HRC/14/31 (31 Mar 2010)

Sepúlveda Carmona M (2013) Report of the special rapporteur on extreme poverty and human rights. UN Doc A/HRC/23/36/Add.2 (30 May 2013)

Slack T, Jensen L (2008) Employment hardship among older workers: does residential and gender inequality extend into older age? J Gerontol B Psychol Sci Soc Sci 63(1):S15–S24

Stanford Center on Longevity (2018) The sightlines project. Special report seeing our way to financial security in the age of increased longevity. https://longevity.stanford.edu/wp-content/uploads/2018/11/Sightlines-Financial-Security-Special-Report-2018.pdf. Accessed 11 Dec 2019

Steen A, Taggart J (2018) Reverse mortgages. In: Field S, Williams K, Sappideen C (eds) Elder law a guide to working with older Australians. Federation Press, Annandale

United Nations (n.d.) Ageing. https://www.un.org/en/sections/issues-depth/ageing/. Accessed 13 Feb 2020

United Nations Department of Economic and Social Affairs Programme on Ageing (n.d.) Income poverty in old age: an emerging development priority. https://www.un.org/esa/socdev/ageing/documents/PovertyIssuePaperAgeing.pdf. 13 Feb 2020

United Nations Secretary-General (2012) Report on follow-up to the second world assembly on ageing. UN Doc A/67/188 (26 July 2012)

UN Women (2015) Progress of the World's Women 2015–2016 Transforming Economies, Realizing Rights. https://www.unwomen.org/-/media/headquarters/attachments/sections/library/publications/2015/poww-2015-2016-en.pdf?la=en&vs=0. Accessed 11 Dec 2019

United Nations Department of Economic and Social Affairs, Population Division (2015) World population ageing. UN Doc ST/ESA/SER.A/390

United Nations Economic and Social Council, Report of the Secretary-General on Review and Appraisal of the Implementation of the Beijing Declaration and Platform for Action and the Outcomes of the Twenty-Third Special Session of the General Assembly. UN Doc E.CN.6/2015/3 (15 Dec 2014)

United Nations Entity for Gender Equality and the Empowerment of Women (2015) Protecting women's income security in old age: toward gender-responsive pension systems, Policy Brief, No. 3. https://www.unwomen.org/en/digital-library/publications/2015/12/women-income-security-in-old-age. Accessed 6 Feb 2020

United Nations Secretary-General (2015) Report on follow-up to the international year of older persons: second world assembly on ageing. UN Doc A/70/185 (24 July 2015)

United Nations Department of Economic and Social Affairs Programme on Ageing (2016) Economic Inequalities in Old Age. https://www.un.org/development/desa/ageing/wp-content/uploads/sites/24/2016/08/Briefing-Paper_Economic-Inequalities_Final.pdf. Accessed 11 Dec 2019

United Nations Department of Economic and Social Affairs, Population Division (2017) World population ageing: highlights. https://www.un.org/en/development/desa/population/publications/pdf/ageing/WPA2017_Highlights.pdf. Accessed 6 Feb 2020

Virick M (2011) Underemployment and older workers. In: Maynard DC, Feldman DC (eds) Underemployment psychological, economic and social challenges. Springer, New York, pp 81–103

Westwood S (2012) I May be older, but i ain't no 'elder': a critique of "elder law. Temple Polit Civ Rights Law Rev 21(2):485–510

Williams J (2003) When i'm sixty four: lawyers, law and old age. Cambrian Law Rev 34:103–119

World Bank (2015) Women, business and the law 2016: Getting to equal. https://documents.worldbank.org/curated/en/455971467992805787/Women-business-and-the-law-2016-getting-to-equal. Accessed 13 Feb 2020

International Instruments

Additional Protocol to the American Convention on Human Rights in the Area of Economic, Social and Cultural Rights (1988) Opened for signature 17 November 1988, entered into force 16 November 1999. OAS Treaty Series No 69

African Charter of Human and Peoples' Rights (1981) Opened for signature 27 June 1981, entered into force 21 October 1986. 21 ILM 58

Committee on Economic, Social and Cultural Rights (1995) *General Comment No. 6: The Economic, Social and Cultural Rights of Older Persons* (8 December 1995)

Committee on Economic, Social and Cultural Rights (2008). *General Comment No. 19: The Right to Social Security (Art. 9)* UN Doc E/C.12/GC/19 (4 February 2008)

Convention on the Elimination of All Forms of Discrimination Against Women (1979) Opened for signature 18 December 1979, entered into force 3 September 1981. 1249 UNTS 13

Convention on the Elimination of All Forms of Racial Discrimination (1965) Opened for signature 21 December 1965, entered into force 4 January 1969. 660 UNTS 195

Convention on the Protection of the Rights of All Migrant Workers and Members of their Families (1990) Opened for signature 18 December 1990, entered into force 1 July 2003. 2220 UNTS 39481

Convention on the Rights of Persons with Disabilities (2006) Opened for signature 13 December 2006, entered into force 3 May 2008. 2515 UNTS 3

European Social Charter (Revised) (1996) Opened for signature 3 May 1996, entered into force 1 July 1999. ETS 163

International Covenant on Civil and Political Rights (1966) Opened for signature 16 December 1966, entered into force 23 March 1976. 999 UNTS 171

International Covenant on Economic, Social and Cultural Rights (1966) Opened for signature 16 December 1966, entered into force 3 January 1976. 993 UNTS 3

International Labour Organization (1952) *Convention 102 (Social Security (Minimum Standards))*

International Labour Organization (1967) *Convention 128 (Invalidity, Old-Age and Survivors' Benefits)*

Madrid International Plan of Action on Ageing (2002) Adopted by second world conference on ageing, Madrid, 8–12 April 2002

United Nations General Assembly (2015) *Transforming Our World: the 2030 Agenda for Sustainable Development*. Resolution 70/1. UN Doc A/RES/70/1 (21 October 2015)

Universal Declaration of Human Rights (1948) Adopted by United Nations General Assembly Resolution 217A (III). UN Doc A/810 (10 December 1948)
Vienna International Plan of Action on Ageing (1982) Adopted by World Assembly on Ageing, Vienna, 26 July–6 Aug 1982. Endorsed by UN General Assembly Resolution 37/51. UN Doc A/RES/37/51 (3 December 1982)

Cases

Deanna Amato v The Commonwealth of Australia (27 Nov 2019) Federal Court of Australia No: VID611/2019
Johnson v Johnson [2009] NSWSC 503
Keech v Sandford (1726) Sel Cas T King 61; 25 ER 223
López Rodríguez v. Spain E/C.12/57/D/1/2013, Committee on Economic, Social and Cultural Rights (4 March 2016)
Marcia Cecilia Trujillo Calero v. Ecuador E/C.12/63/D/10/2015, Committee on Economic, Social and Cultural Rights (14 November 2018)
Nelson v Nelson (1995) 184 CLR 538
Public Trustee of Queensland as Administrator of the Estate of KH, Deceased [2017] QSC 48

Legislation

Family Law Act 1975 (Cth)
Succession Act 2006 (NSW)
Trusts Act 1973 (Qld)

Chapter 9
Accommodation

A person's ability to access safe and affordable accommodation fundamentally influences their enjoyment of a range of rights and, ultimately, whether they enjoy a life of dignity. The right to housing is guaranteed as part of the right to an adequate standard of living. Ageing, however, may necessitate changes to housing and various accommodation alternatives have been developed for older people worldwide. Two broad categories of accommodation for older people are examined in this chapter. First, accommodation for people who decide to age in their own home and who may require additional supports ('ageing in place'). Second, accommodation for those who decide to relocate to an age-specific housing development ('retirement village living'). Whilst the nomenclature of these villages varies from country to country, the developments are similar in that they generally provide accommodation in a secure (often gated) community with shared leisure facilities and varying levels of in-home support. A third category consists of aged care for those people who require a higher level of long-term care, and these are examined in Chap. 10 in the context of healthcare. This chapter argues that segregating people on the basis of age is not representative of a human rights-based approach to housing for older persons. Instead, it must be acknowledged that the accommodation option which best suits older individuals will be determined by a number of factors including their health needs, financial security, and the overarching legal, regulatory and design frameworks. This chapter will detail the human rights principles which ought to guide law and policy around accommodation for older persons, noting the fundamental interdependence between the right to housing and other rights. It will then examine select housing options, as well as the benefits and criticisms of dedicated retirement living, including dispute resolution mechanisms.

© Springer Nature Singapore Pte Ltd. 2020
B. Lewis et al., *The Human Rights of Older Persons*,
https://doi.org/10.1007/978-981-15-6735-3_9

9.1 Introduction

The extent to which older persons can enjoy their human rights depends in no small part on their living arrangements. Health, social participation, cultural connectedness and financial independence, on the one hand, and mistreatment, abuse and neglect, on the other, are all inherently connected with factors of location, physical space, and habitation—either sole or shared occupancy. Not only is the right to housing guaranteed under human rights law as part of the right to an adequate standard of living, a person's ability to access safe accommodation fundamentally influences their enjoyment of a range of other rights and, ultimately, whether they enjoy a life of dignity.[1]

There are increasing accommodation options available for older persons in most countries. Their ability to choose freely between these alternatives is, however, shaped by the financial, cultural and social conditions linked to the enjoyment of other human rights, such as social security (discussed in Chap. 8) and the highest attainable standard of health (Chap. 10).[2] Ageism (Chap. 5), capacity (Chap. 6) and elder abuse (Chap. 7) are also matters underpinning protection of the right to housing and an adequate standard of living.

What is clear is that new trends are emerging based around whether people remain in their homes and the level of support and care they need.[3] It is perhaps useful to highlight again the close connection between accommodation and long-term aged care. While there are clear intersections between the two, accommodation for older people does not necessarily involve long-term care, and to assume otherwise is an example of ageist attitudes. In fact, as discussed in Chaps. 1 and 10, older people are not a homogenous group and their health needs vary greatly, as does their financial ability to access healthcare options. Thus, the accommodation needs of older people also substantially differ. We have therefore chosen to separate long-term residential aged care from accommodation and have instead included discussion of issues arising in relation to residential aged care in Chap. 10 when exploring the right to health. This is a relatively arbitrary delineation given, as has been seen, the fundamentally intermingled nature of accommodation, aged care and health.

For our purposes here, therefore, we focus on two main categories of accommodation for older people, 'ageing in place' and 'retirement village living'. 'Ageing in place' has been defined as living independently in the community distinct from residential care.[4] The level of independence will depend upon the care needs of the person.[5] 'Ageing in place' gained prominence largely because it is viewed as supporting older people to remain in their own homes, thus promoting autonomy, independence, dignity and social connectedness with family and friends, although

[1] *International Covenant on Economic, Social and Cultural Rights* (1966), art 11 ('*ICESCR*').

[2] Ibid., arts 9 and 12.

[3] Wiles et al. (2011), 357; World Health Organization (2007).

[4] Wiles et al. (2011, 357.

[5] Ibid.

this varies among different cultural and ethnic groups.[6] It also has a financial component given that ageing in place helps to avoid the expense and potential inconvenience (if, for example, the family home has to be sold) of having to enter into institutional care.[7] Consequently, it is an accommodation (and care) model favoured not only by governments and policy-makers but also often by older people themselves.[8] This has, in turn, led to the proliferation of care services able to be performed in the home rather than via an institution as was the historical approach.[9] This has also led to a growth of independent living alternatives specifically directed at people aged over sixty-five years.[10]

We have defined 'retirement village living' as village or communal living for older people who are no longer in paid employment wherein they receive accommodation and services, other than services provided in an aged care facility, and for which a contract was entered into with (generally) an initial contribution being paid that was not rent.[11] We acknowledge however, that retirement village schemes differ greatly from one another, and vary even further depending upon local jurisdictional requirements. This is therefore designed to be a general definition only, and the extent to which a given scheme adheres to human rights standards will vary according to design, services and support offered, legal status and financial arrangements.

The range of retirement village accommodation options for older people throughout the world has increased. It has been suggested that this wide variety of independent, but supported, living alternatives for older people is a property developer's solution to the desire of governments around the world for older people to 'age in place'.[12] Moreover, developers have seen the ageing population as a business opportunity to market exclusive 'resort style living' to people aged fifty years and over. Other developments target people aged sixty-five years and over with a combination of independent and supported living accommodation options, often with a high care residential facility within the same complex. Countries such as the United States of America market exclusive 5-star resorts to potential residents while in other countries entire towns and communities have been designed for older people, such as the 'dementia villages' in the Netherlands.[13]

The rise of retirement village living has been driven by a number of factors, especially increased life expectancy and shifting attitudes within society about old age.[14] There are a number of potential benefits which older people are seeking when considering a move to a retirement village. These include the opportunity to meet

---

[6]Ibid., 357–8; Keeling (1999).

[7]Wiles et al. (2011), 357.

[8]Ibid., World Health Organization (2007).

[9]Wiles et al. (2011), 357.

[10]Bevin (2018); Hu et al. (2017a).

[11]Consumer Affairs Victoria (2019).

[12]Petersen and Warburton (2012).

[13]Although offering long-term care, these villages are designed to maximise independence, and so we have chosen to include them with the accommodation options rather than in Chap. 10.

[14]See, for example: Petersen et al. (2017); Hu et al. (2017a).

new people and live in a community of similarly aged people. These factors can be particularly powerful following the death of a spouse, or where the older person has weak family ties or a fear of social isolation.[15] A further element which has affected the growth of retirement villages is the declining rate of older people being cared for by family. The social expectation of caring for older parents has changed in most developed countries due, in part, to smaller family sizes, an increase in the rates of divorce and the increase of women in the workforce.[16]

As always, there are also disadvantages to residing in a retirement village, some of which are not so obvious to potential residents. The transaction itself generally requires significant financial and legal literacy to navigate often complex and lengthy contracts.[17] Once in the retirement community, residents can experience social isolation and ageism because of the multipurpose nature of many modern villages.[18] The cost to enter into a retirement village is also often high and generally requires residents to sell their homes to finance their entry into the village. Residents who had been homeowners therefore generally have a greater financially-based choice of age friendly housing options distinct from non-homeowners who may have had a history of insecure housing and financial uncertainty.[19] Irrespective of the disadvantages however, retirement villages, including those offering ‘assisted living’, often represent an attractive option for those who are seeking a middle ground between living independently in the community and residential aged care.

All of these factors represent challenges for the protection and realisation of older persons’ human rights. This chapter will begin by detailing the specific human rights and principles which ought to guide law and policy around accommodation for older persons, noting the interdependence between the right to housing and other rights, and the international guidelines that support a human rights-based approach in this space. It will then outline the common types of accommodation which older persons typically rely on. As will be seen, these fall along a continuum from independent living in a person’s own home (‘ageing in place’) through to retirement village living and then the high and long term-care residential facilities discussed in Chap. 10. Which option best suits an individual older person will often be determined by their health needs, although this is not the only consideration, with financial security and the relevant legal, regulatory and design frameworks also playing significant roles in the enjoyment of human rights in the context of accommodation. The benefits and criticisms of dedicated retirement living will then be examined followed by a discussion of dispute resolution mechanisms in this context.

[15]Stimson and McCrea (2004).

[16]Doron and Lightman (2003).

[17]Petersen et al. (2017).

[18]Nielson et al. (2019); Gray and Worlledge (2018).

[19]Bridge et al. (2011).

9.2 Accommodation and Human Rights

Article 11 of the *International Covenant on Economic, Social and Cultural Rights* (ICESCR) guarantees to all people the right to housing as part of an adequate standard of living.[20] The obligations contained in article 11 include a range of elements particularly relevant to older people:

1. ensuring accessibility and usability, taking into account capacity decline;
2. ensuring access to transport, healthcare services and social facilities;
3. guaranteeing physical safety;
4. being sufficiently affordable; and
5. respecting the resident's cultural identity.[21]

It is recognised that what adequate housing looks like is different for each individual and is determined by a range of socio-economic factors.[22] For some, this will be living with family and, as noted in Chap. 2, the prospect of multi-generational families living together is contemplated within article 11 of the ICESCR. Article 10 elaborates on the importance of the family and in line with this, paragraph 1 and recommendation 29 of the *Madrid International Plan of Action on Ageing* ('Madrid Plan of Action') encourage governments and non-government organisations to create support services for families whose elderly members live with them at home.[23] In addition to this, it is recommended that specific measures be implemented to financially support families who are on low incomes to keep their elderly family members at home, something that aligns with the right to social security found in article 9 of the ICESCR. For others, 'adequate housing' will be living independently with the necessary assistance and support to be able to do so; in a retirement village community, including all of the options encompassed within this concept; in a long-term residential care environment; or some combination of all of these accommodation options. The importance of providing options for older persons to choose where they live is recognised in priority direction 3 of the Madrid Plan of Action, which also recognised that appropriate and suitable housing can promote good health and well-being.[24] Fundamentally, the right to housing demands not only that all persons have a place to live, but that they are able 'to live somewhere in security, peace and dignity.'[25]

The Madrid Plan of Action further elaborates on the concept of adequate housing found in the ICESCR.[26] Recommendations 19 and 24 clearly state that housing should not only be considered as 'shelter' but should also be afforded with an

---

[20] ICESCR, art 11.

[21] Office of the High Commissioner for Human Rights (2014); Committee on Economic, Social and Cultural Rights (1991); Farha (2018).

[22] World Health Organization (2015), 161.

[23] *Madrid International Plan of Action on Ageing* (2002) ('*Madrid Plan of Action*').

[24] Ibid., para 95.

[25] Committee on Economic, Social and Cultural Rights (1991), para 7. See also Farha (2018).

[26] Madrid Plan of Action.

understanding of the psychological and social significance that such housing offers. Recommendation 20 highlights the need for effective urban planning to consider the needs of an ageing population, in particular to ensure the continued integration and interaction of older persons with general society.[27]

Housing, or 'adequate' housing, depends on a person's ability to finance it (financial security is discussed further in Chap. 8). Many older people, particularly those with no income or wealth security and/or who live in countries where there is no aged pension, find it very difficult to meet their basic need for housing.[28] As people age, the cost of housing becomes one of their biggest expenses, particularly absent home ownership. In many developing, as well as developed, countries this can determine, for example, what healthcare is available, how much food older people can afford or whether the heating can be turned on, with obvious implications for the enjoyment of other human rights.[29]

Beyond financial considerations, older persons seek accommodation that enables them to be both safe and comfortable, and which aligns with their cultural, religious and other needs. For some this may mean remaining in their own home and within their community however, for a range of reasons, this may not be possible for others. Health issues or disability may demand higher levels of support, though it must not be assumed that a person with disability is incapable of living independently, provided that they receive the right level of assistance. Article 19 of the *Convention on the Rights of Persons with Disabilities* (CRPD) recognises the right of all persons with disabilities to live in their chosen community, to choose where and with whom they live and in whatever living arrangements are suitable for their needs.[30] Article 19 also provides for a range of in-home residential services to support a person's continued living and their inclusion within a community.[31] For older persons with disabilities this provides a more cost-effective option at times as opposed to relocating in order to receive care.[32]

Even without disability, however, the core values of autonomy, liberty and dignity implied by article 19 of the CRPD will still apply to demand that the will and preferences of older persons are respected and that accommodation options are facilitated which allow for the fullest possible enjoyment of human rights. This is essentially what is required within a human rights-based approach to accommodation policy, funding and related legal arrangements. The following sections will consider ageing in place and particular varieties of accommodation before then identifying challenges which need to be overcome to meaningfully implement a human rights-based approach in this context.

---

[27] Committee on Economic, Social and Cultural Rights (1995), para 113.

[28] See, for example: Petersen and Jones (2011).

[29] Office of the High Commissioner for Human Rights (2014).

[30] *Convention on the Rights of Persons with Disabilities* (2006), Annex II.

[31] Ibid.

[32] Wiles et al. (2011).

9.3 Ageing in Place

Since the 1990s, the policy of 'ageing in place' has been an integral part of many governments' strategies to deal with their ageing populations. The policy is effectively a commitment by government to support people to age in their own homes rather than relocate to a formal aged care facility.[33] In 1994 the Ministers of Health and Social Policy of OECD countries agreed on taking active steps to ensure older people can remain in their own homes as they age.[34] However, research has identified that not all countries have taken up the policy and from a sample of thirty-two countries, fifteen were found not to have any policies in relation to 'ageing in place'.[35] Countries located in the Asia–Pacific, Latin America, Europe and the Caribbean had some policies enabling older people to remain in their own home. However, in the Arab States, Africa (except South Africa) and Indonesia there was no evidence demonstrating progression towards a policy of 'ageing in place'.[36]

What though is 'ageing in place'? Although we have defined the term here to mean living independently in the community distinct from residential care,[37] the imprecise nature of the term must be acknowledged and the problems arising from that imprecision recognised. Ageing in place is a complex paradigm which extends beyond attachment to the traditional 'family home' and increasingly looks to broader communities and embedded support systems such as the availability of 'in-home care'.[38] It can also involve an older person repeatedly having to 'reintegrate' themselves with new environments. This can include establishing their identity in new settings as well as navigating new systems, processes and people in a dynamic of social, cultural, political and technological change.[39] Interestingly, although readily adopted by researchers, governments and policy-makers, the concept actually means little to older people themselves who are instead focused on 'staying put' or staying at 'home'.[40] Recognising the importance of this in policy development requires that we understand the various factors that make a place 'home'. These can extend beyond just the physical building and surrounding environment to include: the symbolic significance to the person including a shared history and sense of community; the local political, health and social structures; the sense of community engendered by a particular area; and proximity to friends and family. For most people a sense of 'home' is an unquantifiable combination of any number of factors, the exact composition of which depends entirely upon the individual.[41] All of these factors contribute to the concept of 'ageing in place'. Thus, as is a recurrent theme throughout this book, more

[33] Ibid.

[34] See OECD (1994).

[35] United Nations Population Fund and Help Age International (2012).

[36] Ibid.

[37] Wiles et al. (2011), 357.

[38] Ibid.; Howden-Chapman et al. (1999).

[39] Wiles et al. (2011), 358.

[40] Ibid., 364.

[41] Ibid., 358, Wiles (2005).

nuanced research is needed exploring the concept of 'home' and what this means to older people, particularly people from different backgrounds.[42] Failure to do this perpetuates the deliberate systemic ignorance of the need to authentically include the views of older people about the structures which they will have to navigate.

The Madrid Plan of Action was one of the earliest international human rights instruments to outline a range of priorities and recommendations for action that encourages what has come to be known as 'ageing in place'. This is because the Plan supports older people to remain living in their own homes for as long as possible in conjunction with affordable housing options. It also promotes the introduction of policies that assist older people to access affordable services within their community.[43]

Fundamental to 'successful' ageing in place, however, is the intersection between capacity and function, both physical and cognitive (discussed in Chap. 6), and the physical environment of the home.[44] This showcases how appropriate supports, including physical, financial, emotional and technological, can augment an older person's independence.[45] Nevertheless, it is important to recognise that the loss of capacity can occur raising the question of human rights and long-term aged care (discussed in more detail in Chap. 10). The issue of elder abuse (Chap. 7) is also pertinent to the situation wherein an older person with impaired capacity is ageing in place as there is a heightened risk of abuse occurring in these circumstances because the older person may more easily be able to be isolated from support networks.

One consequence of the 'ageing in place' policy is the rapid growth in the development of retirement villages. As discussed above, on the continuum of accommodation options available, retirement villages offer an almost midway point between independent living (the 'ideal' form of 'ageing in place') and long-term residential care. These villages enable older people to remain in their own communities but in a pre-designed retirement environment—a type of 'ageing in place', albeit in a varied form of 'home'.

9.4 Retirement Living Options

As the world's population ages, the retirement village sector is experiencing dramatic growth. Although the data are somewhat inconsistent (particularly in terms of currency) they demonstrate a consistent trend across various parts of the world indicating increased uptake of retirement village living. For instance, in the United States of America, the 'continuing care retirement communities' (CCRCs) are one

---

[42] For research including older participant perspectives see, for example: World Health Organization (2007).

[43] Ibid., para 98; Madrid Plan of Action.

[44] Wiles et al. (2011), 357–8.

[45] Ibid.

of the largest developments containing on average 330 units.[46] CCRCs are residential facilities specifically designed for older people which generally consist of three levels of care: independent living, assisted living (that is, where older people require some assistance with activities of daily living but do not need constant care) and constant care. All three levels are generally situated within the one complex. In 2010 there were approximately 640,000 residents living in approximately 1900 CCRCs in the United States of America.[47] The majority of these CCRCs are not-for-profit, and half are affiliated with religious organisations.[48] In Australia, as of 2014, there were 2,272 retirement villages housing approximately 184,000 people over the age of sixty-five years—or 5.7% of the over-sixty-five population. This number is projected to increase to 7.5% by 2025.[49] Not-for-profit organisations operate 912 (40.1%) of the retirement village developments and 1,360 (59.9%) are for-profit operations. In 2018 there were 31,500 retirement village units in New Zealand in which 13% of the country's over-seventy-five population reside. This represents a 0.4% increase from the previous year (2017).[50] Although retirement villages are a relatively new concept in the United Kingdom (the majority of people aged over sixty-five have historically tended to remain in their own home),[51] as of 2018 there were approximately 730,000 units,[52] a figure which is expected to increase.[53]

The trend of older people relocating to retirement villages is not, however, limited to developed nations. Developing nations, such as Malaysia, are also recognising the potential for growth in the sector. Some analysts have suggested that there is a golden opportunity for the development of retirement villages based on the increased number of older people living at home which increased from 5.1% in 2004 to 9% in 2014.[54]

Given the growth of the sector generally, this section will now briefly explore the design and operation of select retirement living schemes adopted worldwide. There has been a notable lack of engagement with, and participation of, older people in the design process of accommodation and care options which has resulted in the design of such developments being overwhelmed by assumptions, stereotypes and generalisations about older people and their needs. In fact, current retirement village design is arguably undertaken to hide the reality of ageing, thus perpetuating ageist and negative attitudes and practices.[55]

Although applicable laws, regulations and design can vary greatly throughout the world, retirement villages are somewhat similar in that, as discussed above, they offer a combination of independent living through to limited and high care

[46] Zebolsky (2014).

[47] Zarem (2010).

[48] Ibid.

[49] Property Council of Australia and Grant Thornton (2014).

[50] Lasalle (2019).

[51] Elderly Accommodation Counsel (n.d.a).

[52] Ibid.

[53] Office for National Statistics (2018).

[54] Lim et al. (2019).

[55] Petersen and Warburton (2012).

options.[56] It is the implementation of this underlying aim of providing specialised accommodation for older people that differs. There are also variations in terminology used globally—differences that further contribute to the difficulties in rigorous data collection.

The nomenclature frequently adopted in, for instance, the United States of America, Israel and Japan is CCRC. Taking the United States of America as an example, older people can access a range of housing options financed through the sale of assets such as the family home, relying on savings, purchasing long-term care insurance or annuities, entering into a reverse mortgage or, if eligible, with the assistance of government financed programs.[57] For those who can afford it, retirement accommodation is also available on a luxurious scale promising flexible and responsive resort style living with a strong social and community focus within their CCRCs.[58] However, entry into a CCRC is expensive and the cost is not covered by Medicare or Medicaid. As a result, the demographic make-up of residents in CCRCs is comprised of predominantly Caucasian, widowed women and middle-to-upper class people with an average age of seventy years.[59] The communities are regulated by states rather than by the federal government, however the level of regulation varies from state to state.[60] Whilst regulators and providers believe the level of oversight is sufficient, some consumer groups have suggested a higher level of oversight is required.[61]

In Australia, New Zealand, the United Kingdom and Europe, however, the general terminology adopted is 'retirement village'. A retirement village generally consists of a purpose built residential and multi-dwelling complex suitable for 'older' residents (ranging from approximately fifty-five years and over) who are capable of independent living.[62] There are a variety of retirement village designs in Australia, for example, where it is becoming increasingly commonplace to incorporate a retirement village and a long-term residential aged care facility on the same property. However, these are often designed with separate entrances, segregation within the development and frequently distinct names in an attempt to distinguish between the 'levels' of care available which is, in reality, an example of 'spatial' ageism.[63]

Traditionally retirement villages in Australia consisted of single story developments with adjoining facilities such as swimming pools, gymnasiums and community spaces in order to encourage 'active ageing'.[64] However, due to restrictions of land availability in Australia's capital cities, multi-story apartment blocks incorporating recreational facilities in addition to some limited support and care services are also

[56]Croucher (2006).

[57]U.S. Government Accountability Office (2010).

[58]Ibid.

[59]Shippee (2012).

[60]U.S. Government Accountability Office (2010).

[61]Ibid.

[62]Petersen and Warburton (2012); Crisp (2015).

[63]Petersen and Warburton (2012).

[64]Petersen et al. (2017); Holt et al. (2015).

increasing in popularity.[65] Australia's approach to the accommodation and aged care needs of older people is based on a combination of federal and state legislation. Aged care services in Australia are funded by federal, state, territory and local governments, non-government organisations and from private contributions made by those receiving care.[66] The government subsidises the cost of care and recipients contribute through fees and payments. In 2017–2018, approximately 28% of total aged care spending by the federal government was spent on home care and support. Expenditure on home care and support has increased by 34% between 2012–2013 and 2017–2018.[67] Retirement villages (distinct from aged care facilities) are legislated by the Australian state and territory governments. These purpose-built villages are increasingly being marketed in a manner similar to those in the United States of America wherein resort style living is offered for 'active' adults.[68] Whilst originally the domain of the not-for-profit sector, since the late 1980s retirement villages have been operated by both for-profit and not-for-profit organisations.[69]

In New Zealand, retirement villages are regulated by the government through the Ministry of Housing and Urban Development. The villages are of a similar design to those in the United States of America and Australia, comprising residential villages surrounded by fences, large security gates with well-tended gardens and extensive landscaping.[70] The demographic is also largely Caucasian, over seventy years old, university educated, and restricted to those who can afford to purchase a lease as well as pay the ongoing and increasing charges.[71]

Turning to the United Kingdom, as stated above, retirement villages are a relatively new product. There are four different types of retirement living. First, age-exclusive housing specifically caters to older people. It includes communal facilities such as lounges and large outdoor shared spaces, however, it does not offer any on-site care or support. Residents must be capable of independent living. Any emergency support can be obtained by the use of an alarm system and is on a user-pays basis.[72] The second variation is housing with support, which is principally independent living for residents aged fifty-five years and over. All accommodation is self-contained and there is access to some communal facilities. There is an on-site manager who runs the site and is available to address any resident concerns or complaints. On average, a housing with support development contains 20–80 units.[73] Third, housing with care or extra care housing (ECH) has become increasingly popular in the United Kingdom since its introduction in the 1990s. ECH is designed for older people who require a higher level of assistance with personal and household chores, and who

---

[65] Howe et al. (2013).

[66] Australian Institute of Health and Welfare (2019).

[67] Ibid.

[68] Howe et al. (2013).

[69] Property Council of Australia and Grant Thornton (2014).

[70] Nielson et al. (2019).

[71] Ibid.

[72] Elderly Accommodation Counsel (n.d.b).

[73] Ibid.

cannot live safely in independent housing. The concept supports the independence of the older person but with a higher level of support stopping just short of living with the care providers as would occur in an aged care facility. There is a variety of onsite care and support. Residents are in self-contained accommodation, referred to as assisted living. Often ECH provides older people with a more attractive alternative than residential aged care because in addition to the twenty-four-hour on-site support and care, there are also communal facilities, such as a restaurant, gymnasium or lounge room where there are organised activities for the residents. In 2016 there were approximately 1,600 ECH developments in the United Kingdom.[74] The developments vary in size from 20 to 200 units.[75] ECH has been considered by the United Kingdom government to be a viable alternative to aged care given that it supports an older person's independence and their choice of where they wish to live as they age.[76] ECH has attracted interest from the United States of America, Australasia and Europe.[77] Finally, closed care housing is relatively new and is a combination of independent units co-located with a residential care home. The independent units can obtain some services, such as cleaning, which is included in their service charge. Other services can be purchased as required from the care home.[78]

Somewhat surprisingly, the trend of privately owned and run retirement villages is relatively new to northern European countries, such as Finland, where accommodation for the aged has largely been the traditional domain of the government.[79] In the Finnish case, the privatisation of aged care accommodation has resulted in the rapid growth in retirement village development and alternative age-friendly accommodation, such as dementia villages (discussed below).[80]

Interestingly, retirement communities are often marketed differently from conventional housing developments. The avoidance of images or other indicators of 'old age' is prevalent in the marketing material. Selective words such as 'upmarket resorts', '5 star', 'hotel or resort style amenities' are used, often in conjunction with younger models than the average age of retirement village residents.[81] This is all designed to entice the older person to enter the community which, when coupled with the often complex and confusing legal and regulatory framework of retirement villages (discussed below), can mean that an older person is not necessarily aware of their legal obligations arising from entering such communities.[82] In Australia, for instance, units in retirement villages are leased but the marketing around them generally indicates that older people will be 'purchasing'—a very different legal outcome, particularly when considering estate planning outcomes and ensuring financial security.

---

74 Buisson (2016).

75 Elderly Accommodation Counsel (n.d.b).

76 Johnson et al. (2019).

77 Howe et al. (2013).

78 Ibid.

79 Lundman (2019).

80 Ibid.

81 Petersen and Warburton (2012), 73.

82 See, for example: Petersen et al. (2017).

Thus, it is clear that although based upon the same premise, that of providing accommodation and care options for older persons ranging from independent living through to high care, the legal frameworks, regulations, design and terminology of retirement village communities are extremely varied worldwide. Furthermore, what also seems to be emerging is the existence of three key elements: provision of supported independent living in self-contained accommodation; provision and availability of communal facilities such as restaurants, lounges, gardens, and laundries; and access to continuous on-site care (if and when necessary) which is flexible enough to quickly adapt to the changing needs of residents.[83] In addition to these common configurations, two specific types of accommodation options for older people warrant attention for their particular design features and potential enhanced enjoyment of human rights: dementia villages and self-managed or co-housing. These will be examined next, before the discussion turns to the efficacy of retirement villages in satisfactorily meeting their underlying human rights purposes.

9.4.1 Dementia Villages

Dementia villages have been the product of a worldwide movement which originated in Japan called 'dementia friendly communities'.[84] A dementia friendly village or community is a place or environment where people with dementia and their carers or family members are included in, and supported by, society.[85] They aim to provide a more meaningful care environment for dementia patients. There are dementia friendly communities or villages in Africa, the United States of America, South America, Australasia, United Kingdom and Europe.[86] The principles of a dementia friendly community champion empowerment, support and inclusivity and reflect the human rights principles of dignity, equality and participation.[87] Although they have a strong aspect of care we have decided to include them here as an example of the different types of living options available, irrespective of any diagnosed mentally disabling condition.

Europe now leads the way with the design of dementia villages, which are becoming increasingly popular with healthcare professionals and relatives of people with dementia as an alternative to high care residential nursing homes. In the Netherlands, Germany and Denmark these villages are specifically designed with the needs of people with memory disorders in mind.[88] In the Netherlands, for example, the Hogeweyk dementia village consists of twenty-three houses for one hundred and fifty-two older dementia patients. The patients manage their own households with

---

[83] Evans et al. (2017); Evans and Vallelly (2007).

[84] Miyamoto et al. (2011).

[85] Alzheimer's Disease International (n.d.).

[86] Ibid.

[87] Ibid.

[88] Glass (2014).

the assistance of support staff. Similar to a retirement village, the facility offers a restaurant and a communal social space.[89] The main difference between this village and a retirement village is that the facilities and grounds of the village are available to be used by the surrounding neighbourhood thus facilitating inclusion and participation. This inclusive attitude therefore aims to remove the potential for age- or disability-based segregation and the negative human rights impacts of this (see Chap. 4 for more on social and cultural rights). Denmark's first dementia village, Bryghuset, opened in 2015 and was established by relatives of older people with dementia and healthcare professionals. The Danish village is based on the Hogeweyk model.[90]

The concept of dementia villages has been subject to criticism with some arguing that the residents are being deluded into thinking that they are still living within a 'real' community when they are, in fact, still in a controlled environment not dissimilar to a nursing home.[91] A method of addressing this is evident in the Dutch village where the relatives of family members with dementia regularly take them out of the village to visit other family members or take them to appointments in an attempt to connect them to the 'real' community. These activities are supported by the village health professionals.[92] This support is a recognition that there is no 'delusion' amongst the family members and the care professionals that the village is offering a real-life experience for people with dementia, but a more meaningful and inclusive environment than a traditional high care aged care facility. Dementia villages are therefore most closely aligned to a human rights-based approach, because they are grounded in a respect for the individual's dignity and autonomy, seeking to promote social and cultural participation to the greatest extent possible. They also represent a rejection of the traditional medical model for housing people with dementia, which has frequently involved violations of human rights through restricting their freedom of movement and privacy or exposing them to cruel or degrading treatment, particularly through the use of physical and chemical restraints.[93]

9.4.2 Self-Managed or Co-housing

An alternative intermediate-level housing model for older people gaining traction is co-housing or self-managed group housing. The design of co-housing generally includes a cluster of small homes or units around a common shared space, which can include a kitchen and dining room of which the residents share the cost.[94] It is a way that older people can live independently but still be a part of a small community of

[89] Hogeweyk (n.d.).
[90] Peoples et al. (2018).
[91] Zwijsen et al. (2011).
[92] Peoples et al. (2018).
[93] Foebel et al. (2016).
[94] Ibid.

people with shared ideals and values and benefit from the security and support that comes from that connectedness.[95] Since the mid-1980s, co-housing communities for older residents have been developed initially in northern European countries such as Denmark and Netherlands, but have now been adopted in other countries such as Germany, the United Kingdom and United States of America.[96] The first American project opened in 2005 with eight units built in Colorado for a group long-time of friends whose average age was eighty-one years. The largest community of twenty-nine units is in Virginia and includes residents who own their own unit and some who rent government subsidised units.[97] A distinguishing feature of this style of accommodation, and significant from a human rights approach, is that the residents are actively involved in the design process and manage the community themselves.

9.5 Is There Value in Retirement Living?

Despite the marketing campaigns promising continued good health and financial prosperity, the question is increasingly being asked about the utility of retirement villages in satisfactorily providing affordable and quality accommodation and care services.[98] The next sections will therefore discuss the benefits and criticisms of retirement living.

9.5.1 Benefits of Retirement Living

The role of a 'community' in an older person's life becomes increasingly important as they experience life post-retirement, the loss of a partner or the loss of mobility and/or capacity. It is during this time that many older people in developed countries look to relocate to a more communal style of living arrangement where costs of maintaining their home can be minimised. Older people are also often looking to either maintain their social connections or to discover new ones. In fact, one of the main benefits of retirement village living is the opportunities provided for social interaction. The importance of social participation was highlighted in Chap. 4, wherein it was demonstrated that inclusion and participation are essential for the enjoyment of a wide range of human rights. The Madrid Plan of Action recognised and supported the need to incorporate affordable housing for older people with opportunities for social interaction.[99]

---

[95]Brenton (2013).

[96]Baldwin et al. (2019).

[97]Glass (2012).

[98]Hu (2017b).

[99]Madrid Plan of Action, para 98.

The village environment provides an opportunity for residents to take part in social events and a range of activities, for instance some villages offer swimming pools and gymnasiums. As discussed in Chap. 10, there are a number of contributing factors to 'healthy ageing' including, for example, a lifestyle that is physically, intellectually and socially active.[100] Given the prevalence of social isolation and loneliness among older people, and the heightened risk that these factors present for elder abuse, for instance, the move to a retirement village can be an attractive alternative for many people.[101] However, whilst the physical environment assists in promoting a setting of community and social interaction, social and cultural contexts are equally as important for a harmonious group life.[102]

The foundations for a successful community, or effective social interaction, include a shared group identity and collective behaviour, and can be affected by factors of class, education and life experience.[103] Within a retirement village context these social 'norms' involve an expectation of sociability and engagement from all residents[104] and depend on them sharing a similar and functional level of physical health[105] and well-being, and sufficient cognitive capacity to be able to contribute.[106] Residents who live in retirement communities may actually gain psychological benefits and improved functionality when moving into a retirement village by being part of an active community.[107] The success of the retirement living environment is, however, predicated upon the residents not only sharing in these 'norms', but in also having a similar base level from which to be able to contribute to the communal dynamic. A study in the United States of America, for example, noted an 'us and them' view between those living independently and those living in either assisted living or long-term aged care.[108] Residents interviewed admitted to losing touch with residents who moved from independent living to assisted living or aged care as the perception was that their lack of engagement with former friends was a personal choice rather than due to cognitive decline.[109]

Retirement living may also offer the reassurance of living in a supportive environment where assistance is close at hand if required. The ability to make an 'easy' transition from independent living into accommodation providing constant care (if and when needed), and thus not becoming a 'burden' on family and friends, is promoted in varying ways in the marketing campaigns of both property developers and not-for-profit organisations.[110] The implied or even subconscious reinforcement

[100]Bennett et al. (2014).

[101]Sutin et al. (2018); Barbosa Neves (2019).

[102]Durkheim (1984).

[103]Van den Hoonaard (2002).

[104]Ibid.

[105]Johnson and Troll (1994).

[106]Yamasaki and Sharf (2011).

[107]Holland et al. (2017).

[108]Shippee (2012), 3.

[109]Ibid., 4.

[110]Petersen and Warburton (2012), 73.

of the notion that older people are 'burdensome', in effect a living liability, is clearly problematic, however. The ability to access safe accommodation options should be a genuine choice for older persons, albeit one inevitably based in financial reality, not one made out of fear of having to one day depend upon others.

From the relatively little data available, residents who enter into a retirement village arrangement seem content overall with their decision, indicating that they perceive the benefits to outweigh any perceived or actual costs.[111] It is also suggested that the non-conventional living arrangement of a retirement community, as opposed to remaining at home (for people who are not used to living in a communal environment), may better meet the increased and diverse needs of older people.[112] As discussed in Chap. 4, human rights law requires that older people who belong to minority groups are supported to maintain those connections, and thus, where a retirement village community makes this a priority, then it will be beneficial for older people who belong to those minority groups.[113] It is noteworthy, however, that the data are frequently collected by the retirement village industry themselves and thus questions are justifiably raised in relation to both the rigour and the reliability of such findings. There are, nevertheless, benefits to retirement living and residents who choose age-exclusive accommodation are ideally seeking a combination of independence and security in addition to opportunities to be socially and physically active.[114] This style of living can therefore be an effective way of ensuring that human rights are protected and promoted as people age. However, as will now be discussed, there are also persistent challenges which can lead to human rights concerns.

9.5.2 Disadvantages of Retirement Living

Traditionally retirement villages and aged care residences were operated by not-for-profit organisations.[115] With the increasing entry into the market of property developers and the subsequent commercialisation of care, an unfortunate result has been a neglect of the real needs of the ageing population within some retirement communities.[116] Furthermore, in many jurisdictions private corporations are not directly bound by human rights obligations.[117] There is a significant risk therefore that retirement villages will fail to adhere to human rights standards. At a minimum, all residents are entitled to an adequate standard of living, access to necessary healthcare and social

[111] Biggs et al. (2000).

[112] Pynoos (2018) as cited in Lundman (2019).

[113] *International Covenant on Civil and Political Rights* (1966), art 27; ICESCR, art 15.

[114] Crisp et al. (2013).

[115] Property Council of Australia and Grant Thornton (2014).

[116] Cutchin (2007).

[117] Under international law, states bear the primary obligation to protect human rights, including through regulating the actions of corporations. Work is underway, however, to adopt a new treaty which would articulate the human rights duties of businesses. See Human Rights Council (2014).

interaction. Inadequate care or social isolation can have further negative impacts on physical and mental health. This points to the need for accommodation systems to be adaptable to the older individuals' changing needs, while at the same time ensuring that their preferences with respect to their living arrangements are respected along with their dignity and right to participate in decisions about those living arrangements (where capable to do so).

However, despite the marketing of retirement living focusing heavily on a secure and thriving living environment promising independence and care, the reality of these purpose-built villages is often an age-segregated and socially isolated existence—in contrast to the data suggesting that occupants are generally happy with their decision.[118] Upon entry to a retirement village, residents are expected to be capable of independent living requiring minimal assistance and support.[119] Many residents however experience a decline in physical and cognitive capacity requiring a move to more supported accommodation. In that regard, and despite promises to the contrary, retirement villages often frequently fail to provide any sense of the certainty and security that attracted the resident in the beginning.[120] Living in a specifically designed and constructed village for older people can also result in residents struggling with self-esteem if they do not fit within the norms outlined above.[121] Further, where the villages have differing levels of accommodation and care, the residents requiring little care may compare themselves with residents requiring a higher level of care. These residents may be perceived as 'less capable' and therefore are often excluded from activities.[122] The remainder of this section will address some of the more common problems arising in the retirement village environment, notably the financial and legal complexity, cost, social isolation and diversity.

9.5.2.1 Financial and Legal Complexity

A prospective retirement village resident is required to understand complex legal and financial obligations. This is because, in most jurisdictions, entry into a retirement village is not a purchase of property nor is it a purchase of care, it is instead a unique form of tenancy.[123] Residents can therefore experience confusion in relation to understanding their rights in a leasehold relationship, especially if transitioning from owning their own home, in addition to concerns about ever-increasing fees and complex residential contracts.[124]

Take for example, a vision impaired resident who was living in her own home. Over time her family were becoming worried that she would be safer in a retirement

[118]Lundman (2019).

[119]Doron and Lightman (2003).

[120]Lundman (2019).

[121]Ibid.

[122]Dobbs (2008).

[123]Petersen et al. (2017).

[124]Ibid.

village. After some persuasion, the older person decided upon a village and negotiated with the management some additional conditions in her retirement village contract. Upon executing the contract, it was discovered that the contract she signed was only a standard version and did not contain the agreed-upon amendments. The village operator refused to acknowledge the prior negotiated terms and the resident had no funds to pursue the operator because she was reliant on a pension as her only income source. Anecdotally at least the inability of a prospective resident to understand and negotiate terms or pursue remedial relief against operators is not uncommon.

Given that a human rights-based approach values older persons' autonomy, seeks to maximise their participation in decision-making, and respects those choices once made, overly complex legal or financial arrangements in retirement village contracts undermine this objective. It is obviously problematic to assume that an older person will not be capable of understanding technical legal language or complex structures, but the frequently unapproachable and impenetrable way in which retirement village arrangements are drafted, and the usual requirement for the incoming resident to obtain (costly) financial and legal advice, can have the effect of not only reinforcing negative and ageist stereotypes but also forcing older persons to obtain sometimes prohibitively expensive legal advice on lengthy contracts when they may not be financial secure. There is therefore a risk that an older person will fail to realise the full implications of the agreement, or will accept a lesser deal because of family, health, financial or other pressures, also remembering the frequently employed 'burdensome' marketing approach. The use of standard-term contracts and the reluctance of operators to negotiate (or abide by) special terms consequently raises questions in terms of equality before the law and access to justice for older persons. More practically, it means that individual residents may be unable to secure the particular arrangements that they need, or prefer, and which may be necessary to support full enjoyment of their human rights.

Legislation can also come into conflict thus failing to protect the rights of older persons. In the United Kingdom, for example, there are approximately 300,000 private care home residents whose rights are not protected by the *Human Rights Act* ('the Act') because it does not extend to private operators.[125] This effectively means that such residents of a private care home cannot rely on the Act to dispute or prevent an eviction from a private care home. Given that 91% of care homes in England and Wales are privately owned, this represents a potential violation of a resident's right to adequate housing.[126] Furthermore, in 2007 the House of Lords ruled that an eighty-four-year-old diagnosed with Alzheimer's disease could not rely on the Act to prevent her from being evicted from a private care home despite her care being funded by the local council.[127] These examples serve to highlight the fact that a meaningful human rights-based approach would ensure that residents are protected from eviction without adequate access to justice. Appropriate protections

[125] Human Rights Act 1998 (UK), s 6; Kendall-Raynor (2007), 14.

[126] Kendall-Raynor (2007), 14.

[127] Ibid.; *YL (by her litigation friend the Official Solicitor) v Birmingham City Council & Ors* (2007) UKHL 27.

ought to be included in contracts to ensure that an older person is not left at risk of homelessness, which may require safety nets to be established more broadly to provide adequate state-funded or subsidised accommodation.

9.5.2.2 Cost of Retirement Living

The models of retirement living discussed have been predominantly focused on a financially secure older person who has the funds and/or assets to buy into these retirement village 'resorts'. Financially insecure older people lacking income and or asset security frequently do not have the ability to cover the increasing daily charges often imposed by multi-purpose retirement villages. There is an acknowledged and increasing demand for safe and affordable housing for older people on low to moderate incomes.[128] Moreover, in the event that a resident is unhappy in the village, then the deposits and fees are often not refundable.[129]

As noted above, often retirement villages expect residents to be able to live independently with minimum care. When their care needs change, they can find themselves needing to move, but the financial arrangements they have entered into often leave them with insufficient funds to move to the higher care facility of their choice. Consequently, they may have to compromise on physical design, location, cultural appropriateness and connection to community, facilities and services, and/or quality of care. This points to the need for a coordinated, human rights-based approach to the aged accommodation sector, which helps to support older persons better transition through the different levels of care in a way which maximises their options, facilitates their participation, and respects their choices and dignity.

9.5.2.3 Social Isolation, Inclusion and Diversity

As discussed above, a fear of social isolation and desire to increase social networks is often the reason older people choose to enter into a retirement community.[130] Therefore, it is more likely that older people with smaller social networks and greater social needs will seek to move to a retirement village than someone with strong social connections and friendship groups. That is, a prospective resident is more likely to be looking for friendship and companionship than an older person with an active social life.[131] Paradoxically, however, the very design of a 'retirement village' is itself a form of exclusion, or at least labelling and separation, from the community. Whilst the residents often choose a retirement village for reasons of safety and security, they are frequently finding themselves in an age-segregated and

[128] Bridge et al. (2011).

[129] Ayalon and Green (2013).

[130] Medical Advisory Secretariat, Health Quality Ontario (2008); Biggs et al. (2000).

[131] Sheehan and Karasik (1995).

often gated community away from general society.[132] Combine this with healthy and unimpaired residents living side by side with residents experiencing a range of age-related impairments, often advanced in nature, and it is suggested that a retirement village actually presents a perfect environment to foster social isolation. Whilst a 'younger' older resident of a retirement village can experience increased levels of social interaction and engagement, it is often the residents who are not as physically active or adept at navigating social situations who find themselves socially isolated.[133] In fact, residents can often feel isolated and are subject to social exclusion on their entry into retirement village living, particularly when their independence or cognitive capacity deteriorates.[134]

As stated, the marketing of a retirement village is based on the ideal of a particular 'lifestyle' and a sense of a ready-made community. However, the reality is often different given that in order to establish and/or maintain a social life, a resident relies upon either an invitation to be part of social groups or a pre-existing connection to residents already living in the village. Retirement village management usually takes no active role in establishing social groups or encouraging the social participation of new residents. Isolation and exclusion are rife.[135] Therefore, the promised lifestyle is generally unattainable unless the older person has established social connections within the village prior to entry.[136] Individual behaviours in a retirement village may also result in social exclusion such as drinking alcohol, not attending funerals of other residents,[137] being too introverted or not socially engaged,[138] behaving inconsiderately to neighbours,[139] and even being widowed (particularly when the other residents are married).[140] In response to such activity, responses within a community take the form of social exclusion and avoidance of people perceived as not 'fitting in'.[141] However, whilst some behaviour may be anti-social in nature, it may also be aggravated by structural and systemic factors including the design of the village and/or the ways in which activities are managed.

As discussed in Chap. 4, social participation is a driver of the full realisation of human rights, while exclusion is a multidimensional, dynamic problem that negatively impacts on a range of human rights across the many domains of a person's life.[142] Exclusion of older persons is influenced by a variety of factors and has been found to be somewhat of a fluid concept. This is because it can be a transitional experience dependent upon the older person's compliance with rules or social norms,

---

132 Bernard (2012).
133 Shippee (2012).
134 Ibid.
135 Ibid.
136 Petersen and Warburton (2012).
137 Carp and Carp (1980).
138 Ibid.
139 McLean (2006).
140 Van den Hoonaard (2002).
141 Shippee (2009).
142 Walsh et al. (2017).

as well as the level of decline in physical as well as cognitive capacity.[143] Further, although it is arguably not possible to guarantee friendships between residents, structural barriers which impede inclusion and participation need to be addressed. This includes ensuring that organised social activities are inclusive of all levels of ability and appeal to people with a diverse range of interests. Physical design of spaces ought to encourage interaction, while still allowing for adequate privacy and enabling individuals to choose the level of interaction that they prefer. Consideration should also be had to cultural factors which may impact on an individual's participation and inclusion, and opportunities for people to remain connected to their own cultural communities and pre-existing social networks need to be supported.

The importance of ensuring that retirement communities are inclusive and supportive of diversity has become particularly pertinent with respect to recognition of LGBTIQA older persons for whom there is a lack of choice of housing and support in the overall retirement community. There are limited statistics, however, of numbers of older LGBTIQA residents living in retirement communities as residents are often reluctant to self-identify in order to avoid discrimination and prejudice.[144] Of the data that have been collected, a recent study in the United Kingdom indicated that LGBTIQA older people do not feel that their rights are considered and that they are missing out on choices available to mainstream sections of the older population.[145]

Other studies have revealed that older LGBTIQA people also often delay entry into aged care or retirement villages due to a fear of stigmatisation and marginalisation.[146] In response to this a scheme has been adopted by the Netherlands National Centre for Social Development called the Pink PassKey to indicate those aged housing and care providers who have LGBTIQA friendly policies.[147] The first Passkey was awarded to six local centres in 2008 and by 2014 there were more than one hundred local care organisations who held a Pink Passkey. The scheme is now operating in Germany and being developed in Austria, Spain and Costa Rica.[148] Housing, retirement and care communities specifically for LGBTIQA older people have also been created in New Mexico, Los Angeles and San Francisco, with studies estimating that there are approximately three million LGBTIQA people aged over sixty-five years in the United States of America.[149] These communities are usually operated by not-for-profit groups with the goal of providing safe and affordable accommodation and care for LGBTIQA older adults in a way that engenders respect.

---

[143] Shippee (2012); Link and Phelan (2001).

[144] Wathern and Green (2017).

[145] Ibid.

[146] Leyerzapf et al. (2018).

[147] COC Netherlands (n.d.).

[148] Ibid.

[149] de Vries (2005/2006).

The retirement village and aged care industries throughout the world struggle with catering for the needs of people not only from the LGBTIQA community but also from other culturally diverse backgrounds.[150] Therefore, there exists an ongoing requirement for further research to ensure that older people from different backgrounds are not discriminated against as they face decisions relating to suitable housing as they age. The importance of respecting diversity among older people is discussed in more detail in Chap. 4.

9.6 Dispute Resolution

Transitioning into a form of communal retirement living results in residents living in close proximity to their neighbours, often for the first time. Research conducted in 2014 reported that residents complained about an overall lack of power in resolving disputes.[151] The variety of dispute resolution processes provided by the village and the ability of village management to solve disputes have been found to influence residents' satisfaction levels and well-being.[152] From a human rights perspective, it is important that residents feel that their concerns are taken seriously and not dismissed out of hand, and that a genuine effort is made to try to resolve the dispute.

One way to achieve this is through embedding elder mediation as a dispute resolution method. Elder mediation has been described as an 'emerging discipline' whose importance in resolving disputes is becoming more relevant as the population ages.[153] It is a preventative process designed to offer a more time- as well as cost-effective alternative to the traditional court system.[154] As an 'alternate' dispute resolution method, elder mediation should therefore be able to better foster respect for the autonomy, independence, participation and control of the parties. The focus on self-determination may also promote access to justice given that mediation is intended to be less intimidating and more cost effective. It also does not prevent access to more formal forms of dispute resolution such as judicial determination. However, as with other forms of dispute resolution, there are drawbacks, such as lack of enforceability and potential power imbalances that have to be taken into account when considering any access to justice pathway.[155]

---

[150] Ibid.

[151] Malta (2018).

[152] Ibid.

[153] Barry (2018).

[154] Law Council of Australia (2017), 11–2.

[155] Barry (2018); Williams (2013).

9.7 Conclusion

As the population ages worldwide, the increasing issue for older people, governments, policy-makers and society more broadly is ensuring access to adequate, safe and age appropriate housing. This is particularly the case given that the social expectation that families care for older people has changed. Older people want, and are encouraged, to maintain their independence and 'age in place', whether that be in their own home or in a specifically designed community.

The retirement village industry throughout the world has consequently undergone exponential growth fuelled by the combination of the ageing population, government policy and the corporatisation or commodification of ageing. Dedicated retirement living is aggressively marketed as a luxury destination in which to spend the remaining years of one's life enjoying facilities such as swimming pools, gymnasiums and organised social activities. There are numerous benefits to retirement living for those who can afford it, which include an active social life, increased opportunities for activity and fitness, along with the added benefit of downsizing to a more manageable property. In addition, there are integrated communities for older people living with dementia, particularly in northern Europe, which provide a secure environment for the older person and their family without segregating them from society. However, this lifestyle is not accessible to all people as they age. The numerous disadvantages include the complexity of the financial and legal obligations surrounding entry, the high cost of entering into these communities (particularly when considered in conjunction with the question of financial security discussed in Chap. 8), and social isolation (the risks of which are discussed in Chap. 4). Residents who are unhappy are often forced to stay as they cannot afford to leave due to complex fees and charges imposed on a departing resident.

There is an increasing demand for adequate and age specific housing for older people on fixed and low incomes. The challenge for governments is to ensure that there is a sufficient supply of appropriate housing for the population as it ages. Currently, older people with financial security can sell the family home and move comfortably into an accommodation option of their choice. However, older people on low or fixed incomes are often faced with long waiting lists for basic housing. As the population ages, along with declining rates of home ownership throughout the world, it is essential that the right to adequate, safe and secure housing is maintained.

Other human rights should similarly guide the design of physical spaces, the legal and financial arrangements entered into, the management of facilities and provision of care and services. More broadly, retirement living should form one part of an integrated, human rights-based network of accommodation options for older persons to ensure that suitable, affordable, culturally appropriate and inclusive accommodation options are available to suit individuals' needs and ease of transition between them. Legal rights relating to tenure and residency must also be adequately protected to prevent exploitation and appropriate safety nets ensuring access to justice established to protect the most vulnerable older persons from human rights violations.

References

Alzheimer's Disease International (n.d.) Principles of a dementia friendly community https://www.alz.co.uk/dementia-friendly-communities/principles. Accessed 20 Sept 2019

Australian Institute of Health and Welfare (2019) Government spending on aged care. https://www.gen-agedcaredata.gov.au/www_aihwgen/media/2018-factsheets/spending_factsheet.pdf. Accessed 14 Feb 2020

Ayalon L, Green V (2013) Social ties in the context of the continuing care retirement community. Qual Health Res 23(3):396–406

Baldwin C, Dendle K, McKinlay A (2019) Initiating senior co-housing: people, place, and long-term security. J Housing Elderly 33(4):358–381

Barbosa Neves B (2019) "It's the worst bloody feeling in the world": Experiences of loneliness and social isolation among older people living in care homes. J Aging Stud 49:74–84

Barry L (2018) Elder mediation: what's in a name? Conflict Res Q 32(4):435–442

Bennett DA et al (2014) Cognitive and social lifestyle: links with neuropathology and cognition in late life. Acta Neuropathol 127(1):137–150

Bernard M et al (2012) Then and now: evolving community in the context of a retirement village. Ageing Soc 32:103–129

Bevin K (2018) Shaping the housing Grey zone: an Australian retirement villages case study. Urban Policy Res 36(2):215–229

Biggs S et al (2000) Lifestyles of belief: narrative and culture in a retirement community. Ageing Soc 20(06):649–672

Brenton M (2013) Senior co-housing communities—an alternative approach for the UK? Joseph Rowntree Foundation Programme Paper. https://www.basw.co.uk/system/files/resources/basw_24212-7_0.pdf Accessed 14 Feb 2020

Bridge C et al (2011) Age specific housing for low to moderate income older people. AHURI Final Report No. 174. https://www.ahuri.edu.au/__data/assets/pdf_file/0011/2216/AHURI_Final_Report_No174_Age-specific_housing_and_care_for_low_to_moderate_income_older_people.pdf. Accessed 14 Feb 2020

Buisson L (2016) Extra care and retirement communities: UK Market Report, 14th edn. LaingBuisson, London

COC Netherlands (n.d.) Netherlands: Pink passkey project. https://international.coc.nl/66-netherlands-pink-passkey-project/. Accessed 17 Feb 2020

Carp FM, Carp A (1980) Person-environment congruence and sociability. Res Aging 2(4):395–415

Crisp DA et al (2013) What are older adults seeking? Factors encouraging or discouraging retirement village living. Australasian J Ageing 32(3):163–170

Crisp DA (2015) Retirement villages. In: Pachana N (ed) Encyclopedia of Geropsychology. Springer, Singapore, pp 1-7

Consumer Affairs Victoria (2019) What is a retirement village? https://www.consumer.vic.gov.au/housing/retirement-villages/choosing-a-retirement-village/what-is-a-retirement-village. Accessed 3 Dec 2019

Croucher K (2006) Making the case for retirement villages. Joseph Rowntree Foundation, York

Cutchin MP (2007) Therapeutic landscapes for older people: care with commodification, liminality, and ambiguity. In: Williams A (ed) Therapeutic landscapes. Ashgate, Aldershot, pp 181–198

de Vries B (2005/2006) Home at the end of the rainbow. Generations 29(4):64–69

Dobbs D et al (2008) An Ethnographic study of Stigma and ageism in residential care or assisted living. Gerontologist 48(4):517–526

Doron I, Lightman E (2003) Assisted-living for older people in Israel: market control or government regulation? Ageing Soc 23(6):779–795

Durkheim E (1984) The division of labor in society (trans: Halls WD) Free Press, New York

Elderly Accommodation Counsel (n.d. a) https://www.eac.org.uk/. Accessed 14 Feb 2020

Elderly Accommodation Counsel (n.d. b) Finding the right retirement housing for you: retirement village factsheet. https://www.housingcare.org/downloads/kbase/3603.pdf. Accessed 19 Sept 2019
Evans S, Vallelly S (2007) Social well-being in extra care housing. Joseph Rowntree Foundation, York
Evans S et al (2017) A community hub approach to older people's housing. Quality Ageing Older Adults 18(1):20–32
Farha L (2018) Report of the special rapporteur on adequate housing on the implementation of effective rights-based housing strategies. UN Doc A/HRC/37/53
Foebel AD et al (2016) Physical restraint and antipsychotic medication use among nursing home residents with dementia. J Am Med Directors Assoc 17(2):184.e9-e14
Glass AP (2012) Elder co-housing in the United States: three case studies. Built Environ 38(3):345–363
Glass AP (2014) Innovative seniors housing and care models: what can we learn from The Netherlands. Social Housing Care J 22(1):74–81
Gray A, Worlledge G (2018) Addressing loneliness and isolation in retirement housing. Ageing Soc 38(3):615–644
Hogeweyk (n.d.) Hogeweyk, living in lifestyles: a mirror image of recognizable lifestyles in our society. https://hogeweyk.dementiavillage.com/en/. Accessed 14 Feb 2020
Holland C et al (2017) Transition from community dwelling to retirement village in older adults: cognitive functioning and psychological health outcomes. Ageing Soc 37(7):1499–1526
Holt A et al (2015) Are retirement villages encouraging active ageing? J Aging Physical Activity 24(3):407–411
Howe AL, Jones AE, Tilse C (2013) What's in a name? Similarities and differences in international terms and meanings for older peoples' housing with services. Ageing Soc 33(4):547–578
Howden-Chapman P, Signal L and Crane J (1999). Housing and health in older people: ageing in place. Soc Policy J New Zealand 13(13):14–30
Hu X et al (2017a) Retirement villages in Australia: a literature review. Pacific Rim Property Res J 23(1):101–122
Hu X et al (2017b) What is a sustainable retirement village? Perceptions of Australian developers. J Clean Prod 164:179–186
Johnson L, Troll E (1994) Constraints and facilitators to friendships in late late life. Gerontologist 34(1):79–87
Johnson EK et al (2019) Ageing in extra care housing: preparation, persistence and self-management at the boundary between the third and fourth age. Ageing Soc 1–21 https://doi.org/10.1017/S0144686X19000849
Keeling S (1999) Ageing in (a New Zealand) place: ethnography, policy and practice. Soc Policy J New Zealand 13:95–114
Kendall-Raynor P (2007) Eviction of elderly is 'Lawful' despite the human rights act. Nursing Standard 21(45):14
Lasalle JL (2019) Retirement villages market review white paper May 2019. https://www.jll.nz/content/dam/jll-com/documents/pdf/research/retirement-village-white-paper-2019-web-final.pdf. Accessed 17 Sept 2019
Law Council of Australia (2017) The justice project introduction and consultation questions. Consultation Paper. https://www.lawcouncil.asn.au/files/web-pdf/Justice%2520Project/Consultation%2520Papers/Introduction%2520and%2520Consulation%2520Questions.pdf. Accessed 11 Feb 2020
Leyerzapf H et al (2018) Gay friendly elderly care: creating space for sexual diversity in residential care by challenging the hetero norm. Ageing Soc 38(2):352–377
Lim XJ, Ng SI, Basha NK (2019) To retire or not to retire: the intention towards concept of retirement villages in Malaysia. Asian J Business Res 9(1):60–80
Link BG, Phelan JC (2001) Conceptualizing stigma. Ann Rev Sociol 27(1):363–385

Lundman R (2019) A spatio-legal approach to the intermediate housing-with-care solutions for older people: exploring the adoption of a retirement village concept in finland. Ageing Soc pp 1–22. https://doi.org/10.1017/S0144686X19000412

Malta S (2018) An ant against an elephant: retirement village residents' experiences of disputes and dispute resolution. Australian J Ageing 37(3):202–209

McLean A (2006) From commodity to community in nursing homes: an impossibility? Ageing Soc 26(6):925–937

Miyamoto M, George DR, Whitehouse PJ (2011) Government, professional and public efforts in Japan to change the designation of Dementia. Dementia 10(4):475–486

Nielson L, Wiles J, Anderson A (2019) Social exclusion and community in an urban retirement village. J Ageing Stud 49:25–30

OECD (1994) Caring for frail elderly people: new directions in care. Organisation for economic co-operation and development, Paris

Office of the High Commissioner for Human Rights (2014) The right to adequate housing (Fact sheet No. 21/Rev. 1) https://www.ohchr.org/Documents/Publications/FS21_rev_1_Housing_en.pdf. Accessed 14 Feb 2020

Office for National Statistics (2018) Living longer: how our population is changing and why it matters. https://www.ons.gov.uk/peoplepopulationandcommunity/birthsdeathsandmarriages/ageing/articles/livinglongerhowourpopulationischangingandwhyitmatters/2018-08-13. Accessed 19 Sept 2019

Peoples H, Pedersen LF and Moestrup L (2018) Creating a meaningful everyday life: perceptions of relatives of people with dementia and healthcare professionals in the context of a Danish Dementia village. Dementia, pp 1–18. https://doi.org/10.1177/1471301218820480journals.sagepub.com/home/dem

Petersen M, Jones A (2011) Homelessness and older Australians: developing a research evidence base. Parity 24(9):31

Petersen M, Warburton J (2012) Residential complexes in Queensland, Australia: a space of segregation and ageism? Ageing Soc 32(1):60–84

Petersen M, Tilse C, Cockburn T (2017) Living in a retirement village: choice, contracts and constraints. J Housing Elderly 31(3):229–242

Property Council of Australia and Grant Thornton (2014) National overview of the retirement village sector. https://www.retirementliving.org.au/wp-content/uploads/2015/03/National-overview-of-the-retirement-village-sector-Grant-Thornton.pdf. Accessed 14 Feb 2020

Pynoos J (2018) The future of housing for the elderly: four strategies that can make a difference. Public Policy Ageing Rep 28(1):35–38

Secretariat MA, Ontario HQ (2008) Social isolation in community-dwelling seniors: an evidence-based analysis. Ontario Health Technology Assessment Series 8(5):1–49

Sheehan NW, Karasik RJ (1995) Decision to move to a continuing care retirement community. J Housing Elderly 11(2):107–122

Shippee TP (2009) "But i am not moving": residents' perspectives on transitions within a continuing care retirement community. Gerontologist 49(3):57–70

Shippee TP (2012) On the edge: balancing health, participation, and autonomy to maintain independent living in two retirement facilities. J Ageing Stud 26(1):1–15

Stimson RJ, McCrea R (2004) A push-pull framework for modelling the relocation of retirees to a retirement village: the Australian experience. Environ Plann 36(8):1451–1470

Sutin AR et al (2018) Loneliness and risk of Dementia. J Gerontology. Series B:1–9

United Nations Population Fund and Help Age International (2012) Overview of Available Policies and Legislation, Data and Research, and Institutional Arrangements Relating to Older Persons: Progress Since Madrid. Report compiled in preparation for The State of the World's Older Persons 2012. https://www.unfpa.org/sites/default/files/pub-pdf/Older_Persons_Report.pdf. Accessed 14 Feb 2020

U.S. Government Accountability Office (2010) Older Americans: continuing care retirement communities can provide benefits, but not without some risk. Doc GAO-10–611, published 21 June 2010, publicly released 21 July 2010
Van den Hoonaard DK (2002) Life on the margins of a florida retirement community: the experience of snowbirds, newcomers, and widowed persons. Res Aging 24(1):50–66
Walsh K, Scharf T, Keating N (2017) Social exclusion of older persons: a scoping review and conceptual framework. Eur J Ageing 14(1):81–98
Wathern T, Green RW (2017) Older LGB&T housing in the UK: challenges and solutions. Housing Care Support 20(3):128–136
Wiles JL (2005) Conceptualising place in the care of older people: the contributions of geographical gerontology. Int J Older People Nurs 14(8B):100–108
Wiles JL et al (2011) The Meaning of 'aging in place' to older people. Gerontologist 52(3):357–366
Williams K (2013) Elder mediation in Australia. Elder Law Rev 7:6
World Health Organization (2007) Global age-friendly cities: a guide https://www.who.int/ageing/publications/Global_age_friendly_cities_Guide_English.pdf. Accessed 3 Dec 2019
World Health Organization (2015) World report on health and ageing. https://apps.who.int/iris/bitstream/handle/10665/186463/9789240694811_eng.pdf;jsessionid=7AF39462821E35F7BDAF3986D2CE9462?sequence=1. Accessed 14 Feb 2020
Yamasaki J and Sharf BF (2011) Opting out while fitting in: how residents make sense of assisted living and cope with community life. J Aging Stud 25(1):13–21
Zarem JE (2010) Today's Continuing Care Retirement Community (CCRC). American seniors housing association CCRC task force. https://willametteview.org/uploads/documents/Todays_CCRC_LeadingAge.pdf. Accessed 14 Feb 2020
Zebolsky G (2014) An introduction to continuing care retirement communities. https://us.milliman.com/insight/2014/An-introduction-to-continuing-care-retirement-communities/. Accessed 20 Sept 2020
Zwijsen SA, Niemeijer AR, Hertogh CMP (2011) Ethics of using assistive technology in the care for community-dwelling elderly people: an overview of the literature. Aging Mental Health 15(4):419–427

International Instruments

Committee on Economic, Social and Cultural Rights (1991) *General Comment No. 4: The right to adequate housing (Art 11 par 1)* (13 December 1991)
Committee on Economic, Social and Cultural Rights (1995) *General Comment No. 6: The Economic, Social and Cultural Rights of Older Persons* (8 December 1995)
Convention on the Rights of Persons with Disabilities (2006) Opened for signature 13 December 2006, entered into force 3 May 2008. 2515 UNTS 3
Human Rights Council (2014) *Elaboration of an international legally binding instrument on transnational corporations and other business enterprises with respect to human rights.* Resolution 26/9. UN Doc No A/HRC/RES/26/9 (14 July 2014)
International Covenant on Civil and Political Rights (1966) Opened for signature 16 December 1966, entered into force 23 March 1976. 999 UNTS 171
International Covenant on Economic, Social and Cultural Rights (1966) Opened for signature 16 December 1966, entered into force 3 January 1976. 993 UNTS 3
Madrid International Plan of Action on Ageing (2002) Adopted by Second World Conference on Ageing, Madrid, 8–12 April 2002

Cases

YL (by her litigation friend the Official Solicitor) v Birmingham City Council & Ors (2007) UKHL 27

Legislation

Human Rights Act 1998 (UK)

Chapter 10
Health and Aged Care

The ability to enjoy human rights more broadly, including economic, social and cultural rights, is influenced heavily by one's state of health. There are numerous international and regional instruments in relation to the right to health, and the literature is extensive. Notwithstanding this, there is often a poor understanding of the interdependence of the right to health with other human rights, the obligations arising from these rights and the standards required to meet them, particularly in relation to older persons. Issues arising in relation to the right to health and other health-related rights are especially significant for older persons given the ageist assumptions and practices prevalent in health and aged care settings. Older persons continue to face diverse challenges in accessing affordable, appropriate and quality health and aged care services. This chapter is designed to complement the existing body of knowledge through demonstrating how fundamental human rights principles ought to guide legal frameworks in facilitating access to affordable and quality care. It argues for the adoption of a multisectoral person-centred approach embedding the core values of respect for dignity, liberty, participation and autonomy, while also delivering specific safeguards where appropriate. A holistic approach to health is necessary, incorporating determinants such as economic, social and environmental factors, while recognising the importance of functional ability and capacity. This chapter specifically considers the global context of the right to health, issues arising in relation to health and aged care systems, and barriers to accessing care.

10.1 Introduction

The ability to participate in social, cultural, educational, recreational or paid activities is heavily influenced by a person's health.[1] Just as financial security (discussed in Chap. 8) underpins the rights of older people generally, so too does having the health

[1] Beard et al. (2016), 163; European Network of National Human Rights Institutions (2017), 1.

© Springer Nature Singapore Pte Ltd. 2020

B. Lewis et al., *The Human Rights of Older Persons*,
https://doi.org/10.1007/978-981-15-6735-3_10

to enjoy them. Efforts have been made since 1946 to integrate a human rights-based approach into policies and practices globally in order to fulfil the right to the highest attainable standard of health ('the right to health').[2] Numerous international and regional instruments have been developed and subsequently adopted by governments and policy-makers worldwide which recognise the right to health.[3] Social theories and concepts about what constitutes good health are increasingly being developed and refined globally,[4] and there is a rich source of literature exploring the role of human rights in promoting health.[5] Human rights language is also increasingly being utilised in the health context to draw attention to older people who are particularly at risk of not being able to access affordable and quality healthcare.[6]

Nevertheless, throughout low, middle and high income countries, older persons face diverse challenges in accessing affordable, quality health and aged care services, both in the short and long term. There is a pervasive fear that the rates of ageing mean that the number of end users will inevitably exceed the capacity of finite care resources.[7] Moreover, despite the commitments arising from the numerous international instruments, there is often a poor understanding of the human rights of older persons, the obligations arising from these rights, the standards necessary to fulfil them, and the interconnectedness of different rights in the context of health.[8]

A narrow construction of health sees it defined solely in terms of absence of disease. This is, however, limiting as it does not recognise the importance of a holistic approach to the complex and interrelated financial, environmental as well as social factors informing both health and ageing. Failure to appreciate the broad, multifaceted issues connected to health undermines effective responses and ultimately impinges upon older persons' rights. Moreover, despite the growing attention being attracted by the ageing population and increasing technological advances, there is actually little evidence that older people currently enjoy better health than their parents did.[9] In fact, although there are varying declines in physical and mental capacity as people age (discussed in Chap. 6), if an older person is generally of good health and resides in an

[2] Baer et al. (2016), 206.

[3] See, for example: *Universal Declaration of Human Rights* (1948), art 25 ('*UDHR*'); *International Covenant on Economic, Social and Cultural Rights* (1966), art 12 ('*ICESCR*'); *Convention on the Elimination of All Forms of Racial Discrimination* (1965), art 5(e)(iv); *Convention on the Elimination of All Forms of Discrimination Against Women* (1979), arts 11(1)(f), 12 and 14(2)(b) ('*CEDAW*'); *Convention on the Rights of the Child* (1989), art 24; *Convention on the Protection of the Rights of All Migrant Workers and Members of their Families* (1990), arts 28, 43(e) and 45(c); *Convention on the Rights of Persons with Disabilities* (2006), art 25 ('*CRPD*'); *African Charter of Human and Peoples' Rights* (1981), art 16; *European Social Charter* (1996), art 11; *Additional Protocol to the American Convention on Human Rights in the Area of Economic, Social and Cultural Rights* (1988), art 10.

[4] Sadana et al. (2016), 181.

[5] See, for example: Gostin et al. (2018); Harrington and Stuttaford (2010); Tobin (2012); Hunt (2016).

[6] Baer et al. (2016), 206.

[7] European Network of National Human Rights Institutions (2017), 2.

[8] Ibid.

[9] World Health Organization (2018); Beard et al. (2016), 163; Chatterji et al. (2015).

environment supportive of their human rights, then their ability to engage in activities that they value will differ relatively little from that of a younger person.[10] Reliable data in relation to 'healthy ageing' and what this looks like are, however, limited. It is therefore difficult to obtain a complete picture of older people's experiences of health, including what 'health' should look like and how this relates to the right to health.[11] This problem has not only arisen because of a lack of research however, although this is an issue in low and middle income countries. The issue in high income countries is rather the inconsistencies in the data that have been collected emerging largely from definitional discrepancies, for instance, how 'older' is being chronologically defined (discussed in Chap. 1).[12]

What is clear is that in health, as in all other areas, it is important to change the ageist assumptions promoting negative stereotypes of older people as being frail, dependent and a burden on society to a positive understanding and appreciation for the opportunities that arise as people age (discussed in Chap. 5).[13] Ageism in healthcare for older people can present as, for instance, workers adopting patronising attitudes, failing to consult the older person in relation to their health and aged care preferences, and restricting access to necessary medical interventions or care rationing.[14] It is only through eliminating ageist assumptions and practices that policies can be developed to authentically foster the right to health. However, in many countries there is little to suggest that the policies and infrastructure necessary for older people and societies more broadly to benefit from the opportunities arising from ageing exist or are even in development.[15] Moreover, there is limited debate on how these opportunities should be maximised.[16]

Global action on the public health implications of the ageing population is therefore desperately needed. The United Nations' Sustainable Development Goal ('SDG') number 3 focuses on good health and wellbeing 'for all at all ages', although, as with the discussion in Chap. 8 in relation to the right to social security and SDG 1, there is no specific mention of what this means for older people. The World Health Organization (WHO) also produced the first *World Report on Ageing and Health* in 2015 which outlines a public health framework based on the concepts of intrinsic capacity and functional ability.[17] It incorporates a broader human rights paradigm by fostering respect for the intrinsic abilities of older individuals and highlights the importance of the environment they inhabit as a setting in which their rights can be protected and fulfilled, for example in aged care.[18] Aged care is defined here as the care provided by other people to older persons experiencing a significant

---

[10] World Health Organization (2018); Beard et al. (2016), 163.

[11] World Health Organization (2015), 49.

[12] For more on this, see: World Health Organization (2015), 49–51.

[13] Royal Commission into Aged Care Quality and Safety (2019).

[14] World Health Organization (2015), 94–5.

[15] Beard et al. (2016), 163.

[16] Ibid.; Lloyd-Sherlock et al. (2012).

[17] World Health Organization (2015), 64–6; Beard et al. (2016), 163.

[18] World Health Organization (2015), 64–6; Beard et al. (2016), 163.

loss of intrinsic capacity and/or functional ability consistent with their human rights and fundamental freedoms to promote dignified living.[19] It is an especially topical issue in, for instance, Australia, where a Royal Commission Inquiry into Aged Care Quality and Safety ('Royal Commission') was launched in 2019.[20]

In fact, there are significant negative implications for society if SDG 3 is not achieved.[21] Ageing populations will see an increased demand for primary healthcare and long-term aged care requiring a bigger and (much) better trained, as well as supported, workforce. The demand for health and aged care services does not have to be an economic drain on society, however, and to assume it will be is an example of the negative assumption that older people are financial burdens. Moreover, not only could ageing be affordable for communities, it could actually benefit both developing and higher income countries through acknowledging the financial and non-financial contributions made by older people.[22]

The health, functional abilities and capacity of older people are, however, complex and diverse, making the challenge of adequately supporting them, while respecting the rights of older persons, unique. Health, perhaps more than any other issue, demonstrates the need for, and appropriateness of, a multidimensional and comprehensive human rights-based approach. Such an approach embeds core values of respect for dignity, liberty and autonomy, while also delivering specific safeguards and outcomes in particular contexts of care. The idea that the right to health incorporates determinants of health, including economic, social and environmental factors, has also long been accepted within human rights law.[23] A human rights-based approach is therefore best able to accommodate the complexity of factors informing good health as a platform for the enjoyment of other human rights. Achieving the goals of public health and human rights agendas is thus a complex and challenging undertaking that will require ambition and progressive realisation.[24]

This chapter provides a human rights analysis of these crucial and, at times, controversial issues. It will consider the global context of the right to health, beginning with consideration of how health should be conceptualised within a human rights-based approach. Issues arising in relation to healthcare systems, long-term care and aged care will then be examined as will barriers to accessing care. Some comments will then be made in relation to palliative care and the role of pain relief at the end of life before making concluding remarks. It should be noted that this chapter is not intended to, nor could it, be a comprehensive analysis of all of these issues because, as stated, there is rich literature delving into the complexities of each. Instead, this chapter will complement that literature by demonstrating how the application of fundamental human rights principles ought to guide legal frameworks in facilitating access to affordable and quality healthcare. In particular, it argues that the right to

[19]Pot et al. (2018).

[20]Royal Commission into Aged Care Quality and Safety (2019).

[21]Beard et al. (2016), 163; Staudinger et al. (2016).

[22]Fried (2016), 167.

[23]Committee on Economic, Social and Cultural Rights (2000).

[24]Baer et al. (2016), 208; World Health Organization (2015), 27.

the highest attainable standard of health not only requires governments to provide adequate health and aged care, as well as related services, it likewise encompasses each individual's functional and decision-making ability in addition to the various factors which operate as determinants of health in older age.

10.2 Health and Ageing

Biologically, ageing occurs as a result of the cellular and molecular damage that accrues during a person's life course causing a decrease in physical and mental capacity before, ultimately, death.[25] In fact, age, and the biological changes associated with 'ageing', are believed to be the most important determining factor in health.[26] This section will discuss what is meant by 'healthy ageing', functional ability and intrinsic capacity, as well as determinants of health in ageing.

10.2.1 Defining Health in the Ageing Context

As with many concepts in the ageing discourse, 'healthy ageing' (discussed in Chap. 4) is the subject of much discussion with no clear consensus emerging as to how to define, measure or analyse it. It is therefore useful to think in terms of the WHO approach of defining 'healthy ageing' in holistic terms. This focuses on the development and maintenance of an individual's functional ability and intrinsic capacity throughout the life course in order to facilitate well-being as they age, particularly in 'old, old' age, and also has reference to other determinants such as financial, social and environmental factors.[27]

Fostering 'healthy ageing' as a concept is designed to go further, therefore, than only protecting the right of older people to access healthcare.[28] In this sense it is similar to the right to health, as found in the *International Covenant on Economic, Social and Cultural Rights* (ICESCR). As discussed below, the Committee on Economic, Social and Cultural Rights has explained that the right to health encompasses the underlying determinants of good health, and that health is also dependent on the realisation of other rights.[29] In order to experience good health whilst ageing, older people need to be able to enjoy the right to social (financial) security ensuring

---

[25] World Health Organization (2018).

[26] United Nations Department of Economic and Social Affairs Programme on Ageing (2018), 1.

[27] World Health Organization (2019); World Health Organization (n.d.).

[28] World Health Organization (n.d.).

[29] Committee on Economic, Social and Cultural Rights (2000).

that they are free from poverty,[30] have access to safe housing,[31] be able to participate in society,[32] and be shown respect for their dignity[33] and autonomy.[34] The highest attainable standard of health in older age therefore needs to be understood as the culmination of many interacting factors throughout an individual's life. It also includes understanding that governments' obligations extend across an individual's life course.

10.2.2 Functional Ability and Intrinsic Capacity

Turning then to the concepts of functional ability and intrinsic capacity. Functional ability here refers to a person having the proficiency, including physical and mental capacity, to both be, and do, what they reasonably value.[35] It comprises health and environmental factors, and the cumulative impact of these on an individual. For instance, an individual's (functional) ability to move around, develop and preserve relationships, meet their own basic needs including activities of daily living (ADLs), learn, grow, make autonomous decisions and participate in activities they value, as well as decision-making.[36]

Research on intrinsic capacity tends to focus on the loss or reduction of abilities to perform ADLs or instrumental activities of daily living (IADLs) (which require more advanced cognitive functioning).[37] These measures have a valuable practical application, particularly in relation to the assessment of capacity, which was the subject of detailed analysis in Chap. 6, and can therefore give some insight into the environmental factors impacting on an older person.[38] They are, however, limited to detecting more serious losses in function, and assessments are often conducted on an ad hoc and unsatisfactory basis placing the human rights of older persons at risk of abuse.[39] It would therefore be useful to be able to see the trajectory that preceded the loss of ADLs and IADLs, particularly any factors that influenced the loss of capacity, and to what extent.[40] It should be noted, however, that the intrinsic capacity discussed here does differ from the legal concept discussed in Chap. 6. Whereas legal

---

[30] UDHR, arts 22 and 25; ICESCR, art 9.

[31] ICESCR, art 11.

[32] United Nations Principles for Older Persons (1991), Principle 7 ('*UN Principles*'); Kornfeld-Matte (2017), para 13.

[33] UDHR, art 25(1).

[34] See, for example, CRPD, art 3.

[35] World Health Organization (2019); World Health Organization (n.d.).

[36] World Health Organization (2015), 28.

[37] Ibid., 65.

[38] Ibid.

[39] Ibid.; Purser and Rosenfeld (2014); Purser (2017).

[40] On measuring and analyzing this trajectory see, for example: World Health Organization (2015), 65–7. See also: Cosco et al. (2014).

capacity focuses on a person's ability to make legally recognised decisions, intrinsic capacity in the health context is more concerned with promoting a person's good health generally and what this looks like practically.

10.2.3 Health Determinants in Ageing

The extent to which physical and mental decline occurs in ageing, however, is not uniform and does not occur according to a standardised timeline. There is no 'typical' older person.[41] Factors such as financial security (including, for instance, the ability to afford health insurance), retirement and work history (for example a desk job distinct from manual labour),[42] safe housing and nutrition can all further impact the rate and extent of decline in ageing.[43] This is why some seventy-five-year-old people, for example, experience good health while others can be quite frail.[44] Chronological age can also be a poor predictor of physical and mental capacity which is why some seventy-five-year-old people can have the capacity of much (chronologically) younger people, whereas others experience significant decline.[45] Assuming that people become physically and/or mentally infirm merely because of age alone is therefore erroneous and ageist (discussed in Chap. 5).

Nevertheless, the risk of experiencing poor health, either acute or chronic, does increase as people age, particularly for people in the 'old, old' cohort, with older age being characterised by the emergence of 'geriatric syndromes'.[46] Geriatric syndromes are a combination of multiple complex health states that do not fit within discrete disease categories and tend to develop only in later life, such as frailty, falls, pressure ulcers and incontinence.[47] Frailty, for instance, is estimated to have a prevalence rate of approximately 10% in people aged sixty-five years and over.[48] These geriatric syndromes can be a better predictor of death than specific diseases but are frequently undetected or ignored, especially in countries that lack a recognised speciality in geriatric medicine.[49] Conditions in older people are also not static, that

---

[41] World Health Organization (2017a).

[42] Staudinger et al. (2016), 281.

[43] World Health Organization (2018); United Nations Department of Economic and Social Affairs Programme on Ageing (2018), 1.

[44] World Health Organization (2018).

[45] Beard et al. (2016), 164.

[46] United Nations Department of Economic and Social Affairs Programme on Ageing (2018), 1; United Nations Department of Economic and Social Affairs, Population Division (2015); World Health Organization (2018); See also: World Health Organization (2015), 62–4.

[47] World Health Organization (2018).

[48] World Health Organization (2015), 26.

[49] World Health Organization (2018).

is, there is no one predetermined pathway that an illness will follow.[50] Consequently, in considering appropriate care, it is necessary to look at not only the nature and stage of the illness but also its effect(s) on both physical and mental functioning.[51]

The natural, physical and social environments of a person, for instance the extent of air pollution in the person's location, are also determining factors of health in ageing.[52] It is likewise affected by a person's exposure throughout the life course to risks including poor diet, inactivity and smoking, as well as social change such as isolation through the loss of loved ones.[53] Genetics play some role, and are estimated to be responsible for approximately 25% of changes in health and function as people age.[54]

Socio-economic status, of both the state and the individual, is another significant factor. Individually this includes occupation, income, education level, gender and membership of ethnic or other minority groups, all of which will have different significance in different communities.[55] Older people in, for instance, lower socio-economic countries experience significantly poorer health outcomes and reduced life expectancies.[56] Furthermore, accumulated disadvantages experienced throughout life are compounded as people age.[57] For example, although the rates of multi-morbidity, that is, the probability of experiencing more than one chronic condition, increase generally as people age, the rates are estimated to be higher in lower to middle income countries.[58] Interestingly, even in higher income countries the multi-morbidity rates increased disproportionately for those in the lower socio-economic groups.[59] It is estimated in Germany, for example, that 24% of people in the seventy to eighty-five year old bracket experience five or more chronic conditions simultaneously.[60] Further, the life expectancy rates discussed in Chap. 1, although increasing

---

[50] World Health Organization (2015), 26.

[51] Ibid.

[52] United Nation Department of Economic and Social Affairs Programme on Ageing (2018), 1; World Health Organization (2018); Beard et al. (2016), 164; Dannefer (2003); Sadana et al. (2016), 179.

[53] United Nation Department of Economic and Social Affairs Programme on Ageing (2018), 1; United Nations Department of Economic and Social Affairs, Population Division (2015); Courtin and Knapp (2017).

[54] Foebel and Pedersen (2016); United Nations Department of Economic and Social Affairs Programme on Ageing (2018), 1.

[55] World Health Organization (2018); United Nations Department of Economic and Social Affairs Programme on Ageing (2018), 1; Beard et al. (2016), 164; Sadana et al. (2016), 179; Dannefer (2003).

[56] World Health Organization (2017b); United Nations Department of Economic and Social Affairs Programme on Ageing (2018), 1.

[57] Sadana et al. (2016), 179; United Nations Department of Economic and Social Affairs Programme on Ageing (2018), 1, 3.

[58] United Nations Department of Economic and Social Affairs Programme on Ageing (2018), 5; World Health Organization (2018); See also in World Health Organization (2015), 58–62.

[59] United Nations Department of Economic and Social Affairs Programme on Ageing (2018), 5. See also St. Sauver et al. (2015).

[60] World Health Organization (2015), 26.

worldwide, do differ.[61] Globally, for instance, the life expectancy of women is greater than that for men, with the greatest difference being in Europe where women live on average four years longer than men.[62] This can be compared to Africa where women only live on average one and a half years longer than men.[63] Within societies, life expectancy can differ significantly for certain minority groups, especially Indigenous people.[64] In relation to the impact of education on life expectancy, persons who have attained the highest education level live approximately six years longer than people with the lowest education level in member states of the Organisation for Economic Co-operation and Development ('OECD').[65]

These factors can also be seen in the patterns of death globally. For example, in low to middle income countries, deaths from non-communicable diseases ('NCDs') occur at earlier ages than deaths from NCDs in high income countries, and significant numbers of deaths result from communicable diseases (at all ages).[66] Worldwide, NCDs are the most common health condition in older people and the primary cause of death across all countries and regions.[67] Ischaemic heart disease, stroke and chronic pulmonary disease lead the conditions responsible for deaths amongst older people, with these diseases having a much greater impact in low to middle income countries.[68] Hearing and vision loss, chronic obstructive pulmonary disease, falls, fractures, dementia, depression, diabetes, back and neck pain as well as osteoarthritis are estimated to be the most significant causes of disability for older people.[69] Hearing and vision loss as well as heart disease in particular have a higher prevalence rate in low to middle income countries partly because quality interventions such as hearing aids, glasses and medication are more readily available (and accessible) in high income countries.[70] The rates of dementia, however, are actually higher in high income countries. This could, however, reflect greater awareness and diagnostic ability.[71]

---

[61] United Nations Department of Economic and Social Affairs Programme on Ageing (2018), 3; United Nations Department of Economic and Social Affairs, Population Division (2015).

[62] United Nations Department of Economic and Social Affairs Programme on Ageing (2018), 3–4.

[63] Ibid.

[64] See, for example: LaVeist (2005).

[65] United Nations Department of Economic and Social Affairs Programme on Ageing (2018), 4.

[66] Ibid.; World Health Organization (2015).

[67] World Health Organization (2017a), United Nations Department of Economic and Social Affairs Programme on Ageing (2018), 4; World Health Organization (2015).

[68] United Nations Department of Economic and Social Affairs Programme on Ageing (2018), 4.

[69] Ibid., 5; World Health Organization (2018); World Health Organization (2015), 53–64.

[70] United Nations Department of Economic and Social Affairs Programme on Ageing (2018), 3; World Health Organization (2015), 54, 57.

[71] World Health Organization (2015), 57; United Nations Department of Economic and Social Affairs Programme on Ageing (2018), 5.

These conditions can all significantly affect an older person's life, although it is important to note that the existence of 'disease' does not necessarily impact function.[72] Take, for instance, hearing loss. Untreated, it negatively impacts communication and may consequently result in isolation.[73] This, in turn, can lead to depression and anxiety.[74] The inability to hear can also be misconstrued as mental incapacity, particularly when coupled with ageist assumptions that older people are incapable.[75] Any or all of these factors can then cause the older person to isolate themselves even further in order to reduce their interactions with other people, and thus significantly affect the enjoyment of their human rights.[76] Yet, in many cases, the condition can be effectively treated or supported, so the impact on a person's life (and their human rights) is determined not by the fact of the disease, but rather by the response to it.

Law, policies and regulation likewise have a significant role to play in how older people experience ageing, especially in relation to fundamental structural determinants like health and aged care services, and social security.[77] Given the ageing population, protecting the right to health assumes a much greater sense of urgency, although the fact that every person experiences ageing differently presents a unique challenge in how to ensure adequate access to affordable, quality services.[78] Furthermore, this must be achieved within a framework of multiple, interrelated actions from governments and policy-makers which all affect the enjoyment of other, interrelated human rights.

10.3 The Right to Health

What then, does the right to health incorporate? As discussed in Chap. 2, older people are entitled to the universal rights and freedoms recognised in the ICESCR. This includes the rights to health and social security,[79] which require that states make available, without discrimination of any kind, good quality health facilities, and goods and services that are accessible and acceptable.[80] Availability here refers

[72] World Health Organization (2015), 57.

[73] Ibid., 55.

[74] Ibid.

[75] Ibid.

[76] Ibid.

[77] United Nations Department of Economic and Social Affairs Programme on Ageing (2018), 1.

[78] Beard et al. (2016), 164.

[79] ICESCR, art 12.

[80] World Health Organization (2015), 14; Baer et al. (2016), 209; Committee on Economic, Social and Cultural Rights (2000).

to non-discrimination, physical and economic accessibility, affordability and user-friendly information.[81] This is particularly pertinent when considering that older people may experience age-based healthcare rationing as well as physical limitations and financial insecurity that can make accessing quality healthcare facilities, goods, services and information difficult.[82]

Inherent in the right to health is the element of acceptability of goods, services and facilities, which should be compliant with ethical principles and culturally appropriate models.[83] They should also be gender-responsive.[84] That is, it is fundamental to recognise that older people are not a homogenous group.[85] The right to health is therefore a right to a set of social structures, norms, institutions and laws, in addition to an environment facilitating the ability for people to experience good health in ageing.[86] This section will explore what a rights-based approach to health incorporates as well as the global response to protecting the right to health, including the role of assistive technologies.

10.3.1 A Human Rights-Based Approach to Health

A human rights-based approach (Chaps. 2 and 3) can help to understand the full range of interconnected rights and principles underpinning health in older age. Former Special Rapporteur on the right to health, Paul Hunt, has explained that a human rights-based approach to health is broader than simply the right to health, acknowledging that it draws on the range of other human rights that underpin enjoyment of health.[87] These rights include the right to life;[88] freedom from torture, violence, abuse or other cruel or degrading treatment;[89] freedom of movement, including freedom from restraint;[90] equal recognition before the law;[91] the right to privacy;[92] rights to participation and social inclusion;[93] freedom of expression, freedom of

[81] The availability of user-friendly information in the Australian aged care system was heavily criticised in Royal Commission into Aged Care Quality and Safety (2019), 2. See also Committee on Economic, Social and Cultural Rights (2000), para 12.

[82] World Health Organization (2015), 14; Baer et al. (2016), 209.

[83] Baer et al. (2016), 211.

[84] Ibid.

[85] Ibid.

[86] Ibid., 208.

[87] Hunt (2016).

[88] International Covenant on Civil and Political Rights (1966), art 6 ('ICCPR').

[89] Ibid., arts 7 and 9.

[90] Ibid., art 12.

[91] Ibid., art 26.

[92] Ibid., art 17.

[93] UN Principles, Principle 7; Kornfeld-Matte (2017), para 13.

thought, conscience, beliefs, culture and religion;[94] and rights to social security and an adequate standard of living.[95] Within a human rights-based approach to health, the right to the highest attainable standard of health forms a cornerstone, however, and international instruments and scholarship have contributed a great deal to our understanding of what this right requires.[96] Together, this body of rights imposes a wide range of obligations on governments to ensure minimum standards of treatment and protections for older persons, discussed in more detail below. To deliver these rights, a commitment to the principles of participation in decision-making and respect for the will and preferences of individuals is essential, as is the meaningful provision of access to justice in situations where rights are infringed. As with all matters relating to older persons, respect for dignity, autonomy and liberty must underpin health policy and delivery of health services. Abiding by these core values is the principal means through which to ensure that other human rights are realised.

The availability of healthcare services highlights the importance of equality and non-discrimination,[97] as well as the interconnectedness of various human rights. A key point to emphasise is that, under a human rights-based approach to health in older age, where possible, individuals are entitled to choose the type of care they receive. This concept has been clearly articulated for persons with disabilities, who are entitled under international law to choose the type of care they prefer, including residential, home or community care.[98] Where an older person experiences disability then this provision obviously extends to them, but even in the absence of disability an older person's choices must still be respected as part of valuing their fundamental dignity and autonomy. This points to the need, identified in Chap. 3, for a dedicated international treaty to protect the human rights of older persons, as it must not be assumed that older persons will fall under the *Convention on the Rights of Persons with Disabilities* (CRPD).

The fact that a person is older, or in poor health, does not result in the loss of any of their interrelated, interdependent and inalienable rights. In fact, a rights-based approach to health and ageing can assist in overcoming the legal, regulatory, structural and social barriers that can prevent older people from accessing affordable and quality healthcare by integrating human rights norms and standards as legally enforceable principles.[99] Furthermore, such an approach can actually make clear the legal responsibilities of state and non-state actors in relation to meeting not only the right to health but to all of the interrelated rights.[100] Consequently, a human

---

94 ICCPR, art 18.

95 ICESCR, art 9. See also: European Network of National Human Rights Institutions (2017), 7.

96 ICESCR, art 12; Committee on Economic, Social and Cultural Rights (2000).

97 Baer et al. (2016), 207.

98 CRPD, arts 3, 19 and 25; Committee on the Rights of Persons with Disabilities (2017); European Network of National Human Rights Institutions (2017), 7.

99 Baer et al. (2016), 206.

100 World Health Organization (2015), 14.

rights-based approach can support the ability of 'duty-bearers' to fulfil their legal obligations as well as empower rights holders to enforce their rights.[101]

While a human rights-based approach therefore extends to a range of interrelated considerations and encompasses the duties of various actors, it can be limited where specific obligations are not translated into domestic law and policy with sufficient precision. For instance, while a human rights-based approach would respect older persons' choices about the type of care they receive, in many jurisdictions older persons are not entitled to choose a particular service-provider or facility.[102] The selective implementation and interpretation of different human rights in domestic jurisdictions leads to a variable application of a human rights-based approach to health.[103] Where human rights are not meaningfully adopted into domestic legislation then a human rights-based approach will be of a more limited effectiveness.

There are three main stages when implementing a human rights-based approach to health and long-term care policies and practices: planning, implementation and continuous monitoring facilitating genuine improvement.[104] Aged care facilities, as demonstrated by the Australian example, frequently have significant difficulty in maintaining a human rights-based approach.[105] This is especially in relation to fulfilling the rights to dignity, participation, privacy and health.[106] Aged care facilities also fail to prevent abuse and uphold the right to access justice.[107] This is not to say that good practices do not exist, however, as research has demonstrated that some workers naturally adopt a person-centred approach.[108]

Adopting a human rights-based approach would also require recognition of the socio-economic factors that are inherent in accessing determinants of good health such as housing and nutrition.[109] Domestic legal, policy and regulatory frameworks also offer vehicles through which to create enforceable obligations to fulfil the right to health. States must adopt the principle of progressive realisation.[110] That is, they must take all steps available in order to devote maximum available resources in order to progress towards fulfilling the right to health of older people, in addition to the other interrelated human rights. A key element in achieving this is ensuring the active participation of older people in the system design, development and implementation.[111] Participation is not only reflective of a rights-based approach, but it is also central to a person-centred approach to health and aged care contrary to,

---

[101] Baer et al. (2016).

[102] European Network of National Human Rights Institutions (2017), 7.

[103] Ibid.

[104] Ibid., ii, 17.

[105] Ibid., ii. See also: Royal Commission into Aged Care Quality and Safety (2019).

[106] European Network of National Human Rights Institutions (2017).

[107] Ibid. On hospital-based elder abuse interventions see, for example: DuMont et al. (2015).

[108] European Network of National Human Rights Institutions (2017), ii, 3.

[109] United Nations Department of Economic and Social Affairs Programme on Ageing (2018), 1.

[110] World Health Organization (2015), 15, 27; Baer et al. (2016), 208. The obligation of progressive realization is set out in article 2 of the ICESCR.

[111] World Health Organization (2015), 15.

for example, the consumer-driven approach currently found to be dominant in the Australian aged care system.[112]

A person-centred model to healthcare, promoting the concepts of love, attachment, comfort, identity, occupation and inclusion started to gain favour in the 1970s and the 1980s.[113] This approach prioritises: valuing the older persons in need of care and those who undertake this role; recognising and responding to people as individuals in a timely, flexible and appropriate manner; viewing the world from the older person's perspective; and promoting a positive and social environment focusing on ensuring equal opportunities to access care in order to improve the older person's quality of life.[114] A person-centred model is reflective of the principles and practices promoted in a human rights-based model, the main difference being that a human rights-based model is grounded in legal obligations and associated rights of redress and remedies.[115] The principles adopted by the person-centred model should therefore also be considered within a rights-based approach to health and ageing—the umbrella of the rights-based approach offering greater options for enforceability.

10.3.2 The Global Response

Given the fundamental importance of the right to health and ensuring access to affordable and quality health and aged care services, the question arises as to what is actually being done to achieve this ambitious goal. As discussed in Chaps. 2 and 4, there are two main international policy instruments that have established a framework to guide the global response to the increasing rates of population ageing since 2002. These are the *Political Declaration and Madrid International Plan of Action on Ageing*[116] and the WHO's *Active Ageing: A Policy Framework.*[117] The framework includes health and social services as one of its six key determinants for achieving 'active ageing', recommending four constituent elements necessary for a sound health policy response designed to facilitate active ageing.[118] These are to: diminish the burden of 'excess disabilities, chronic disease and premature mortality' on society; promote good health throughout the entire life course and decrease the risk of major diseases; foster a range of affordable, accessible and high quality healthcare and social services that are age-friendly recognising the right to dignified

[112] Ibid.; Royal Commission into Aged Care Quality and Safety (2019), 79.

[113] European Network of National Human Rights Institutions (2017), 13.

[114] Ibid.

[115] Ibid., 15–6.

[116] Madrid International Plan of Action on Ageing (2002).

[117] World Health Organization (2002).

[118] The others being economic, behavioural, personal, social, and the physical environment: World Health Organization (2015), 5.

and autonomous living; and to provide effective education and training to health, long-term and aged care sector workers.[119]

These two documents sit within an overarching legal framework informed by international human rights law. They document large scale aims in relation to ensuring security as people age, including the importance of health, both on its own but also as the basis for the protection of a number of other rights, including participation in society.[120] Significantly, however, they do not address the systemic transformations that are necessary to meet their stated goals and fulfil the right to health, as well as the interrelated rights.[121] Moreover, a recent review covering the progress of more than one hundred and thirty countries worldwide since 2002 concluded that the demographic changes resulting from the rates of ageing are not attracting the attention of policy-makers and that it is a 'low priority' area.[122] Further, there is little training for health professionals involved with older persons, despite the global rates of ageing, and that supporting caregivers, although necessary, is not a focus of state action on ageing.[123]

The lack of progress is significant. This is not only because of what it means in relation to the right to health, but also because of the relationship between health and the other rights, as well as in the effectiveness (or otherwise) of the SDGs. Absent clear progress on health, the wider agenda set out in the SDGs becomes increasingly unreachable.[124] In fact, in recognition of the interconnected nature of the right to health with, especially, the right to social security, SDG 3 attempts to address what it means to age 'well' by promoting 'healthy lives and well-being for all at all ages'.[125] That is, implicit in SDG 3 is the idea that the right to health should provide access to at least essential healthcare, including long-term care.[126] It also serves as an economic framework for financing health systems.[127]

In addition to SDG 3, there are a number of other SDGs that are relevant to the right to health. SDG 5, for instance, includes in its targets the aim of ceasing 'all forms of discrimination against all women and girls everywhere', including in relation to accessing adequate healthcare.[128] Goal 10 incorporates in its targets the goal to 'ensure equal opportunity and reduce inequalities of outcome', including through addressing the effects of discrimination, and to 'empower and promote the social, economic and political inclusion of all, irrespective of age, sex, disability,

[119] Ibid.

[120] Ibid., 4.

[121] Ibid.

[122] Ibid.

[123] Ibid.

[124] Ibid.

[125] United Nations General Assembly (2015) Goal 3 ('*UNSDGs*'); United Nations Department of Economic and Social Affairs Programme on Ageing (2018), 1.

[126] World Health Organization (n.d.); International Labour Organization (2017), 101.

[127] Gostin et al. (2018), 2733–4.

[128] UNSDGs Goal 5; United Nations Department of Economic and Social Affairs Programme on Ageing (2018), 1.

race, ethnicity, origin, religion or economic or other status.'[129] That is, addressing the inequities present when accessing health, long-term and aged care when developing person-centred policies.[130]

There are, however, currently large gaps in coverage worldwide resulting in millions of persons, including a significant number of older persons as well as people residing in rural areas, being left without any coverage or access to quality health and aged care.[131] Population wide (distinct from focusing exclusively on older people), the discrepancies in access to quality care between rural and urban areas is astounding with 56% of the global rural population lacking health coverage compared to 22% of people in urban areas.[132] Workforce issues are compounded in rural areas with shortages of skilled health workers available to deliver quality care.[133] This shortfall, for example, is estimated to be seven million people in rural areas contrasted to three million in urban areas.[134] Healthcare spending per capita is also twice as high in urban areas than rural areas thus making where a person lives a significant factor in whether that person lives or dies and the healthcare available.[135] Older people in particular are at risk of experiencing gaps in coverage and access to healthcare, especially long-term care, as a majority of countries globally do not provide long-term care protection.[136] Consequently over 48% of persons are not covered, with women the most affected.[137] A further 46.3% of older persons are excluded from long-term care because of strict means-testing.[138] In fact, only an estimated 5.6% of persons globally reside in countries that provide legislatively mandated health coverage for all.[139]

The WHO has also developed a *Global Strategy and Action Plan on Ageing and Health* (2016–2020)[140] calling for transformative change that was adopted by 194 countries globally.[141] It utilised evidence from the 2015 *World Report on Ageing and Health*[142] and aligns to the SDGs.[143] Five priority areas were developed.[144] First, a commitment to 'healthy ageing' which necessitates an appreciation for the

[129] UNSDGs Goal 10; United Nations Department of Economic and Social Affairs Programme on Ageing (2018), 1.

[130] World Health Organization (n.d.).

[131] Ibid.; International Labour Organization (2017), 101; Burholt and Dobbs (2012).

[132] International Labour Organization (2017), 101.

[133] Ibid.

[134] Ibid.

[135] Ibid.

[136] Ibid.

[137] Ibid.

[138] Ibid.

[139] Ibid.

[140] World Health Organization (2017b).

[141] World Health Organization (2019).

[142] World Health Organization (2015), 49.

[143] World Health Organization (2019).

[144] World Health Organization (2018).

worth of the concept and what it entails, including an unbroken commitment to developing evidence-based policies and practices to support older people.[145] Second, a recognition that health systems should be better aligned to the needs and preferences of older people.[146] Such realignment should acknowledge the abilities of older people through the adoption of a person-centred approach whilst being better integrated across health and aged care settings, as well as providers, in order to strengthen the right to access universal healthcare. Third, better systems, including governance systems, infrastructure and workforce capability, need to be developed to provide quality long-term, including palliative, care in all countries that adequately meet the needs of older people.[147] Fourth, age-friendly environments need to be created which address ageism and associated practices whilst facilitating autonomy and 'healthy ageing' more broadly.[148] Finally, improvements must be made in measuring, monitoring and understanding, first of all, what it is to age and, subsequently, what 'healthy' ageing looks like.[149]

In addition to this, and in response to the SDGs, the WHO is facilitating the *Decade of Healthy Ageing* from 2020–2030 in which there is to be a concerted effort to achieve the transformative change called for in the *Global Strategy and Action Plan on Ageing and Health*. This is designed to be an opportunity for united multisectoral action. It will therefore include a diverse range of stakeholders such as governments, international bodies, private sector, academia and the media to engage in 'concerted, catalytic and collaborative action' to better the lives of older people as well as their families, local communities and society more broadly.[150] This is because it is clear that the current approaches are not working, the human rights of older persons are being abused and systemic change is needed.[151] Consequently, in order to meet SDG 3 and to protect the right to health, as well as associated rights, a paradigm shift is required from a disease-centric model to a person-centred rights-based approach.[152]

The adoption of a person-centred approach in integrated care models would help to not only realise the right to health but also the associated rights and fundamental freedoms of older people enshrined in international law.[153] Government and policy responses need to be multisectoral, recognising that healthcare systems are currently better able to deal with acute individual health conditions rather than

---

145 Ibid.

146 Ibid.; World Health Organization (2017a, b).

147 World Health Organization (2018).

148 Ibid.

149 Ibid.

150 World Health Organization (2019).

151 World Health Organization (2015), 6; Beard et al. (2016), 164. On the abuse of older people see, for example: Pillemer et al. (2016).

152 World Health Organization (n.d.).

153 Beard et al. (2016), 164; Baer et al. (2016).

the chronic requirements more often associated with old age, including the geriatric syndromes.[154] It is also necessary to ensure that any response is gender positive in light of the particular challenges experienced by older women when both giving and receiving care, especially long-term care.[155] Equity in access to quality health and aged care, and reliable financing are likewise fundamental, as is ensuring decent working conditions.[156] The fact that health, long-term and aged care systems frequently operate independently of one another also produces poor health outcomes as well as being economically questionable with inefficient use of services and workforce members as well as engaging in cost-shifting.[157] Significantly, care workers are often unsure of what human rights obligations are even owed to older persons let alone how to best fulfil them.[158] Better education, training and accreditation of workforce members is therefore needed, including how to respect individual autonomy and foster the participation of older persons in decision-making about their own healthcare whilst protecting them from harm.[159]

10.3.3 The Role of Assistive Technologies

An emerging area which has the potential to both support and hinder older persons' enjoyment of their human rights is assistive technology.[160] Assistive technologies range from basic and well-known aids such as walking canes to newer and emerging technologies such as 'memory and communication aids, safety devices, GPS tracking, companion robots, and technology for so-called smart homes'.[161] Their purpose is 'to maintain or improve an individual's functioning and independence, to facilitate participation, and enhancement of overall well-being and quality of life'.[162] This objective clearly correlates with a human rights approach, since it focuses on improved autonomy, liberty and participation, yet the design and use of assistive technologies also raises potential risks for human rights which need to be addressed in policy and regulation to ensure that the potential benefits can be realised.

To begin with, the benefits of assistive technology need to be accessible and affordable to all. This involves addressing not only the financial or geographical barriers

---

[154] World Health Organization (n.d.).

[155] UN Women (2017), 3.

[156] International Labour Organization (2017), 102.

[157] World Health Organization (n.d.).

[158] European Network of National Human Rights Institutions (2017), ii, 3.

[159] Ibid.

[160] World Health Organization (2015), 137; Royal Commission into Aged Care Quality and Safety (2019), 134–5. For more on assistive technologies for older people see, for example: Garçon et al. (2016).

[161] Bennett et al. (2017), 749.

[162] Garçon et al. (2016), 293, 295.

which might prevent an older person from accessing technology, but also recognising the educational, social and cultural factors which can influence a person's take-up of technology and their level of confidence and comfort if they decide to use it.[163] Issues around affordability and accessibility of technology might also exacerbate existing inequalities, for example for people living in rural, regional and remote areas and/or people on lower-incomes.[164] Policies on the use of assistive technologies must therefore foster an equitable and non-discriminatory approach.

New technologies potentially have the capability to assist older people with ADLs, decision-making and independent living, but it is essential that these are deployed in a way which respects their preferences and decision-making autonomy to the greatest possible extent. For example, technology could be useful for supported-decision making for older people with impaired capacity, but should not be used to substitute their decision-making, a process which requires decisions to made in a person's best interests on a case by case basis whilst respecting their individual will and preferences. For people with dementia, for instance, assistive technologies may prolong independent living and improve quality of life, which supports the right to health. Technologies such as communication aids, tracking, and smart-home technology may provide an alternative to traditional restrictive practices. However, their use raises a potential threat to the right to privacy if data is not managed carefully, and to freedom of movement if the technology can constrain this.[165] The use of technology must, therefore, be guided by human rights principles as discussed in Chaps. 2, 3 and 6, and appropriate safeguards need to be in place to ensure the technology does not create a heightened risk of isolation and/or elder abuse (discussed below).

New technologies also have great potential to enhance older people's inclusion and connectedness, and to thereby combat the negative impacts of isolation and loneliness which can occur as people age.[166] These connections could be facilitated by, for example, digital platforms, virtual communities, online shopping, accessing government and financial services, or experiences with social robots.[167] The increased social connection has obvious human rights and health benefits, as discussed in more detail in Chap. 4. However, assistive technologies may also lead to marginalisation and loneliness, and thus a heightened risk of abuse, among older persons if not properly managed. For example, as assistive technologies become more affordable and more widely available, and as pressures on health and aged care systems increase, we may see these technologies, in particular robots, replace care workers in some settings. This transition to technology-provided care may reduce the amount of human contact experienced by older people and have negative impacts on their well-being.[168] It may

---

[163] Francis et al. (2019); Australian Digital Inclusion Index.

[164] Lewis et al. (2018), 17.

[165] Bennett et al. (2017).

[166] Waycott et al. (2019); Baez et al. (2019).

[167] Kornfeld-Matte (2017), para 13.

[168] Ibid., para 14; Lewis et al. (2018), 17; Batti and ul Haq (2017).

give rise to perceptions of objectification and loss of control,[169] with consequent impacts on dignity if a person feels that they are being treated as an object and not a person. While there is disagreement as to how well robots might be able to substitute for human company, a human rights-based approach would, at a minimum, stipulate that an individual's choices as to the role of robots in their lives must be respected.[170]

It is also critical that we properly manage the implementation of new technologies to ensure that they do not turn into a tool for elder abuse (discussed in Chap. 7). Access to technologies can heighten the risk to older people of financial abuse through romance or online financial scams, for example.[171] As another illustration, a proposal by the Australian Law Reform Commission to create an online register of Enduring Power of Attorney instruments may have benefits in terms of enabling verification of delegated powers,[172] but may also have the undesired effect of exposing older people to abuse if, for example, a family member becomes aware that the older person has an enduring power of attorney which appoints a different person as attorney. The family member may then pressure the older individual to appoint them. Further, the right to privacy should be respected unless there is a legitimate reason to make information about appointments public.

The use of assistive technologies clearly has great potential to enhance the quality of life of older people by increasing independence and social connectedness. However, the risks identified above demonstrate the importance of a human rights-based approach in this area. It is essential that law and policy relating to assistive technologies are guided by core values of respect for autonomy and dignity, and that rights such as freedom of movement and the right to privacy are protected.

10.4 Healthcare Systems

Given that the right to health requires a holistic approach, effective responses need to be both systemic and systematic. A break-down in one area can create health and human rights risks elsewhere. These, in turn, can be further compounded given the individualised nature of 'health' for each and every older person, the complexity of having multiple actors and different timeframes and the lack of communication that can exist in the provision of care. The next two sections will therefore discuss some of the major issues arising in health and aged care systems in relation to the right to health, followed by a discussion about some of the main barriers to accessing quality care.

High income countries are arguably better equipped to address the challenges of designing, developing and implementing transformative systemic approaches to facilitate good health in ageing. However, the systems in these countries tend to

---

[169] Sharkey (2012).

[170] Kornfeld-Matte (2017), 10.

[171] Button and Cross (2017).

[172] Australian Law Reform Commission (2017), Recommendation 5.3.

be acute disease specific rather than focused on maximising function and capacity throughout the life course with a view to minimizing the effects of chronic conditions as people age.[173] Moreover, acute services are often used to deal with chronic conditions, especially in long-term care. This not only fails to promote the functional ability of older people, but it can also result in excessive wastage of resources and is therefore unnecessarily costly.[174]

Treatment can involve numerous healthcare professionals, particularly in high income countries where medical specialists are more readily available. In Germany, for example, 39% of older individuals reported seeing at least four doctors in one year.[175] To ensure continuity of care which is safe and effective, interventions need extensive coordination across providers, treatment levels and settings.[176] When this coordination is not in place or falters the burden falls on the older person and/or their family members to be aware of, understand, and effectively communicate this information to health and medical professionals when necessary—often in stressful and/or emergency medical situations.[177] Such a system breakdown can therefore result in older people experiencing gaps in their treatment programs.[178]

Furthermore, the development of healthcare systems in discipline silos results in disparate methodologies which can then negatively impact the effectiveness of the care regime resulting in, for instance, unnecessary and sometimes invasive interventions and polypharmacy which can lead to further negative impacts.[179] For example, in 2015 it was estimated that in one of France's biggest hospitals, 20% of all older patients were substantially worse at completing ADLs when they were discharged than when they were admitted, with the presenting condition accounting for the decrease in ability in under a half of the cases.[180] Moreover, the problem was avoidable in an estimated 80% of cases through the use of, for instance, approaches that better facilitated the mobility of the older person or responding to incontinence issues.[181]

Healthcare systems in lower and mid-income countries face a different problem, where access to healthcare services is often inadequate.[182] In Mexico, for example, it is estimated that only 21% of patients have access to effective health coverage.[183]

---

[173] World Health Organization (2015), 6.

[174] This can lead to costly acute services being used to meet chronic care needs and a failure to fully foster the functioning of older people receiving long-term care: World Health Organization (2015) 23, 25.

[175] Ibid., 93.

[176] Ibid.

[177] Ibid.

[178] Ibid.

[179] Ibid., 6.

[180] Ibid.

[181] Ibid.

[182] World Health Organization (2015), 6.

[183] Ibid.

Moreover, healthcare workers may receive limited training and education in relation to caring for older patients and the chronic conditions that can commonly occur.[184] However, as demonstrated by the Royal Commission in Australia, this is not a problem experienced solely by lower or mid-income countries, and high-income states also experience systemic problems.[185] Dementia, in particular, does not receive adequate attention.[186] Further, conditions such as heart disease and stroke can be missed or misdiagnosed, negatively impacting the effectiveness of the treatment plan.[187] The lack of access to quality healthcare results in higher rates of older people who have limited function and thus reduced ability to undertake the ADLs, thus increasing their need for support.[188] Given that the healthcare and social services systems are often ill-equipped to provide long term care in low to mid-income countries, the burden of this care therefore often falls to family members, mainly women, which can in turn result in them having to limit their work to care for their older family members.[189] This then leads to the risk of increased financial insecurity (discussed in Chap. 8).[190]

10.5 Aged Care Systems

Decline in functional ability, both physical and mental, can result in older people transitioning their living environment either to include more in-home supports or to move to an environment offering increased supports, such as an aged care facility.[191] Aged care can therefore be loosely divided into two main categories, community-based care and residential or institutional care. Residential or institutional care involves a person receiving personal care, such as meals and cleaning services, and/or nursing care in addition to accommodation. Community-based care has been defined as all modes of care that do not necessitate the older person permanently living in an institutional or residential care environment.[192] It includes care provided for older people in their own homes or short-term in community centres.[193] Consequently, community-based care can be used as a strategy to support ageing in place (discussed in Chap. 9) because it can assist older people to remain in their own homes thus postponing or

[184] Ibid., 93.

[185] Ibid., 10; Royal Commission into Aged Care Quality and Safety (2019).

[186] Morris (2010).

[187] World Health Organization (2015), 6.

[188] Ibid.

[189] Ibid.

[190] Ibid.

[191] As discussed in Chap. 9, an arbitrary distinction has been made mainly for manageability between health and aged care, discussed here, and accommodation options, discussed in that chapter.

[192] World Health Organization (2015), 129.

[193] Ibid.

even eliminating the need for admission to an aged care facility.[194] It is growing in popularity and availability, with between one half to three quarters of older people in need of long-term care receiving it in their homes.[195] It also has the potential to minimise the time spent in hospital.[196] Even more significantly, community-based care and ageing in place have the potential to improve the quality of life of an older person because care is being provided in a familiar environment and established supports can continue to be accessible.[197]

Residential aged care is mainly delivered in assisted-living and aged care facilities. The human rights implications of residential aged care are numerous and interconnected. Changing living environments can have a significant impact on an older person's sense of self and their autonomy as they may have a strong sense of connection, familiarity and therefore security with their current home and local community.[198] Losing connections with existing communities and support networks can have significant negative impacts on older persons' well-being, and many residents of aged care facilities report feeling lonely and isolated.[199] This can be particularly damaging for older people from culturally and linguistically diverse backgrounds, as the aged care setting can present structural as well as cultural and linguistic barriers to their ongoing participation, with associated human rights consequences as discussed in Chap. 4.[200]

The provision of residential care has traditionally been based on the medical model of service delivery and was therefore designed to be more akin to a hospital than a home in both layout and provision of short and long-term care.[201] The impersonal feel of the care environment can be dehumanising for residents. When coupled with a lack of workforce education about human rights obligations towards older people, care facilities regularly fail to meet the physical, psychological and/or socialisation needs of older persons.[202]

In fact, aged care facilities not only frequently fail to meet the needs of their residents, in certain circumstances the facilities are actually causing harm, as recently demonstrated by the Australian Royal Commission, which has provided a contemporary, comprehensive investigation into aged care in Australia.[203] The Royal Commission highlighted many issues in the delivery of aged care which, unfortunately, are not unique to Australia. Despite particularities of Australia's funding models, as well as legal and regulatory landscape, the problems identified by the Royal Commission are,

---

[194] Ibid.

[195] Ibid.

[196] Ibid.

[197] Ibid.; Low and Fletcher (2015); Szanton et al. (2015).

[198] World Health Organization (2015), 36.

[199] Barbosa Neves et al. (2019); Australian Institute of Health and Welfare (2019).

[200] See, for example: Australian Government Department of Health, Aged Care Sector Committee Diversity Sub-Group (2019); World Health Organization (2015), 36.

[201] European Network of National Human Rights Institutions (2017), 12.

[202] Ibid.

[203] See, for example: Royal Commission into Aged Care Quality and Safety (2019).

at heart, the same dilemmas facing many jurisdictions: fundamental lack of respect for dignity, autonomy, liberty and human rights; inadequate funding, regulation and monitoring; and lack of access to justice. These could all be addressed by greater adherence to a human rights-based approach. In this way Australia illustrates the problems and recommendations for aged care more broadly.

In its interim report titled 'Neglect', the Royal Commission noted that the Australian aged care system is comprised of 'deficiencies' and 'outright failures', and that significant reforms and transformations are needed to address the broken system.[204] This includes rectifying the human rights abuses currently being perpetrated against older Australians. Moreover, the Royal Commission identified an apathetic culture resulting in sub-standard care that is reinforced by a focus on funding.[205] It also pointed to a lack of curiosity generally in relation to exploring the possibility of aged care as a restorative and positive care environment for older Australians.[206]

In fact, the Royal Commission characterised the Australian system as 'cruel and discriminatory' resulting in the neglect of older people in both home care and residential aged care facilities.[207] The following 'shameful' quality and safety issues were identified: insufficient wound prevention and management, sometimes resulting in death; inadequate continence management, including the rationing of continence pads which would leave older residents sitting in urine and/or faeces; appalling nutrition and hydration with poor care for oral health; a high rate of abuse perpetrated by staff against residents with the use of physical and chemical (through the use of over-prescribing medications)[208] restraints common in order to render them easier to manage (distinct from being necessary for their care); and fragmented and unpredictable palliative care.[209] Moreover, it was noted that the aged care sector in Australia does not listen to the voices of older people. They are unable to participate in bettering a system they are forced to endure—a system that is 'woefully inadequate' and disdainful of their basic human rights.[210]

The utilisation of restrictive practices in particular was highlighted and described as 'inhumane, abusive and unjustified' in the Australian system.[211] Restrictive practices are defined as interventions of activities which may be pharmacological or physical that restrict a person's movement or decision-making ability. They, by very definition, infringe upon a person's human rights, most obviously rights to liberty of the person and freedom of movement, and more generally are incompatible with basic respect for dignity, autonomy and liberty.[212] Despite the use

---

204 Ibid., 12.

205 Ibid., 1, 5.

206 Ibid.

207 Ibid., 3–4.

208 Ibid., 10.

209 Ibid., 6.

210 Ibid., 7, 8, 12.

211 Ibid., 193.

212 ICCPR, arts 7, 9 and 12; Royal Commission into Aged Care Quality and Safety (2019), 194.

of restrains often resulting in serious psychological and physical harm, they are commonly employed in aged care facilities throughout Australia.[213] One aged care facility, Earle Haven, acknowledged to the Royal Commission that 71% of residents were given psychotropic medication and 50% of people were physically restrained as recently as June 2019.[214]

The overuse of chemical restraints has been identified as being in particular need of urgent action.[215] While they are frequently justified as being necessary to control 'difficult' behaviours and/or to prevent falls,[216] medications should only be given in circumstances where there is therapeutic merit—not as a form of restraint.[217] The types of drugs used include, for example, antipsychotic drugs, benzodiazepines, opioid analgesics and sedative-hypnotic drugs or sedatives.[218] As seen in the Australian context, the use of these chemical 'protective' mechanisms can so very easily go terribly wrong and their use is instead symptomatic of neglect in providing adequate care to and engaging with older people.[219] This is especially the case when considering the psychological and physical harm (potentially even death) that can result from restraints, coupled with the significant fact that they are actually of questionable efficacy in managing 'challenging' behaviours.[220] Further difficulties arise in managing the use of restraints because there is currently a lack of consensus as to what constitutes 'chemical restraint' given that there are difficulties in determining whether the prescribing physician's aim is to treat a person's symptoms or to modify their behaviour.[221]

The use of physical and chemical restraints are just two examples of the human rights abuses which can occur in aged care settings. The evidence before the Royal Commission discussed above reveals multiple breaches of human rights, including freedom from cruel and inhuman treatment, freedom of movement, rights to privacy and an adequate standard of living, in addition to the right to health.[222] These breaches flow from pervasive failures to respect older individuals' dignity, autonomy and liberty, and a system which frequently does not support participation of older persons in decisions which affect them or respect their preferences.[223]

Interestingly, it has been suggested that the older person may not be the only victim because of the risk of burnout in carers as a result of onerous working conditions, indicating they too may be vulnerable to human rights abuses.[224] Understanding the

[213] Royal Commission into Aged Care Quality and Safety (2019), 193.

[214] Ibid; Australian Government (2019).

[215] Royal Commission into Aged Care Quality and Safety (2019), 10.

[216] World Health Organization (2015), 133; Meyer et al. (2009); Feng et al. (2009).

[217] Human Rights Watch (2019).

[218] Ibid.

[219] Royal Commission into Aged Care Quality and Safety (2019), 216.

[220] Ibid., 201; Human Rights Watch (2019).

[221] Royal Commission into Aged Care Quality and Safety (2019), 195.

[222] Ibid., 5, 111, 114–5.

[223] Ibid., 73ff; Carnell and Paterson (2017).

[224] World Health Organization (2015), 133; Johannesen and LoGiudice (2013).

situation of carers is important to understanding the causes of neglect and abuse (discussed in Chap. 7), so as to develop suitable responses. However, this should not be viewed as condoning or accepting sub-standard treatment or elder abuse, and a human rights approach mandates that those who have the responsibility of caring for vulnerable older people are supported and trained to perform this role consistently with human rights obligations.

Given the potential for widespread human rights violations like those uncovered in Australia, alternative residential care models have therefore been gaining attention (discussed in more detail in Chap. 9). In Germany, Japan, the Netherlands, Sweden and the United States of America, for example, the traditional medical model is being replaced by healthcare design focusing on the delivery of smaller group homes and the provision of constant care in a home-like environment.[225] This model is more representative of a person-centred, non-discriminatory approach in keeping with a human rights model to healthcare and ageing.[226] High income countries in particular are examining the efficacy of these new care models, including the promotion of ageing in place where possible, and engaging in healthcare system redesign as a result.[227] These new care models are promising with respect to improving the quality of care provided to older people and thereby promoting their human rights, and are an innovative way for governments to meet their associated obligations under human rights law.[228]

A human rights-based approach to aged care would therefore foreground respect for dignity, autonomy and liberty of each individual, and ensure that both regulation of the sector and management of individual facilities are guided by principles of participation and non-discrimination. Specific human rights, such as the rights to health, privacy, freedom of movement and freedom from cruel and inhuman treatment, provide essential standards for care. A human rights-based approach should also guarantee that in any situation where an older person's rights have been neglected or abused, they have appropriate and timely access to justice. Furthermore, the objective of progressive realisation articulated within the ICESCR should encourage governments to continue to strive for better quality of care and greater participation and inclusion of older persons. The adoption of a person-centred approach would align well with a human rights-based framework, promoting the development of integrated long-term community-based and residential aged care facilities which would assist in protecting the right to health and associated rights. This will necessarily involve a transformation in thinking about how these systems operate, with the focus being on the creation and implementation of a system wherein functional ability and intrinsic capacity are promoted in order to respect an individual's dignity, autonomy and right to participate.[229]

---

[225] World Health Organization (2015), 129; Pot (2013).

[226] World Health Organization (2015), 129; Pot (2013).

[227] World Health Organization (2015), 129.

[228] Ibid.; De Bruin et al. (2010); Verbeek et al. (2012); Willemse et al. (2014); Smit et al. (2014); te Boekhorst et al. (2009).

[229] World Health Organization (2017a).

10.6 Barriers to Care

In addition to the systemic issues faced by older people in realising their right to health, barriers can also exist to accessing affordable, appropriate and quality health and aged care, with a significant number of older persons lacking access to care worldwide.[230] In a survey undertaken in 2010, for example, 63% (out of 1265 respondents aged sixty years and over from across thirty-two countries throughout Africa, Asia, Europe and the Caribbean) said that it was difficult to access healthcare when needed.[231] Interestingly, however, there also appears to be a disconnect between the need for healthcare services and utilisation of those services.[232] This occurs not only in low to middle income countries but also in disadvantaged subgroups in high income countries.[233] Health and aged care utilisation appears to increase as people age in high income countries with the trend reducing in middle income countries and disappearing in low income countries.[234] The patterns of care utilisation are important when considering the barriers that exist to accessing affordable and quality services. There are, thus, two main questions in relation to the provision of health and aged care services. The first is accessibility, including affordability. Provided that health and aged care is even accessible, the second question then relates to the quality of care available.

10.6.1 Accessibility of Care

Considering the first question, that of accessibility, chief amongst the barriers preventing access to adequate and quality health and aged care systems is affordability.[235] There is a symbiotic relationship between good health and financial security (discussed in Chap. 8) with financial security as people age being reliant, in part, on good health, that is, not experiencing any significant illnesses requiring extensive, and expensive, medical interventions.[236] Conversely, good health is also dependent on having the financial resources, both income and wealth security, to afford quality health and aged care services. However, as discussed in Chap. 8, individuals often lack both the income and wealth security necessary to be able to pay for health and aged care, especially long-term care. Moreover, families are becoming increasingly unable to offer support to older relatives.[237] The provision of long-term care is particularly significant because it can vary in scope and intensity, while care environments

[230]United Nations Department of Economic and Social Affairs Programme on Ageing (2018), 1, 2.

[231]Ibid., 2; HelpAge International (2011).

[232]World Health Organization (2015), 89–90.

[233]Ibid.

[234]Ibid.

[235]United Nations Department of Economic and Social Affairs Programme on Ageing (2018), 2.

[236]United Nations Department of Economic and Social Affairs Programme on Ageing (n.d.), 4.

[237]United Nations Department of Economic and Social Affairs Programme on Ageing (2018), 2; United Nations Secretary—General (2012).

can range from the older person's home to hospitals, assisted living facilities, aged care and other health facilities.[238]

Absent government funded or subsidised health and aged care systems, the out-of-pocket costs for formal care can be tremendous and, in worst case scenarios, ruinous.[239] In Europe, for example, the out-of-pocket expenses for healthcare total on average approximately 9.6% of an older person's income.[240] In 2014 just under 5% of people aged sixty-five years and over had difficulty accessing healthcare because of cost in France, Sweden and Norway.[241] This figure rose to 5% in the United Kingdom, and between 5 and 10% in the Netherlands, Switzerland, Germany and Australia.[242] The figure in New Zealand during the same period was 10%, and just under 20% in the United States of America.[243] Equitable access to state-funded services therefore serves as a safeguard against the risk of poverty, a risk that can significantly increase in the event of substantial out of pocket healthcare costs.[244]

Some states, however, compel older persons to use their savings and assets (that is, their wealth security) before they can access health and aged care that is government funded or subsidised.[245] This obviously has the effect of potentially exhausting the wealth security of an older person, again increasing the risk of poverty (and the other attendant threats to human rights). Older people are thus often placed in a position of having to choose between their health and aged care needs and other fundamentals necessary to survive, for instance safe housing and food, which can create further health issues when the inability to access nutritional food leads to, in extreme cases, malnutrition.[246] That is, older people are being forced to rank the importance of their internationally recognised basic human rights and then choose between them. In such situations, older people tend to forego any preventative healthcare believed to be 'unnecessary', and sometimes even 'necessary' treatment and/or medications, in order to afford, for example, housing—which can often be seen, understandably, as a more pressing concern.[247]

Consequently, if an individual lacks the income and/or wealth security necessary to fund health and aged care costs, and there is an absence of publicly available state funded health and aged care schemes, this can jeopardise the rights of older persons, particularly to long-term healthcare, and results in the lack of access to,

[238] World Health Organization (2015), 129.

[239] United Nations Department of Economic and Social Affairs Programme on Ageing (n.d.), 4.

[240] Ibid; World Health Organization (2015).

[241] United Nations Department of Economic and Social Affairs Programme on Ageing (n.d.), 4; World Health Organization (2015).

[242] United Nations Department of Economic and Social Affairs Programme on Ageing (n.d.), 4; World Health Organization (2015).

[243] United Nations Department of Economic and Social Affairs Programme on Ageing (n.d), 4; World Health Organization (2015) Fig. 4.3.

[244] World Health Organization (2015); United Nations Department of Economic and Social Affairs Programme on Ageing (2018), 4.

[245] United Nations Department of Economic and Social Affairs Programme on Ageing (n.d.), 4.

[246] Ibid., 2; Cox (2015), 46.

[247] United Nations Department of Economic and Social Affairs Programme on Ageing (2018), 2.

sometimes life-saving, health interventions.[248] In addition to the international obligation on governments to protect the right to health, this further emphasises the need for governments to protect the right the social security to facilitate (at a minimum) the income security necessary for older people to be able to access vital health and aged care services.[249]

As discussed above, location is another significant issue affecting accessibility of quality health and aged care services.[250] Rural, regional and remote areas experience significant shortages of skilled health professionals as well as a limited availability of services which can negatively affect an older person's ability to be able to access health and aged care.[251] This is particularly the case when considering issues in relation to limited mobility that can restrict an older person's movement as well as the often limited transportation infrastructure in rural, regional and remote areas.[252] This last issue is especially significant in countries where large distances may need to be covered in order to be able to access specialist healthcare, such as in Australia.[253] This also forces the older person from their home, a familiar environment, into an unfamiliar and often intimidating situation lacking their usual support systems in order to seek medical assistance, when they are already experiencing poor health—a situation which can result in fear and anxiety and which can, in turn, adversely affect health outcomes.[254]

Barriers can also exist when accessing health and aged care because of cultural or ethnic background, discrimination and ageism, gender, socio-economic status and education level.[255] A distinction was noted, for example, between education levels of older people attending upon medical specialists and dentists in a study across twelve European countries.[256] Older people in lower socio-economic groups in the United Kingdom also experience hurdles in accessing healthcare services such as vaccinations, eye and dental exams, and heart surgery.[257] In fact, it is often the most vulnerable who confront the greatest challenges in accessing health and aged care services.[258]

Older women in particular experience disadvantages in accessing healthcare, including long-term care, which is significant given that they live longer and are therefore at a higher risk of experiencing chronic conditions.[259] Mammography screening

---

[248] United Nations Department of Economic and Social Affairs Programme on Ageing (n.d.), 4.

[249] Ibid.

[250] Sadana et al. (2016), 179.

[251] United Nations Department of Economic and Social Affairs Programme on Ageing (2018), 2.

[252] Ibid.

[253] Ibid.; United Nations Secretary-General (2012); HelpAge International (2011).

[254] World Health Organization (2015), 129; Low and Fletcher (2015); Szanton et al. (2015).

[255] United Nations Department of Economic and Social Affairs Programme on Ageing (2018), 3.

[256] Ibid.; Terraneo (2015).

[257] United Nations Department of Economic and Social Affairs Programme on Ageing (2018), 3; Equality and Human Rights Commission, Age Concern and Help the Aged (2009).

[258] United Nations Department of Economic and Social Affairs Programme on Ageing (2018), 3.

[259] UN Women (2017).

is another service that older women in lower-socio economic groups are often disadvantaged in accessing. This demonstrates not only the cumulative and intersectional nature that these barriers can have, but also the need for widely available services that are culturally appropriate and gender sensitive.[260]

10.6.2 Quality of Care

Presuming then that care can be accessed, the second question in relation to quality arises. Whether care can be considered 'quality' care depends not only on the nature of the care but also the system(s) in which it is provided. To ensure streamlined and effective care that is responsive to patients' needs, as well as their will and preferences, health and aged care services need to be integrated, with effective regulatory standards.[261] Above all, the nature of the care must meet the health needs of the individual and not cause them harm. While this seems an obvious statement, the Australian Royal Commission found that, in fact, the prevalence of abuse of older people within aged care systems, for example, is alarmingly common.[262] Various factors contribute to this lack of quality care, which will now be discussed.

The current anachronistic health and aged care delivery paradigm reinforces the flawed idea of older people as passive recipients of care wherein the focus is on service providers rather than the needs and preferences of the older person as required in a human rights model.[263] The provision of quality care can also be hindered by health and aged care professionals who have limited understanding of the distinct health issues facing older people which can, in turn, be compounded by age-inappropriate systems.[264] As mentioned, this can be a particular issue in rural, regional and remote areas, particularly in developing countries, given the shortage of skilled health professionals.[265]

Workforce issues are one of the most pressing challenges facing the health, and especially the aged care, systems globally.[266] Older people receive care from a wide range of diverse workers with varying skill and knowledge levels, and who are in receipt of varying levels of remuneration.[267] At one end of the range are informal, untrained and unpaid caregivers who may receive little to no outside assistance,

---

[260] United Nations Department of Economic and Social Affairs Programme on Ageing (2018), 3; Equality and Human Rights Commission Age Concern and Help the Aged (2009).

[261] World Health Organization (2015), 133.

[262] Royal Commission into Aged Care Quality and Safety (2019), 6, 60; Carnell and Paterson (2017). See also World Health Organization (2015), 133.

[263] World Health Organization (2015), 133.

[264] United Nations Secretary—General (2015); United Nations Department of Economic and Social Affairs Programme on Ageing (2018), 3.

[265] United Nations Department of Economic and Social Affairs Programme on Ageing (2018), 2.

[266] World Health Organization (2015), 129. For a recent example see: Royal Commission into Aged Care Quality and Safety (2019) 9, 10.

[267] World Health Organization (2015), 120–130.

who are often family members and friends, while highly trained health professionals reside at the opposite end.[268] The majority of family caregivers are women which, as discussed in Chap. 8, can result in them having time away from the workforce and thus potentially endangering their own financial security as they age (particularly through limited or no contributions to pension schemes when not in paid employment).[269]

Significantly, however, especially when coupled with the expected decline in informal familial-based carer availability, the rates of ageing mean that there will be a substantially increased demand for paid carers (estimated to at least double by 2050).[270] As discussed though, there are a substantial number of paid carers who lack adequate training to care for older people, particularly when considering the chronic nature of some conditions, and thus are not prepared for these roles which can see them caring for some of the most vulnerable members of society.[271] A lack of understanding about human rights principles and the associated obligations borne by all levels of care staff can be problematic.[272] Remuneration levels, the low status of being a care-giver and/or poor employment conditions (perceived or otherwise) can all combine to make the recruitment of appropriate people difficult.[273] In fact, working conditions may not meet human rights standards either with respect to, for instance, labour rights and occupational health and safety, even when these are legislatively mandated.[274] A system of elder laws and policies which was truly based on a human rights framework would ensure that the rights of aged-care workers, residents and even families would be fully guaranteed.

Another concern is that of the fiscal sustainability of the current models. This is especially important given that the common focus on acute disease management rather than chronic conditions can lead to unnecessary clinical interventions, as discussed above. Health and aged care funding is generally divided between public systems (government funded), private systems (requiring contributions from individuals) or a combination of the two.[275] Government spending on long-term care across OECD member states increased on average by 4.8% annually between 2005 and 2011, with the rate of expenditure estimated to at least double in European Union member states by 2060.[276] Nevertheless, the specifics of the nature of expenditure vary wildly between countries because of variations in the approach of governments to aged and long-term care services, such as whether there is a means test for accessing government assistance and the extent to which out-of-pocket expenses occur.[277] In Europe, for instance, the out-of-pocket payments for older people comprise, on

[268] Ibid., 129–130.

[269] Ibid., 130.

[270] Ibid., 130–1.

[271] Ibid., 131.

[272] European Network of National Human Rights Institutions (2017), 32.

[273] World Health Organization (2015), 130–131; International Labour Organization (2017), 101.

[274] International Labour Organization (2017), 101.

[275] For more on this, see: World Health Organization (2015), 131–3.

[276] Ibid., 131.

[277] Ibid.

average, 9.6% of their household income and can be as high as 25%. Again, the importance of income and wealth security is evident. The most vulnerable, such as the poor, women and those in the 'old, old' cohort are the most at risk from these costs.[278] Significantly, however, informal care is generally excluded from all estimates.[279] It should be noted though, that age is not necessarily determinative of increased healthcare costs with expenditure believed to peak around the age of seventy and then decline in higher income countries. In fact, time to death may be a more useful determinant of healthcare services expenditure than age.[280] Furthermore, increased healthcare expenditure is believed to be more closely aligned to factors such as technological advancements and changes in clinical practice than to population ageing.[281] Thus, the idea of the aged population as an economic drain jeopardising the fiscal sustainability of healthcare systems is not necessarily accurate.

In addition to acting as a barrier to accessing care, discrimination and the negative stigma or stereotypes attached to ageing can also serve as a barrier to receiving quality care. Health and aged care professionals can bring ageist attitudes and practices to their care for older people, for instance, believing that all older people are frail, 'senile', or that caring for older people is easy.[282] Consider, for example, dementia. Ageist assumptions and practices are believed to be a significant factor in the discrepancy between prevalence estimates for Alzheimer's disease (a type of dementia) and the diagnosis rates for this condition, where actual prevalence rates are lower than commonly assumed.[283] This is particularly significant in light of the impact an incorrect diagnosis of a dementing illness can have upon the perception of a person's capacity (discussed in Chap. 6). Depression is another condition which can significantly affect older persons, including their capacity to make legally recognised decisions, and yet is often misdiagnosed or completely overlooked because of the ageist belief that depressive symptoms are 'normal' in the ageing process.[284] Such misdiagnosis has clear implications for the delivery of quality care.

Furthermore, age discrimination and the ageist attitudes prevalent amongst some health and aged care workers can result in care rationing wherein care is limited or withheld solely on the basis of the age of the patient, irrespective of the expected health outcomes of the treatment.[285] A 2009 survey of two hundred members of the British Geriatrics Society, for instance, demonstrated that 72% of members believed it

---

[278] Ibid.

[279] Ibid.

[280] United Nations Department of Economic and Social Affairs Programme on Ageing (2018), 5.

[281] Ibid.

[282] World Health Organization (2015), 130–1.

[283] Alzheimer's Disease International (2012), cited in United Nations Secretary-General (2013); United Nations Department of Economic and Social Affairs Programme on Ageing (2018), 2.

[284] United Nations Department of Economic and Social Affairs Programme on Ageing (2018), 2.

[285] Ibid.

would be less likely that older people would be referred for surgery or chemotherapy, while 66% of respondents indicated that they felt that the symptoms exhibited by older people were less likely to be investigated.[286] The prevalence of negative attitudes may also result in society reinforcing the view to an older person that because they are 'older' then resources and the time of health professionals should not be 'wasted' on them when there are other people more 'deserving' of care (discussed in Chap. 5).[287]

Finally, the lack of available data in relation to the health status and outcomes of older persons not only serves as a barrier to accessing quality care, it also has repercussions for monitoring and evaluating health outcomes and developments, as well as developing and implementing targeted systemic improvements.[288] Significantly, research gaps are reinforced by common health indicators which often exclude older people. This can be seen, for instance, in the frequency rates for HIV that are collected for people aged from fifteen to forty-nine years inclusive.[289] Drug and treatment trials also generally exclude older people or, if they are included, it is generally only on a limited basis.[290] For instance, although older people comprise over 60% of cancer patients in the United States of America, they only account for 25% of participants in relevant clinical treatment trials.[291]

Thus, there are significant difficulties in not only accessing health and aged care, but also in ensuring the quality of that care which may result in inequities in health outcomes. Under international human rights law, states are obliged to work towards progressive realisation of economic, social and cultural rights, including the right to health.[292] This recognises that states' actions will be inevitably curtailed by available resources, but the expectation is that they will continue to make improvements in the provision of health and aged care services, without discrimination, and 'move as expeditiously and effectively as possible towards the full realization of article 12'.[293] States must therefore seek to address demographic changes and attendant health and aged care demands in a way which avoids retrogressive measures and makes the best use of the state's available resources. Ensuring accessibility of quality care is an essential component of a human rights-based approach to health and aged care.

---

[286] See BBC News (2009), cited in United Nations Secretary-General (2012); United Nations Department of Economic and Social Affairs Programme on Ageing (2018), 2.

[287] United Nations Department of Economic and Social Affairs Programme on Ageing (2018), 2.

[288] Ibid.; Beard et al. (2016), 165.

[289] United Nations Department of Economic and Social Affairs Programme on Ageing (2018), 2; United Nations Secretary-General (2015).

[290] United Nations Department of Economic and Social Affairs Programme on Ageing (2018), 2.

[291] Ibid., 2–3; Weaver Williams (2009).

[292] See ICESCR, art 2 for the general obligations of states.

[293] Committee on Economic, Social and Cultural Rights (2000), para 31.

10.7 Palliative Care and End of Life

Despite medical and technological advances, one ultimate outcome remains inevitable, that of death. This highlights the issues of palliative care at the end of life and the ability to be free from pain at that time.[294] A detailed discussion about the ethics and regulation of palliative care, freedom from pain and euthanasia[295] is outside the scope of the discussion here. It is, however, necessary to raise two issues which are pertinent to a human rights-based approach. The first is in relation to freedom from pain as people age. The second is the importance of advance care planning, which is part of estate or future planning (discussed in Chap. 8).

The Committee on Economic, Social and Cultural Rights has noted that, particularly with respect to older persons, the right to health must involve 'attention and care for chronically and terminally ill persons, sparing them avoidable pain and enabling them to die with dignity.'[296] While human rights law provides the right to life in article 6 of the ICCPR, the core human rights values of respect for dignity and autonomy when applied in conjunction with the right to health indicate that, in cases where a person is experiencing unavoidable pain, they are entitled to the option of choosing to die. Rather than being contrary to the right to life, this is in fact consistent with fundamental human rights principles and, indeed, can be thought of as part of a nuanced understanding of the right to life which recognises that this is not just a right to *any* kind of life, but to a life lived in dignity. Requiring a person to live with unavoidable, untreatable pain could in fact be viewed as a violation of their right to be free from cruel, inhuman or degrading treatment.[297] A human rights-based approach therefore supports palliative care and euthanasia, and access to these is consistent with states' obligations under the right to health.[298]

Notwithstanding the clear framing of adequate pain relief as a human rights issue, and the importance of palliative care as part of the continuum of healthcare, the accessibility of pain-relieving medications is limited in many countries.[299] This is because they are deemed to be controlled substances, and progress to strengthen access to such medications where they are needed has been slow.[300] Uganda, for example, was one of the first African countries to facilitate the availability of medications to relieve pain, including liquid morphine, in an attempt to strengthen the overall health system

---

[294] Brennan et al. (2019).

[295] This is incredibly topical in Australia, for example, with Victorian legislation legalising voluntary assisted suicide commencing in that state in 2019. Queensland and Western Australia are currently debating introducing legislation as well. On the history of voluntary assisted suicide and physician assisted suicide in Australia see, for example: McGee et al. (2018); Willmott et al. (2016); White et al. (2019a, b).

[296] Committee on Economic, Social and Cultural Rights (2000), para 25.

[297] Nowak (2009), para 72; Mendez (2013), para 56.

[298] Committee on Economic, Social and Cultural Rights (2000), para 25. For more on the terminology adopted in this context see, for example: McGee et al. (2018), 1371–2 and Willmott et al. (2016), 6–7.

[299] Baer (2016), 209.

[300] Ibid.

as well as palliative care in particular.[301] Even in higher income countries, where there is a better acceptance of pain-relieving medications, these may not be accessible or may be highly restricted.[302] There is therefore a global need to better understand the access to pain-relieving medication as fundamental to a person's dignity.

One practical way to facilitate a human rights-based approach to palliative care and end of life is to support advance care directives as a way for older persons to participate in decision-making in anticipation of a loss of capacity to make health and personal decisions themselves. As discussed, they generally form part of the estate planning process and play an important role in planning for the future, including by indicating the individual's wishes about how they wish to be cared for in the event that they lose capacity. It is not the intention here, however, to discuss the intricacies involved in the legality and/or implementation of advance care directives other than to note their role in facilitating a human rights-based approach to healthcare, particularly towards the end of life.[303] Advance care directives can remove the risk that a person will continue to live in pain, or an otherwise undignified life, when they would have chosen otherwise but for the lack of capacity to express such a choice.

10.8 Conclusion

To best protect the rights of older people it is necessary to recognise and understand that ageing is not an inevitable downward spiral resulting in the complete loss of physical and mental function and capacity.[304] In fact, to assume this is to perpetuate unhelpful and damaging ageist attitudes. Nor is it possible to assume that ageing looks the same for each older person. As the factors relevant to the ageing process discussed above demonstrate, ageing is not the same for everyone and does not occur in a linear or homogenous fashion.[305] Instead, it is important to appreciate that ageing offers opportunities for change and growth—some of which have not previously been realised given the unprecedented global trends in ageing and the advances in assistive technologies.

The question therefore arises as to what can be done in relation to protecting the right to health given the unique challenges facing older people. International obligations are helpful in establishing core minimum standards which are complemented by supervision and enforcement mechanisms under human rights treaties. The effectiveness and enforceability of international human rights standards depend a great deal on domestic implementation, however. As can be seen from the above discussion in relation to the aged care system in Australia, the existence of international

---

301 Ibid.; Human Rights Watch (2011).

302 Royal Commission into Aged Care Quality and Safety (2019), 117.

303 There is a wealth of literature on advance health directives, see, for example: Detering et al. (2019); White et al. (2019a, b); Willmott et al. (2010).

304 World Health Organization (2018).

305 Ibid.

obligations through ratification of treaties does not always give full effect to human rights principles and responsibilities. This is particularly problematic where the state is either prioritising protection of the individual or fiscal concerns over individual dignity and autonomy, or attempting to achieve a balance between these concepts but failing. Detailed implementing legislation and policies are required to ensure a meaningful shift away from paternalistic, medicalised models of care towards a human rights-based approach which advances dignity, autonomy and liberty.

In order to ensure compliance with a human rights-based approach it is therefore necessary to: make affordable and quality health and aged care services available; ensure accessible, affordable and quality healthcare and aged care design that is inclusive of age; and improve the collection, analysis and usage of research as well as prevalence data about all facets of health and ageing; ensure that health and aged care organisations as a whole understand the obligations arising from a human rights-based approach and what these obligations mean in practice; incorporate the human rights-based approach into the organisation's strategic objectives; review policies and procedures to ensure compliance with legislative requirements but also human rights principles including addressing age-based discrimination and ageist practices in the accessibility and delivery of health and aged care services; develop and implement action plans with various responsibilities clearly defined; monitor progress and implement improvements accordingly; and facilitate effective communication, not only to staff about the human rights of older persons but also to the older person themselves, as well as to their family and support networks, so they can understand their rights and access justice where necessary.[306] A key theme emerging is therefore the need for (better) workforce and community education, training and, importantly, accreditation in relation to improving the understanding of all involved with the delivery of health and aged care about human rights principles, the obligations arising from them and how they translate into best practice.[307] Information programs can also be useful in increasing awareness and understanding around accessing health and aged care services, and what systems may be in place to facilitate this in order to protect the right to health. However, this all costs money—not only to develop these information, education and training programs but to also ensure that the quality health and aged care systems exist.

Multiple sectors, most notably governments and policy-makers, must therefore act. Policy responses have a significant role to play, particularly with respect to addressing the link between abuses of an older person's right to health and the right to social security.[308] Policies can promote the development of more effective education programs which can, in turn, help influence health determinants more broadly given that how people age is influenced by many factors including their physical environment as well as social and financial considerations.[309] Thus far, policy development has been haphazard with cost containment being the focus over acknowledging the

---

[306] Beard et al. (2016), 165; European Network of National Human Rights Institutions (2017), 18.
[307] Beard et al. (2016), 165; European Network of National Human Rights Institutions (2017), 18.
[308] United Nations Department of Economic and Social Affairs Programme on Ageing (2018), 5.
[309] Ibid.; World Health Organization (2015), 6.

investment needed to fulfil the right to health for older people. Furthermore, policy tends to be divergent in approach with older people being represented as either frail and dependent and therefore in need of support, or as a cohort who have the (financial) ability to contribute to society, for instance, the 'baby boomer' generation.[310] This is not to deny either classification but rather, especially when recognising the individual nature of ageing, to recognise that these depictions are two opposite ends of a 'continuum of diversity' for older people.[311] Correcting policies so they better address the individualised nature of ageing will require the evidence-based development of an integrated and person-centred approach prioritising the function of older people over disease-focused services that are reactive in nature.[312] Moreover, a coordinated response is necessary to create and maintain age-friendly environments including, housing, employment, financial security, transport, and social protection, in addition to integrated person-centred health, long term and aged care.[313]

Underpinning this, however, is the need to accept that a global public health response is both essential and urgent.[314] Such a response will necessarily be a transformative alignment implementing systemic change to adequately meet the needs of older persons worldwide, including emphasising continued function, both physical and mental, throughout the life course.[315] This will require a shift away from a disease specific focus to one emphasising functional ability and intrinsic capacity, and how to maintain an optimum trajectory to promote health in ageing.[316] Systems must therefore meet the needs of older people which are often complex and chronic, traversing all aspects of physical and mental functioning.[317] They must also be able to meet long-term care needs recognising two key factors: first, that even where older people experience a significant loss of functioning they are still entitled to respect and to live a dignified life; and second, that as capacity is not static, steps may be able to be taken to support autonomous decision-making processes.[318]

This will, in turn, require attention to not only the role of the individual in ageing but also building supportive and enabling environments, both social and physical.[319] The role of the family is therefore significant and informal caregivers must be supported.[320] In order to protect the right to health and associated rights, it is fundamental that the voices of older people are represented in any policy and system design, implementation, review and/or change. The only way to adequately fulfil the

---

310 Beard et al. (2016), 165.

311 Ibid.; World Health Organization (2015), 8.

312 Beard et al. (2016), 164.

313 Ibid.

314 Ibid., 165.

315 World Health Organization (2015), 6.

316 Beard et al. (2016), 165. For hypothesized trajectories, see: World Health Organization (2015), 31. See also: Cosco et al. (2014).

317 World Health Organization (2015), 93.

318 Ibid., 128.

319 Beard et al. (2016), 165.

320 Ibid.

human rights of older people is to first understand how they are being violated—an investigation that must include older people themselves.[321] It is only then that governments and policy-makers will truly be able to implement measures to effect real change with a view to protecting the rights of older people—measures which must acknowledge the interrelated nature of human rights.

References

Alzheimer's Disease International (2012) World Alzheimer report 2012: Overcoming the stigma of dementia. https://www.alz.co.uk/research/world-report-2012. Accessed 14 Feb 2020

Australian Digital Inclusion Index. https://digitalinclusionindex.org.au. Accessed 14 Feb 2020

Australian Government Department of Health, Aged Care Sector Committee Diversity Sub-Group (2019) Actions to support older culturally and linguistically diverse people. A guide for aged care providers. https://apo.org.au/sites/default/files/resource-files/2019/02/apo-nid221041-1333696.pdf. Accessed 14 Feb 2020

Australian Institute of Health and Welfare (2019) Social isolation and loneliness. https://www.aihw.gov.au/reports/australias-welfare/social-isolation-and-loneliness. Accessed 14 Feb 2020

Baer B et al (2016) The right to health of older people. The Gerontologist 56(S2):S206-217

Baez M et al (2019) Technologies for promoting social participation in later life. In: Barbosa Neves B, Vetere F (eds) Ageing and digital technology. Springer, Singapore

Barbosa Neves B, Sanders A, Kokanovic R (2019) "It's the worst bloody feeling in the world": experiences of loneliness and social isolation among older people living in care homes. J Aging Stud 49:74–84

Batti A, ul Haq A (2017) The pathophysiology of perceived social isolation: effects on health and mortality. Cureus 9(1):e994

BBC News (2009) NHS discrimination 'common' (27 Jan 2009) https://news.bbc.co.uk/2/hi/health/7850881.stm. Accessed 14 Feb 2020

Beard JR, Officer AM, Cassels AK (2016) The world report on ageing and health. The Gerontologist 56(S2):S163-166

Bennett B et al (2017) Assistive technologies for people with Dementia: ethical considerations. Bull World Health Organ 95(11):749–755

Brennan F, Lohman D, Gwyther L (2019) Access to pain management as a human right. Am J Public Health 109(1):61–65

Burholt V, Dobbs C (2012) Research on rural ageing: where have we got to and where are we going in Europe? J Rural Stud 28:432–446

Button M, Cross C (2017) Cyber frauds, scams and their victims. Routledge, Abingdon

Carnell K, Paterson R (2017) Review of national aged care quality regulatory processes. https://www.health.gov.au/sites/default/files/review-of-national-aged-care-quality-regulatory-processes-report.pdf. Accessed 14 Feb 2020

Chatterji S et al (2015) Health, functioning, and disability in older adults—Present status and future implications. The Lancet 385:563–575

Cosco TD et al (2014) Successful ageing, education, and functional trajectories in later life: a longitudinal latent variable modelling analysis. The Lancet 384(S2):S9

Courtin E, Knapp M (2017) Social isolation, loneliness, and health in old age: a scoping review. Health and Social Care 25(3):799–812

Cox CB (2015) Social policy for an aging society: a human rights perspective. Springer, New York

Dannefer D (2003) Cumulative advantage/disadvantage and the life course: cross-fertilizing age and social science theory. J Gerontol Ser B Psychol Sci Soc Sci 58:S327–S337

[321] European Network of National Human Rights Institutions (2017) 18.

De Bruin SR et al (2010) Day care at green care farms: a novel way to stimulate dietary intake of community-dwelling older people with dementia? J Nutr Health Aging 14(5):352–357
Detering KM et al (2019) Prospective multicentre cross-sectional audit among older Australians accessing health and residential aged care services: protocol for a national advance care directive prevalence study. BMJ Open 9(10):e031691
DuMont J et al (2015) Development of a comprehensive hospital-based elder abuse intervention: an initial systematic scoping review. PLoSONE 10(5):e0125105
European Network of National Human Rights Institutions (2017) Applying a human rights based approach to long-term care for older persons: a toolkit for care providers. https://ennhri.org/wp-content/uploads/2020/01/Applying-a-HRBA-to-Long-Term-Care-for-Older-Persons-A-Toolkit-for-Care-Providers.pdf. Accessed 17 Feb 2020
Equality and Human Rights Commission, Age Concern and Help the Aged (2009) Just ageing? Fairness, equality and the life course: socio-economic inequalities in older people's access to and use of public services. https://www.hscbusiness.hscni.net/pdf/EHRC-Just_ageing_final_report_2009_pdf.pdf. Accessed 14 Feb 2020
Feng Z et al (2009) Use of physical restraints and antipsychotic medications in nursing homes: a cross-national study. Int J Geriatr Psychiatry 24(10):1110–1118
Foebel AD, Pedersen NL (2016) Genetic influences on functional capacities in aging. The Gerontologist 56:S218–S229
Francis J et al (2019) Aging in the digital age: conceptualizing technology adoption and digital inequalities. In: Barbosa Neves B, Vetere F (eds) Ageing and digital technology. Springer, Singapore
Fried LP (2016) Investing in health to create a third demographic dividend. The Gerontologist 56(S2):S167–S177
Garçon L et al (2016) Medical and assistive health technology: meeting the needs of aging populations. The Gerontologist 56(S2):S293–S302
Gostin LO et al (2018) 70 years of human rights in global health: drawing on a contentious past to secure a hopeful future. Lancet 392:2733–2734
Harrington J and Stuttaford M (eds) (2010) Global health and human rights: legal and philosophical perspective. Routledge, New York
HelpAge International (2011) The right to health in old age: unavailable, inaccessible and unacceptable. https://www.helpage.org/silo/files/the-right-to-health-in-old-age-unavailable-inaccessible-and-unacceptable.doc. Accessed 14 Feb 2020
Human Rights Watch (2011) Global state of pain treatment. Human Rights Watch, London
Human Rights Watch (2019) Fading away. How aged care facilities in Australia chemically restrain older people with dementia. https://www.hrw.org/report/2019/10/15/fading-away/how-aged-care-facilities-australia-chemically-restrain-older-people. Accessed 14 Feb 2020
Hunt P (2016) Interpreting the international right to health in a human rights-based approach to health. Health and human rights journal blog post. https://www.hhrjournal.org/2016/12/interpreting-the-international-right-to-health-in-a-human-rights-based-approach-to-health/. Accessed 2 Feb 2020
International Labour Organization (2017) World social protection report 2017–2019: universal social protection to achieve the sustainable development goals. https://www.ilo.org/wcmsp5/groups/public/---dgreports/---dcomm/---publ/documents/publication/wcms_604882.pdf. Accessed 14 Feb 2020
Johannesen M, LoGiudice D (2013) Elder abuse: a systematic review of risk factors in community-dwelling elders. Age Ageing 42(3):292–298
Kornfeld-Matte R (2017) Report of the independent expert on the enjoyment of all human rights by older persons. UN Doc A/HRC/36/48
LaVeist TA (2005) Minority populations and health: an introduction to health disparities in the United States. Jossey-Bass, United States

Lewis B et al (2018) Submission by the Australian centre for health law research to the Australian human rights commission human rights and technology issues paper. https://eprints.qut.edu.au/122938/. Accessed 7 Feb 2020
Lloyd-Sherlock P et al (2012) Population ageing and health. The Lancet 379:1295–1296
Low LF, Fletcher J (2015) Models of home care services for PERSONS with dementia: a narrative review. Int Psychogeriatr 27(10):1593–1600
McGee A et al (2018) Informing the euthanasia debate: perceptions of Australian politicians. Univ of New South Wales Law J 41(4):1368–1417
Mendez JE (2013) Report of the special rapporteur on torture and other cruel, inhuman or degrading treatment or punishment. UN Doc A/HRC/22/53
Meyer G et al (2009) Restraint use among nursing home residents: cross-sectional study and prospective cohort study. J Clin Nurs 18(7):981–990
Morris J (2010) human rights and healthcare: changing the culture. Age Ageing 39:525–527
Nowak M (2009) Report of the special rapporteur on torture and other cruel, inhuman or degrading treatment or punishment. UN Doc A/HRC/10/44
Pillemer K et al (2016) Elder abuse: global situation, risk factors, and prevention strategies. The Gerontologist 56(S2):S194–S205
Pot AM (2013) Improving nursing home care for dementia: is the environment the answer? Aging Mental Health 17(7):785–787
Pot AM, Briggs AM, Beard JR (2018) The sustainable development agenda needs to include long-term care. JAMDA 19:725–727
Purser K (2017) Capacity assessment and the law: problems and solutions. Springer International, Switzerland
Purser K, Rosenfeld T (2014) Evaluation of legal capacity by doctors and lawyers: the need for collaborative assessment. Med J Aust 201(8):483–485
Royal Commission into Aged Care Quality and Safety (2019) Interim report: neglect, vol 1. https://agedcare.royalcommission.gov.au/publications/Documents/interim-report/interim-report-volume-1.pdf. Accessed 10 Dec 2019
Sadana R et al (2016) Healthy ageing: raising awareness of inequalities, determinants, and what could be done to improve health equity. The Gerontologist 56(S2):S178–S193
Sharkey A (2012) Granny and the robots: ethical issues in robot care for the elderly. Ethics Inf Technol 14(1):27–40
Smit D et al (2014) Wellbeing-enhancing occupation and organizational and environmental contributors in long-term dementia care facilities: an explorative study. Int Psychogeriatr 26(1):69–80
St. Sauver JL et al (2015) Risk of developing multimorbidity across all ages in an historical cohort study: differences by sex and ethnicity. BMJ Open 5(2):e006413
Staudinger UM et al (2016) A global view on the effects of work on health in later life. The Gerontologist 56(S2):S281–S292
Szanton SL et al (2015) Preliminary data from community aging in place, advancing better living for elders, a patient-directed, team-based intervention to improve physical function and decrease nursing home utilization: the first 100 individuals to complete a centers for medicare and medicaid services innovation project. J Am Geriatr Soc 63(2):371–374
te Boekhorst S et al (2009) The effects of group living homes on older people with dementia: a comparison with traditional nursing home care. Int J Geriatric Psychiatry 24(9):970–978
Terraneo M (2015) Inequities in health care utilization by people aged 50+: evidence from 12 European countries. Soc Sci Med 126:154–163
Tobin T (2012) The right to health in international law. Oxford University Press, New York
United Nations Department of Economic and Social Affairs, Population Division (2015) World population ageing. UN Doc ST/ESA/SER.A/390
United Nations Department of Economic and Social Affairs Programme on Ageing (2018) Health inequalities in old age. https://www.un.org/development/desa/ageing/wp-content/uploads/sites/24/2018/04/Health-Inequalities-in-Old-Age.pdf. Accessed 6 Feb 2020

United Nations Department of Economic and Social Affairs Programme on Ageing (n.d.) Income poverty in old age: an emerging development priority. https://www.un.org/esa/socdev/ageing/documents/PovertyIssuePaperAgeing.pdf
United Nations Secretary-General (2012) Report on follow-up to the second world assembly on ageing. UN Doc A/67/188 (26 July 2012)
United Nations Secretary-General (2013) Report on follow-up to the second world assembly on ageing. UN Doc A/68/167 (19 July 2013)
United Nations Secretary-General (2015) Report on follow-up to the second world assembly on ageing. UN Doc A/70/185 (24 July 2015)
UN Women (2017) Long-term care for older people: a new global gender priority. Policy Brief No 9. https://www.unwomen.org/-/media/headquarters/attachments/sections/library/publications/2017/un-women-policy-brief-09-long-term-care-for-older-people-en.pdf?la=en&vs=1608. Accessed 14 Feb 2020
Verbeek H et al (2012) Small-scale, homelike facilities in dementia care: a process evaluation into the experiences of family caregivers and nursing staff. Int J Nurs Stud 49(1):21–29
Waycott J, Vetere F, Ozanne E (2019) Building social connections: a framework for enriching older adults' social connectedness through information and communication technologies. In: Barbosa Neves B, Vetere F (eds) Ageing and digital technology. Springer, Singapore
Weaver Williams P (2009) Age discrimination in the delivery of health care services to our elders. Marquette Elder's Advisor 11(1), Article 3
White BP et al (2019) Prevalence of advance care directives in the community: a telephone survey of three Australian states. Int Med J 49(10):1261–1267
White BP, Willmott L, Close E (2019) Victoria's voluntary assisted dying law: clinical implementation as the next challenge. Med J Aust 210(5):207–209
Willemse BM et al (2014) The relationship between small-scale nursing home care for people with dementia and staff's perceived job characteristics. Int Psychogeriatr 26(5):805–816
Willmott L, White BP, Mathews BP (2010) Law autonomy and advance directives. J Law Med 18:366–389
Willmott L et al (2016) (Failed) voluntary euthanasia law reform in Australia: two decades of trends, models and politics. Univ New South Wales Law J 39(1):1–46
World Health Organization (n.d.) Ageing and life-course. Healthy ageing and the sustainable development goals. https://www.who.int/ageing/sdgs/en/. Accessed 14 Feb 2020
World Health Organization (2002) Active ageing: a policy framework. https://www.who.int/ageing/publications/active_ageing/en/. Accessed 7 Feb 2020
World Health Organization (2015) World report on ageing and health. https://www.who.int/ageing/events/world-report-2015-launch/en/. Accessed 19 Dec 2019
World Health Organization (2017a) 10 facts on ageing and health. https://www.who.int/features/factfiles/ageing/en/. Accessed 14 Feb 2020
World Health Organization (2017b) Global strategy and action plan on ageing and health (2016–2020). https://www.who.int/ageing/WHO-GSAP-2017.pdf?ua=1. Accessed 14 Feb 2020
World Health Organization (2018) Ageing and health. https://www.who.int/news-room/fact-sheets/detail/ageing-and-health. Accessed 30 Dec 2019
World Health Organization (2019) Decade of healthy ageing 2020–2030. https://www.who.int/ageing/decade-of-healthy-ageing. Accessed 10 Dec 2019

International Instruments

Additional Protocol to the American Convention on Human Rights in the Area of Economic, Social and Cultural Rights (1988) Opened for signature 17 November 1988, entered into force 16 November 1999. OAS Treaty Series No 69

African Charter of Human and Peoples' Rights (1981) Opened for signature 27 June 1981, entered into force 21 October 1986. 21 ILM 58

Committee on Economic, Social and Cultural Rights (2000) *General Comment No. 14: The right to the highest attainable standard of health (Art 12)* UN Doc E/C.12/2000/4 (11 August 2000)

Committee on the Rights of Persons with Disabilities (2017) *General comment No. 5 on Living Independently and Being Included in the Community.* UN Doc CRPD/C/GC/5 (27 October 2017)

Convention on the Elimination of All Forms of Discrimination Against Women (1979) Opened for signature 18 December 1979, entered into force 3 September 1981. 1249 UNTS 13

Convention on the Elimination of All Forms of Racial Discrimination (1965) Opened for signature 21 December 1965, entered into force 4 January 1969. 660 UNTS 195

Convention on the Protection of the Rights of All Migrant Workers and Members of their Families (1990) Opened for signature 18 December 1990, entered into force 1 July 2003, 2220 UNTS 39481

Convention on the Rights of the Child (1989) Opened for signature 20 November 1989, entered into force 2 September 1990. 1577 UNTS 3

Convention on the Rights of Persons with Disabilities (2006) Opened for signature 13 December 2006, entered into force 3 May 2008. 2515 UNTS 3

European Social Charter (Revised) (1996) Opened for signature 3 May 1996, entered into force 1 July 1999. ETS 163

International Covenant on Civil and Political Rights (1966) Opened for signature 16 December 1966, entered into force 23 March 1976. 999 UNTS 171

International Covenant on Economic, Social and Cultural Rights (1966) Opened for signature 16 December 1966, entered into force 3 January 1976. 993 UNTS 3

Madrid International Plan of Action on Ageing (2002) Adopted by Second World Conference on Ageing, Madrid, 8–12 April 2002

United Nations General Assembly (2015) *Transforming our world: the 2030 agenda for sustainable development.* General Assembly Resolution 70/1. UN Doc A/RES/70/1 (21 October 2015)

Universal Declaration of Human Rights (1948) Adopted by United Nations General Assembly Resolution 217A (III). UN Doc A/810 (10 December 1948)

United Nations Principles for Older Persons (1991) Adopted by United Nations General Assembly Resolution 46/91. UN Doc A/RES/46/91 (16 December 1991)

Chapter 11
Conclusion

Several key themes have emerged through the analysis presented in this book and these are summarised in this concluding chapter. In doing so, this chapter highlights the key contributions the book has made to the literature on elder law, principally by exploring the content and value of a human rights framework which can be used to improve law and access to justice in a number of interrelated fields. Recommendations are also presented, drawing together suggestions which have been discussed in more detail in previous chapters, and highlighting areas for future research and analysis. This chapter concludes with the overarching claim, introduced throughout the preceding chapters, that a human rights-based approach to elder law in fact demands a commitment to human rights across the whole life course. Stronger protections of human rights for people of all ages guard against poverty, marginalisation, ill-health and isolation—all factors which contribute to violations of rights and denial of justice in older age. A genuine conviction in support of human rights can also help to realise the crucial paradigm shift which is required in our treatment of older persons, away from positioning them as dependent recipients of care and a burden on economies, towards recognition as active rights-holders entitled to respect for dignity and autonomy.

11.1 Introduction

Ageing populations are recognised the world over as giving rise to major social and economic challenges that have never before been seen. There is a risk that, in attempting to respond to the challenges associated with these shifting demographics, governments will prioritise economic rationalities over the human rights of older persons. Neo-liberal ideals of productivity, consumerism and privatisation feed into ageist assumptions about the capabilities and value of older persons, with potentially serious and deleterious impacts for their human rights.

© Springer Nature Singapore Pte Ltd. 2020

B. Lewis et al., *The Human Rights of Older Persons*,
https://doi.org/10.1007/978-981-15-6735-3_11

This book has argued that a paradigm shift is needed in the way that society thinks about and approaches ageing. There is a need to move away from the ageist attitudes which position older persons as incapable, dependent, vulnerable and a burden on the economy, towards an idea of older persons as valuable members of the community, entitled to choose for themselves how they wish to live their lives while being supported to live out those choices. This would be a transition away from medical and charity-based models which have shaped law and policy in the past, and instead position older persons as active holders of rights recognised as enforceable legal entitlements. Given the fundamental connection between human rights and the core values of dignity, autonomy, liberty and equality, we have argued throughout this book that a human rights-based approach is a promising way of achieving this necessary paradigm shift.

The law is the medium through which many human rights threats are embodied for older persons, while simultaneously offering a means for reducing and remedying those threats and realising a more human rights-supportive way of operating. This book has defined 'elder law' as a broad range of domains of law and policy which shape the realities of older people's lives, frequently in interconnected and complex ways. The fields of financial security, accommodation, and health and aged care canvassed in this book represent key legal areas which older persons are likely to encounter. Issues of ageism, capacity and elder abuse intersect with the law in ways which have particular impacts on the lives of older people. We have analysed these to show the potential human rights risks and offered suggestions for how human rights could be better incorporated into these contexts.

By way of conclusion, this chapter draws together a selection of the key themes which emerged from the novel analysis undertaken throughout the book and summarises the main recommendations that we see as providing a pathway for legal and policy reform into the future. To this end, it also identifies areas where further research is required. We recognise, however, that meaningfully altering the ageist assumptions and practices pervasive in the current policy approaches to ageing will require detailed and considered input from a variety of stakeholders, including not only legal actors but also, for example, governments, policy-makers, insurers, advocacy groups, financial bodies and community organisations. Principal amongst the voices who should be heard are older people themselves—participation is key to ensure respect for their rights, will and preferences. The following suggestions are therefore made with a view to contributing to the dialogue on what is a fundamentally important measure of our society as a whole—how we treat older people.

11.2 Key Themes and Contributions

The preceding chapters have examined a wide range of legal areas and cross-cutting issues, demonstrating the varied and complex ways in which the law can impact on the human rights of older people. This analysis has uncovered a number of key themes, the discussion of which we hope will make a meaningful contribution to the dialogue on a human rights-based approach to elder law. These are summarised below.

11.2.1 A Human Rights Framework

The central argument of this book, that a human rights-based approach can achieve a necessary paradigm shift in the way we contemplate, design and implement 'elder law', is advanced through a comprehensive and interconnected human rights framework. Drawing on rights that are guaranteed within international and regional human rights law, principally within the *International Covenant on Civil and Political Rights* (ICCPR)[1] and the *International Covenant on Economic, Social and Cultural Rights* (ICESCR)[2] (outlined in Chap. 2), the book articulates a novel, multifaceted framework (set out in Chap. 3) which has been used to analyse the specific areas of capacity, elder abuse, financial security, accommodation, as well as health and aged care, as common areas of law that older people frequently interact with. The framework is also informed by human rights theory and scholarship, as well as the work of various intergovernmental and non-government agencies seeking to advance the rights of older persons.

The framework is structured as three interrelated layers (see Fig. 11.1), beginning with the core values which underpin all human rights law: respect for dignity, autonomy, liberty and equality. They provide a fundamental set of criteria against which to test the components of any set of 'elder laws'. Any law which does not respect the dignity, autonomy, liberty and equality of older persons could not be said to be consistent with human rights. Given the prevalence of elder abuse, age-based discrimination and other affronts to dignity reported the world over, these core values provide an indispensable framework for improving elder law through fostering respect for the rights of older persons.

Building on the core values, the framework then incorporates the specific rights found within human rights law, as well as the corresponding duties which support them. Older people are entitled to the full range of human rights on an equal basis with all other people, and these rights are understood as being interdependent and indivisible. There are select rights which are more directly related to areas of elder law, such as rights to social security, health, an adequate standard of living, housing, freedom from cruel or inhuman treatment, privacy, freedom of movement, and equality before the law. Accompanying these rights are three levels of obligation under human rights law: the duties to respect, protect and fulfil human rights. By understanding this tripartite system as it applies to specific rights, it is possible to flesh out what a human rights-based approach ought to mean for particular areas of elder law.

---

[1] *International Covenant on Civil and Political Rights* (1996) ('*ICCPR*').

[2] *International Covenant on Economic, Social and Cultural Rights* (1966) ('*ICESCR*').

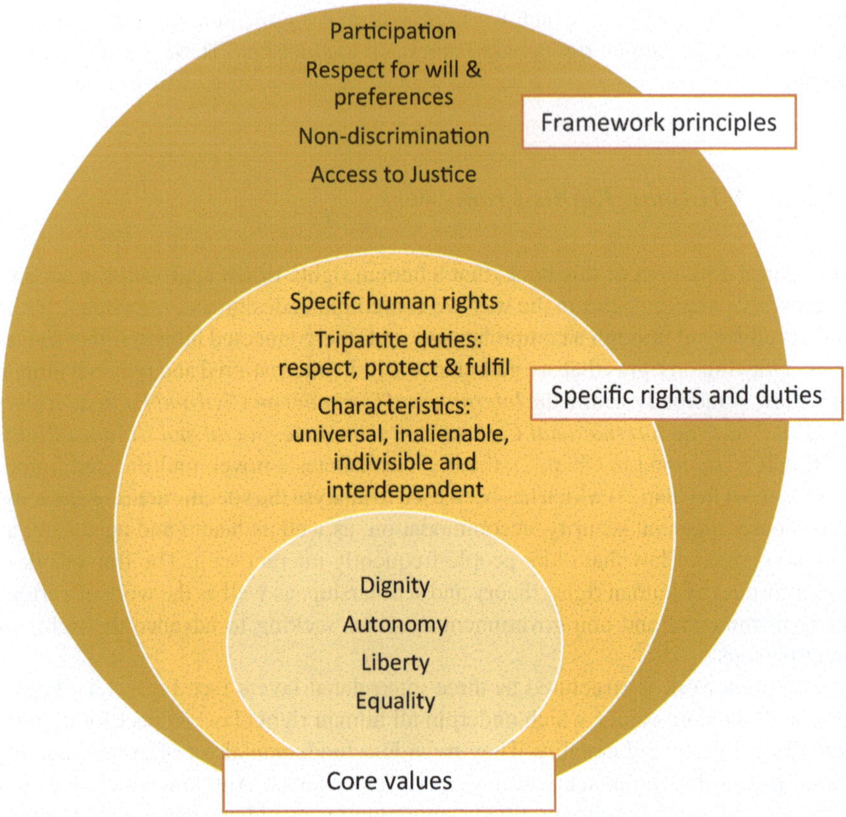

Fig. 11.1 Human rights framework

The third level of the framework sets out a number of principles which can help to operationalise the values, rights and duties and provide more practical guidance for reform of elder law. These principles include participation, non-discrimination, respect for will and preferences and access to justice. Taken together, the values, rights and principles of the human rights framework have enabled a detailed and deep analysis of select areas of elder law and cross-cutting issues. The framework can be used as the basis for analysis of other issues, or to help develop new laws and policies which are oriented towards the promotion and protection of human rights.

11.2.2 Importance of Participation

One of the framework principles set out in Chap. 3 is participation, and it arises as a key theme throughout the discussion in the book. Participation has substantive and procedural dimensions as part of a human rights-based approach to elder law.

As explained in Chap. 4, older people are entitled to participate in the civil, political, economic, social and cultural lives of their communities.[3] Social inclusion has multiple benefits for the well-being of older people and the enjoyment of their other rights, while exclusion and marginalisation can lead to various negative impacts, and can even operate as risk factors for abuse as well as significant health issues such as dementia and depression.[4] A human rights-based approach to elder law is therefore one which supports older people to participate in their communities in the ways in which they choose.

At the same time, participation is an important means of ensuring that elder law meaningfully addresses human rights issues. By authentically including older people in the design and development of policies we can help to ensure that their lived experiences are given due regard—which is useful in improving the quality of those policies and an important way of respecting older people's opinions, knowledge and, ultimately, their dignity. This extends to enhancing access to justice for older people when their rights are threatened or violated through improving understanding of their experiences of legal systems, including interactions with legal actors. Participation ought therefore to be a principle which guides all work in the elder law space.

11.2.3 Ageism

One of the challenges which cuts across all areas of elder law and creates barriers for the enjoyment of human rights is ageism. As discussed in Chap. 5, ageism comprises the various assumptions and stereotypes (usually negative) that are made about older people, for instance that older people are frail, incapable, dependent, and an economic burden. It is a socially-constructed phenomenon—we participate in ageist practices and learn ageist attitudes from society, including through portrayals of older people in the media and popular culture, and it is so ingrained that even older people themselves can come to internalise negative opinions about ageing and being 'old'.

These assumptions have come to influence policy and can be observed in various sectors with which older people must interact. For instance, age discrimination has long been reported across the spectrum of employment, from job advertisements and recruitment, through training and promotion, and ultimately to the cessation of employment through retirement or redundancy. Ageist ideas can also be seen in the design of accommodation options for older people, with assumptions being made about the sort of housing that older people prefer or what would be beneficial for 'them'. Ageism can also be detected in the healthcare sector, with studies showing that healthcare professionals adopt ageist language and tone in their communication with older people, and that certain treatments and diagnostic processes can be made unavailable to people over a certain age.[5] At its worst, ageism can lead to incorrect

---

[3]ICCPR, art 27; ICESCR, art 15.

[4]Barbosa Neves et al. (2019).

[5]See, for example: Wyman et al. (2018); Adelman et al. (2000).

determinations of incapacity, elder abuse and/or exploitation, because it undermines respect for dignity and equal treatment, and allows the 'othering' of older people.

Ageism is so widespread and ingrained that it has emerged as a prevalent issue in all the areas of law that are considered in this book. This also means that, in order to address ageism, legal changes will need to be accompanied by socio-cultural and attitudinal shifts. Recommendations for achieving this will be discussed below.

11.2.4 Balancing Rights and the Dignity of Risk

A perennial problem in implementing human rights is how to balance competing rights when they come into conflict, and these challenges have emerged throughout this book. In the context of elder law, they are most obvious when the level of support which an older person requires to safely undertake the range of activities of daily life makes some limitation of their rights inevitable. For example, an older person with a diagnosis of advanced dementia may require ongoing supervision or even restrictions on movement in order to ensure they are safe and do not present a danger to themselves or to other people. At the less extreme end of the spectrum, an older person living alone may have monitoring or security devices installed to ensure they can access help if required, but these may entail some interferences with privacy. These issues can become particularly problematic if an older person's family, friends, doctor, allied health professional or other carer feel that a particular intervention is in their best interests, but the older individual themselves is opposed to it on the basis that it interferes with their liberty, autonomy, dignity, or other rights. The challenge of resolving these issues may be exacerbated further when there are questions about the individual's decision-making capacity, their ability to consent and/or the existence of any undue influence.

Human rights law and theory provide some guidance here. In general terms, limitations on an individual's rights should only be permitted where it is necessary to protect the rights of others, or is in the interests of public safety or another legitimate aim.[6] Generally, protecting an individual's own safety would not be sufficient justification, though frequently this occurs in the context of legitimate public interests. Even where a legitimate purpose exists, the limitations on rights should be necessary and proportionate to that objective. In the context of elder law, the individual's preferences ought to carry great weight, even where they present some risk to their own health and safety or seem arbitrary or capricious to others. This is embodied in the concept of the 'dignity of risk', which recognises that there is value in supporting an individual's choices, even where others might view those choices as irrational

[6]See, for example: ICCPR, art 12(3), which states that the freedom of movement guaranteed in article 12 'shall not be subject to any restrictions except those which are provided by law, are necessary to protect national security, public order (ordre public), public health or morals or the rights and freedoms of others, and are consistent with the other rights recognized in the present Covenant'.

or even dangerous.[7] This extends not just to material issues of housing, health and aged care, but also to, for example, financial and legal issues, including decisions regarding estate planning.

Taken together, these ideas can help to shape elder law to ensure that an individual's preferences are given the greatest possible respect, that older people are given meaningful opportunities to be consulted and have their opinions heard, and that their rights ought only to be infringed upon when absolutely necessary. The concept of the dignity of risk ought to be recognised and facilitated within relevant laws and policies, as will be recommended below.

11.2.5 Capacity

A common challenge in ensuring that older people are heard and that their wishes are respected arises where an older person has impaired or lost capacity. The question of when a person ought to be viewed as lacking capacity such that their decisions are not legally recognised is extremely complex, as explained in Chap. 6. That chapter outlined the different factors which affect the assessment of capacity, and the ways in which an incorrect assessment can affect an individual's human rights. Capacity is therefore another cross-cutting issue which potentially impacts on a range of elder law topics—any area where an individual may be required to make legally binding decisions (for example, financial matters or in relation to housing, health or aged care) raises the potential for questions of legal capacity to arise.

A key consideration in determining a human rights-based approach to capacity is the notion of substitute decision-making. In most cases where an individual has impaired capacity it will be preferable to implement some form of supported decision-making, if possible. This will enable their own preferences to be given effect to the greatest extent practicable, and thereby minimise the risk of their rights being violated. However, it is inevitable that some older people will lose the capacity necessary to make a decision. In such cases, the question arises: what is the most human rights-compatible approach?

It could be argued that substituting one person's preferences and decisions for another's is incompatible with human rights, particularly with the core values of respect for autonomy and dignity. Yet, in cases where the individual has lost capacity, the use of supported decision-making may create other risks, for instance exploitation, neglect or abuse because, if capacity has been lost, then a 'supported' decision is not possible and thus any decision made will actually be a form of substituted decision-making. This is not to deny the issues that exist with substitute decision-making, most notably the risk of abuse. Rather, it is important to recognise that there are situations in which a person will lose capacity (distinct from impaired or diminished capacity) and thus not be able to be supported in making a decision. As will be recommended below, a system which allows for substitute decision-making will

[7] Nay (2002); Morgan (2004); Ibrahim and Davis (2013); Mitchell (1994).

then be the most appropriate, provided that it is accompanied by strong human rights guidelines and standards to ensure that decisions are made in the best interests of the principal, and accompanied by appropriate causes of action or other avenues through which to seek remedial relief and achieve justice in the event of a breach.

11.2.6 Access to Justice

In order to deliver meaningful protection of human rights across the various fields of elder law, it is necessary to recognise human rights and to enact legal reforms which are genuinely informed by human rights principles as well as seeking to implement these in practice. However, these steps on their own are insufficient if barriers remain for older individuals to access justice. The analysis in preceding chapters has identified various formal and substantive barriers to justice which can prevent older people from accessing legal remedies or other suitable relief, even where legal processes exist which are designed to assist them. Such barriers may be financial, geographical, cultural or linguistic, or connected to issues of capacity. Chap. 7 highlighted the serious problem of ensuring access to justice in the context of elder abuse and pointed to the need for a range of formal and informal processes to provide appropriate pathways and support. Further recommendations on improving access to justice are set out below.

11.3 Recommendations

Through an application of a human rights framework and having regard to the cross-cutting issues identified throughout the book, a number of recommendations can be made for legal reforms, policy approaches and areas for further research. At the outset, it is important to remember that elder law, despite its common features, can vary markedly in detail across different geographical and legal jurisdictions as well as between different substantive legal areas. The analysis presented in the preceding chapters has drawn on examples from a number of jurisdictions, but ultimately the conclusions must be generalised to some extent as the book has not attempted a comprehensive detailed comparison of all legal systems, instead focusing on the application of the overarching human rights-based framework to a unique combination of legal topics. The recommendations presented below are similarly simplified, pointing to general approaches which could (or should) be pursued, highlighting best practice and articulating the core elements which need to be embedded in a human rights-based approach to elder law to enable the subsequent adoption of the framework within jurisdictionally specific legal systems.

11.3.1 Strengthen Human Rights Law

As a starting point, we have recommended that human rights law needs to be strengthened at both the international and national levels. We have employed a human rights framework which draws heavily on existing international human rights law and soft-law instruments but, while the current law provides many useful principles, rights and duties, there remain gaps and deficiencies which need to be addressed.

11.3.1.1 Adopt a Convention on the Rights of Older Persons

A key recommendation presented in this book is therefore that the international community should continue working towards the adoption of a dedicated treaty on the rights of older persons. While existing human rights law applies equally to older people as it does to all people, there are particular ways in which the rights of older persons are impacted in their lived experiences which are not adequately addressed by existing human rights law.[8] In particular, the pervasive problem of ageism threatens a range of human rights and requires specific action, and the application of core values of autonomy and dignity to specific issues, such as aged care, capacity and participation, warrants clarification.

The project of adopting a convention on the rights of older persons (CROP) has made significant strides in recent years, with the establishment of the Open-ended Working Group on Ageing (OEWGA) which has been tasked with identifying the core elements of a new instrument.[9] In delivering on this mandate the OEWGA has conducted annual discussions on select topics including non-discrimination and equality and freedom from violence, abuse and neglect,[10] autonomy and independence, long-term care and palliative care,[11] social protection and social security and education and lifelong learning,[12] and the right to work and access to justice.[13]

While the OEWGA has thus made a significant contribution to fleshing out the scope and content of particular human rights in older age with a view to settling the content of a detailed international instrument, it has not yet recommended the particular form that such an instrument would take. There remains the possibility that, rather than pursue a legally binding treaty, the OEWGA may recommend a non-binding, 'soft-law' instrument, for instance a set of guiding principles. Given that the previous series of principles and plans of action have been unable to address ageism, elder abuse and systemic violations of older people's rights in the past, it is argued that the time has come for a legally binding treaty on which enforceable domestic obligations can be based.

---

[8]Mégret (2011), 68ff.

[9]United Nations General Assembly (2013).

[10]Open-ended Working Group on Ageing (2017, 2018).

[11]Open-ended Working Group on Ageing (2018).

[12]Open-ended Working Group on Ageing (2019a).

[13]Open-ended Working Group on Ageing (2019b).

The chief advantage of a treaty is that it would create international legal obligations for states to respect, protect and fulfil the rights contained within it. While enforcement of human rights treaties has always been a challenge, this could be assisted by the establishment of national and international supervisory, reporting and complaints processes. Furthermore, recognising older persons' human rights in a dedicated convention would signify the level of importance which the international community attaches to these issues, making it clear that addressing older people's rights is not an optional approach or aspirational target—it is a legally enshrined duty. States which sign up to the new treaty can then be held accountable for delivering on that undertaking. The adoption of a CROP would also provide impetus for increased education and understanding of the human rights of older people across all sectors, helping to improve the quality of service and care and helping to combat the effects of ageism in society more broadly.

The analysis presented in this book can supplement the work of the OEWGA in clarifying how particular issues ought to be addressed within a human rights treaty. Some of these areas will be discussed below in relation to select recommendations, but in general terms a new instrument needs to:

i. Recognise that older persons are entitled to all human rights on an equal basis with all other people, and that those rights are interdependent and indivisible. This would also recognise that ageing is an ongoing and universal experience throughout the entire life course and, rather than segregate older people into a distinct category with different rights, the treaty could reinforce the global commitment to protecting and promoting the human rights of all people, regardless of their age.
ii. Clarify how particular human rights apply in older age, addressing the most relevant issues (including capacity, ageism, participation, elder abuse and aged care). Importantly, a new instrument would not in fact be creating any new obligations for states. Instead, just like the *Convention on the Rights of Persons with Disabilities*, a CROP would clarify and unpack how existing human rights apply to older people. States ought therefore to be encouraged to adopt the treaty on the basis that their substantive obligations will not change or expand, but they will receive greater clarification in how to ensure those obligations are met.
iii. Articulate of the nature of states' obligations with respect to those rights, including in regulating the actions of private entities. The standard of obligation could allow for some form of progressive realisation for some rights where this is appropriate, for example rights to the highest attainable standard of health and social security, having regard to what is required of states to fulfil the right and their individual circumstances. At the same time, the treaty should specify that for other rights, for example freedom from cruel, inhuman or degrading treatment, states are expected to take the necessary steps without delay.
iv. Recognise the importance of international cooperation in fulfilling the human rights of older people, particularly for developing countries relying on progressive realisation.

v. Provide for the establishment of national bodies tasked with implementing the new convention, monitoring compliance, gathering data and resolving complaints of non-compliance.
vi. Establish an independent international body, with a mandate for both monitoring states' compliance and hearing complaints from individuals who allege their rights have been breached. Following the model of other human rights treaties, the international complaints process could be made contingent on the exhaustion of local remedies such that a complaint to the international body would only be open where no meaningful resolution can be achieved at the domestic level. This would not only facilitate the creation of enforcement procedures that are tailored to the context of each national jurisdiction, but would also help manage the workload of the international body, and achieve timely and cost-effective dispute resolution. Perhaps most importantly, this approach encourages states to implement meaningful domestic pathways for older persons to enforce their rights, thus enhancing access to justice.
vii. In creating an independent body to enforce the obligations contained in the new treaty it will be important to clarify issues of standing to ensure access to justice for older persons whose capacity may be impaired or lost. For example, the treaty body could grant standing to guardians or other representatives existing within national law to bring complaints on behalf of an individual with impaired or lost capacity. In the absence of a national representative, an international guardian or other similar office could be created to ensure all older individuals can approach the treaty body when their rights have been violated.

11.3.1.2 Strengthen Domestic Human Rights Legislation

In addition to the substantive recommendations presented in this book for reforms to select areas of elder law, general human rights protections must also be strengthened in domestic legislation, especially in countries which do not already have strong human rights laws in the form of a charter or bill of rights. The standards found in international human rights must be embedded into domestic law, accompanied by meaningful complaints mechanisms, allowing for greater enforcement of those rights by older persons.

Human rights legislation should include provisions for causes of action and standing that enable an individual to seek relief in the event of a breach of their human rights. This could take the form of a complaints pathway to a dedicated body, like a Human Rights Commission, or by empowering courts to hear human rights-based claims. An independent cause of action or complaints mechanism for human rights breaches could help to overcome barriers to justice which may exist within other, more specialised remedial processes (such as anti-discrimination legislation or equity), as well as ensuring that all incidents of human rights breaches are adjudged according to consistent principles and processes. Ideally such a claim would be available against both state and non-state actors—an important feature given that in many jurisdictions financial, accommodation, health and aged care services are delivered by private entities. Claims should also be possible to challenge law or policy on human rights grounds, even in the absence of an individual claimant who has suffered harm. A constitutionally-enshrined bill of rights would be preferable in this respect, as it would be most likely to lead to invalidation of laws which are incompatible with human rights.

Genuine commitment to human rights at the domestic level ought also to promote a stronger human rights culture within society and, in turn, this should support transition towards more rights-focused, less ageist, attitudes. Furthermore, enhancing and implementing domestic law as a part of adopting a human rights-based approach could serve as an important interim step towards the enactment of a dedicated covenant. Once states are committed to a human rights-based approach and have integrated it into their laws and policies, the move to a dedicated treaty at the international level ought to be seen as a less onerous undertaking.

An important contribution of a human rights-based approach is that it addresses issues throughout the life course which can manifest as vulnerabilities or challenges in older age. While this book has examined a number of specific legal areas that older people encounter most commonly, many of the human rights issues that have been identified can in fact be traced back much earlier in their lives. Vulnerability in relation to financial insecurity, health, decision-making, safe housing, education, social connectedness and employment is frequently the product of various factors which begin early in life and build up over time. This points to the need for a human rights approach which is broader than just 'elder law', and which meaningfully supports the rights of people of all ages. By protecting and promoting human rights to education, employment, social security, health and accommodation across the life course through strong domestic human rights laws we can minimise vulnerability in older age.

11.3.2 Address Ageism

As noted above, a recurring theme in our analysis was the significant negative impacts of ageism on the human rights of older persons. We have identified a two-way relationship between human rights and ageism. Ageism is fundamentally a violation of human rights, not only of the right to equality before the law and to be free from discrimination, but also in the sense that ageism operates as a barrier to the full enjoyment of all human rights. A meaningful commitment to human rights has the potential to combat ageism, however, by helping to identify areas where older people are vulnerable to discrimination and, perhaps most importantly, by emphasising the inherent dignity, autonomy and worth of each older person, thereby combatting the negative stereotypes upon which ageist attitudes are based. We have therefore argued that a human rights-based approach to elder law will contribute to a paradigm shift in the fundamental way we think about older persons but also about ageing, with the ultimate objective of eliminating ageism and resultant discriminatory practices.

However, waiting for human rights to bring about this change is not our only option—more concrete, short-term reforms can be implemented to address ageism (and in turn to further bolster respect for human rights).

An obvious first step is to review all anti-discrimination legislation to ensure that age is adequately covered, along with other heads of discrimination such as sex, race or religion. The spheres in which discrimination is prohibited also need to be broad enough to cover the main problem areas where older people experience ageism, including not only employment but also education, health and aged care, and housing. Governments and other policy-makers should also commit to meaningful participation of older people in the design of policies and processes, both for addressing ageism specifically and for other areas of elder law. Not only is participation essential to respecting older people's experiences and opinions, it is also key to developing strategies that will actually work. For example, including older people in the design of housing and urban spaces will ensure that facilities and infrastructure can truly meet their needs and preferences, without making assumptions about what they would want to see or what would be 'good for them'.[14] With respect to ageism, older persons' voices must be included in the development and implementation of laws and policies in order to genuinely understand and begin to address the repercussions of ageism. Participation also needs to be ensured at the individual level. In many cases, ageist views are given space to flourish when older individuals are excluded from decisions that affect them. Ensuring meaningful consultation, supported where necessary, enables older people to express their views and have them heard, which can help correct ageist assumptions.

Policy approaches should also move away from narratives of 'successful ageing', 'productive ageing' and 'active ageing', as discussed in Chap. 4. Many of these are based on assumptions that are essentially ageist, for example, that the older population is an economic burden which needs to be minimised by encouraging older people to stay in the workforce for longer. Even where not originally based on ageist ideas, these narratives can reinforce ageist views by seemingly valuing older persons' contributions to society more than their intrinsic worth, dignity and autonomy. Under such an approach, an older person who is not 'productive' or 'successful' can easily be ignored, disrespected or be at a heightened risk of human rights abuse. Narratives of 'healthy ageing', discussed in Chap. 10, can be a more useful framing for policy as they are more closely connected to individual human rights, but even here we must be careful to recognise that health is a product of numerous, interconnected factors throughout the entire life course, some of which are structural or systemic and may themselves be influenced by ageist ideas.

Given that ageism is a socially constructed phenomenon, steps need to be taken, however, to shift attitudes and combat negative stereotypes if policy reforms are to endure. One recommendation to achieve this is to promote authentic intergenerational interaction. There are numerous ways that this can be achieved, and communities ought to be encouraged to devise creative opportunities for older and younger generations to integrate. One promising option is through more integrated education

[14]For example, see: Harding (2006), 3; Adams (2009), 88.

opportunities which encourage older people to participate and share their knowledge, while also supporting their learning needs, as discussed in Chap. 4. Not only does this benefit the older individual in terms of developing skills and facilitating social inclusion, it also benefits younger generations who are able to learn from older people's experiences and knowledge.[15]

Design of physical spaces can also help to promote greater inclusion and integration, as exemplified by the "8 80 Cities" initiative.[16] By designing cities, urban spaces and buildings in ways that are accessible to people of all ages, and supporting the mobility of older people and those with disability, barriers to inclusion can be removed. Over time it is hoped that greater intergenerational interaction will break down ageist stereotypes, as well as promoting human rights more generally. It may also begin to address abusive attitudes and behaviours demonstrated towards older people, for example, the 'inheritance expectation' displayed by some members of younger generations wherein they believe they are entitled to the assets of older people (discussed in Chap. 7).

11.3.3 Improve Access to Justice

One of the framework principles of a human rights-based approach is access to justice. Human rights law obliges governments to ensure that any person whose rights have been violated has access to a suitable remedy, in keeping with the broader rights of equality before the law and the right to a fair trial.[17] It is recognised that older people often experience difficulties accessing justice, for example because of financial, geographical, cultural, gender, health or other factors, and especially where a number of these factors intersect. This creates potential for further human rights violations, for example where elder abuse is allowed to continue, where appropriate redress is not given, or where legal rights relating to health or housing are not able to be enforced.

A key recommendation for a human rights-based approach to elder law is therefore to improve access to justice for older persons. There are three aspects to this:

i. ensure human rights protections are accompanied by appropriate enforcement processes and remedial relief, including against state and non-state actors;
ii. ensure diverse dispute resolution methods, appropriate causes of action and rules of standing are accessible and affordable within specific fields relevant to elder law to enable older people to obtain a suitable remedy; and
iii. address social, economic, physical and/or other barriers to justice so that older people can have meaningful access to the legal and regulatory processes listed

---

15 *Madrid International Plan of Action on Ageing* (2002), Objective 2 (para 42).

16 8 80 Cities (n.d.).

17 ICCPR, arts 2, 14, 26.

above, including education and training programs for older people and all professionals who work with them to ensure an adequate understanding of relevant human rights and how to best protect them.

The need for enforcement processes within international and domestic human rights law was discussed above, but some brief comments will now be made in relation to the other two dimensions.

11.3.3.1 Availability of Remedies

Legally enforceable human rights are necessary for the reasons outlined above but, in order to fully ensure access to justice, the substantive areas contributing to 'elder law' must all be supported by appropriate enforcement processes through which remedial relief is available. This involves ensuring that causes of action exist which can be actioned by an affected individual and that they are accompanied by appropriate remedies. Some suggestions for further development are set out below.

Equitable remedies should be further explored as vehicles through which to protect the rights of older persons, notably fiduciary duties, unconscionable conduct and undue influence, as well as estoppel. Fiduciary obligations may arise in a setting where an older person is in a relationship of trust with other individuals or agencies, and where that other party has agreed to act in the older person's interests even to the extent of placing that person's interests above their own if there is a conflict. Breaches of fiduciary obligations have a significant role to play in, for example, financial elder abuse arising as a result of the erroneous use of substitute decision-making power under an enduring power of attorney. In fact, discussion throughout the book often unintentionally lends itself to fiduciary-based concepts, suggesting the potential for equitable remedies to play a greater role in supporting human rights.

Unconscionable conduct and undue influence are also frequently present in elder abuse cases where the defendant has (allegedly) engaged in conduct that is either taking advantage of a weaker party (unconscionable conduct) or overbearing their will (undue influence). Similarly, equitable estoppel is concerned with addressing unconscionable behaviour where a defendant has made a promise to another person that they have relied upon to their detriment. Causes of action pleading unconscionable conduct, undue influence and/or estoppel are frequently occurring in, for example, assets for care cases (discussed in Chap. 8) wherein younger people are failing to perform promises to care for older people in return for some financial gain (often a transfer of the older person's house).

As with any legal action, however, equitable remedies can be difficult to access. They are both legally complex and, consequently, potentially unaffordable given the need (generally) to engage legal representation. This burden can be particularly challenging where the older person may have experienced, for example, financial abuse thus depleting the income and/or wealth security they need to access equitable relief. Therefore, equitable causes of action should be the subject of further research in order to not only assess their efficacy in protecting the rights of older persons throughout 'elder' law more broadly, but in also exploring ways in which to make accessing them more cost effective and readily available.

Civil actions and criminal offences for elder abuse should be established or expanded, coupled with appropriate remedies which are capable of achieving meaningful justice for the older person. Alternative dispute resolution processes, for example elder mediation, should be also developed. This could be particularly fruitful in certain settings where disputes with service provides can arise (for example, retirement villages, aged care, healthcare) and where there needs to be an ongoing relationship between the disputing parties, for example in familial settings.

Where necessary, rules of standing also need to be examined to ensure that they adequately enable older people to commence a claim, or allow another person to claim on their behalf, for instance where the older individual lacks or has impaired legal capacity. As noted above, if an international treaty body is established through the work of the OEWGA then standing issues need to be addressed at both the national and international level to ensure all individuals have adequate access to all processes designed to protect their rights.

11.3.3.2 Removing Barriers to Justice

Even where the legal infrastructure exists to support older people in upholding their rights, there are many other reasons why they may be unable to access justice. It is essential that these barriers are addressed. This is a complex area and nuanced responses will need to be developed in each jurisdiction, but in general we recommend:

i. Providing adequate information and education about relevant human rights and what they require. This should be provided for both service providers, to ensure they understand their obligations, but also for older persons, their families, carers and support workers (including, for example, community legal centres, health practitioners etc.) so that they are able to identify when rights have been violated and are equipped with the necessary language and strategies to advocate for change.
ii. Adequately funding community legal centres and other advocates to be able to support older persons whose rights have been violated. This is necessary to ensure ready and affordable access to legal and relational support. Such funding must be available to pursue the full range of remedies available, including not just human rights claims but other legal and equitable causes of action.
iii. Peer to peer programs wherein older people can approach their peers within their local community if they feel that their rights are being violated or if they have any questions.
iv. Including training about human rights in accreditation mechanisms for financial, health and aged care services, particularly in relation to impaired and lost capacity.

v. Establishing informal and formal support mechanisms to assist with accessibility, be they physical, emotional, educational or cultural. Significantly in the case of elder abuse, this can also include the provision of services such as housing if the older person lives with their abuser.

11.3.4 Elder Abuse

As discussed in Chap. 7, elder abuse is a significant threat to human rights which is prevalent across many jurisdictions. It must therefore be addressed in all its forms if the human rights of older people are to be fully respected and promoted. Chapter 7 outlined a number of recommendations in more detail, but these can be summarised here:

i. Reporting guidelines need to be developed to ensure that elder abuse can be identified and adequately responded to. Critical to this is determining when reporting should be mandated and when it should merely be encouraged. For example, mandatory reporting might be appropriate where the older person has lost capacity, provided the person required to report (usually a professional in the legal, health or financial sector) has adequate training about how to identify abuse and is supported through the reporting process. Mandatory reporting might not necessarily be appropriate for other situations where there may be other negative consequences, for instance impacts on privacy and/or family relationships. Nuanced approaches are therefore required to ensure meaningful consideration of all human rights impacts.
ii. Further research is required to fully understand the prevalence and causes of elder abuse in order to develop appropriate strategies for prevention. This includes research into violence against older women or other cohorts, and evaluation of strategies for responding to this.
iii. Legal, health, financial and other professions could implement professional standards and/or training as well as accreditation programs (where appropriate) to facilitate the effective identification and prevention of abuse and implement appropriate responses.[18] Areas for training and associated standards could include being aware of the signs of elder abuse, understanding the situation as a potential form of abuse, determining whether action is within the scope of the professional's responsibility, knowing strategies for action and then deciding whether to act in the specific circumstances, including responding to any mandatory reporting requirements as discussed above.
iv. Given the links between elder abuse and impaired or lost capacity, guidelines need to be developed to ensure capacity is accurately assessed so that vulnerability can be identified. Best practice guidelines for assessing capacity are discussed in more detail in the next section.

---

[18] Ries (2018), 29. On screening measures for health professionals see, for example: Ries and Mansfield (2018).

11.3.5 Capacity

As noted above, issues of capacity cut across the various areas of elder law discussed in this book. Given that incorrect assessments of capacity can contribute to various violations of human rights, and that respect for autonomy is a core part of a human rights-based approach, it is essential that law and policy supports accurate assessments of capacity. Further, these assessment processes need to adhere to human rights standards, including respecting older individuals' dignity, privacy and liberty to the greatest extent possible, as well as their ability to participate in and be informed about the assessment process. More detailed recommendations are included in Chap. 6, but these can be summarised here:

i. In keeping with respect for autonomy and dignity, the law ought to adopt a presumption of capacity unless a loss/impairment of capacity can be proven.
ii. Context specific best practice guidelines should be developed to improve assessment processes bearing in mind the audience, that is, whether the guidelines are for legally trained assessors, those in the health or allied health disciplines, the financial industry or lay people. These guidelines should include, for instance: private face-to-face meetings with the client asking open-ended questions; carefully documenting the instructions as well as the scope of the retainer; refusing to act for more than one client, noting the relevant legal, professional and ethical rules; suggesting independent legal advice where appropriate; considering issues in relation to diminished or lost capacity, including the necessity of an assessment, who should be involved and when they should become involved; and considering speaking to health and allied health professionals, particularly where the client is in aged care or a hospital setting.[19]
iii. Individuals should be provided with appropriate supports and education about the process throughout to maximise their participation in the assessment and to produce an 'optimal' outcome. Optimal here does not necessarily mean a finding of capacity, instead referring to the process and the 'correct' capacity determination—be it one of capacity or incapacity.
iv. Avenues of review need to be available so that an older individual or their family/carer can challenge an assessment which they justifiably believe to be incorrect.
v. The law ought to facilitate supported decision-making wherever possible, as this helps to maximise an older individual's participation in decisions which affect them. However, substitute decision-making should also be available when absolutely necessary *provided* that it is accompanied by strong human rights principles to ensure that the decision-maker complies with relevant protections and standards. This is necessary to recognise the vulnerability which exists

[19]Cockburn and Hamilton (2009); Hamilton and Cockburn (2008); Purser et al. (2020); Lonie and Purser (2017); Purser and Rosenfeld (2014).

where a person lacks capacity and the potential for both a supported and substitute decision-making arrangement to be exploited or inadvertently used in ways which are not consistent with human rights.

vi. Education and training need to be implemented for all professions involved with capacity, particularly those in the legal, health, allied health and financial as well as banking industries.

vii. Education should also occur for lay people about what capacity is, its significance, what happens when a person experiences diminished or lost capacity, and what avenues are available to facilitate decision-making after the loss of capacity.

11.3.6 Housing and Urban Design

Repeatedly throughout this book issues of place and physical environments have emerged as factors which influence the enjoyment of human rights. This is most noticeable in relation to accommodation and aged care (discussed in Chaps. 9 and 10) but it also connects to older persons' experiences of social inclusion, vulnerability, cultural safety and sense of place. Select recommendations are given here which could be implemented to help improve protections of human rights in relation to physical spaces which older persons experience.

i. As noted above, and discussed in detail in Chap. 4, participation is key to ensuring that design of accommodation, public buildings and urban spaces suits the needs and preferences of older users. Laws and policies ought to support maximum consultation with older persons, not only in the design of physical spaces, but also in the design of those laws and policies themselves.

ii. In relation to housing, the emerging trend of 'ageing in place' was noted in Chap. 9. Supporting older people to remain in their homes for as long as they wish is a positive way of respecting their rights and wishes, but is contingent on appropriate supports being accessible when required. This includes understanding the different experiences of older people who rent rather than own their own homes. Reforms to tenancy laws and increased funding should support older tenants to make necessary modifications to facilitate ageing in place.[20]

iii. Chap. 9 also identified the often-complex nature of legal and financial arrangements surrounding retirement villages and other accommodation options for older persons. This complexity can mean that older people are unable to negotiate terms which suit their personal needs, and the financial investment involved can create further vulnerabilities should they find themselves needing to move to a higher level of care—provided they have the required financial security to enter the particular accommodation option in the first place. The law should ensure that these arrangements are flexible enough to cater to individuals' needs, and that their rights are protected in the event of any dispute or other issue arising.

[20] Stewart and Stein (2019); Power (2016).

11.3.7 Financial Security

As was noted in Chap. 7, financial security underpins the enjoyment of a wide range of human rights, while poverty operates as a significant risk factor for human rights abuse. Essential to a human rights-based approach to elder law is ensuring that all older people have access to adequate social protection, including through the provision of non-contributory pension schemes for those who lack independent means. Concepts like universal basic income or the social protection floor offer promising developments in this regard and warrant further research.

Education is, again, also a significant factor. People need to be educated throughout the entire life course as to the significance of financial security post-paid employment. Ensuring adequate financial literacy is particularly important for women and other groups who are more vulnerable to poverty-related human rights impacts in later life. In addition to education, appropriate processes should be available to identify vulnerability and support those who are experiencing or at risk of financial insecurity.

A particular question which warrants greater research is how the concept of dignity of risk ought to apply to financial and other legal matters in the context of a human rights-based approach. As noted above, the dignity of risk has emerged as an important principle in the health and aged care sectors, where it is increasingly recognised that the ability to make one's own decisions, even where those decisions might be considered poor ones, is an important exercise of autonomy and preferable to overly-paternalistic approaches which limit free decision-making.[21] Similar work has been lacking in other areas of elder law, however. More careful thought is required, for instance, on how the dignity of risk can be embedded within a human rights-based approach to financial and estate planning, and what reforms may be required to the legal frameworks which support decision-making. More broadly, further work on the interaction between dignity of risk and human rights would be beneficial for identifying appropriate limits on individual decision-making where risks to personal safety or well-being are present.

11.4 Conclusion

There are clearly a number of ways in which a human rights-based approach to elder law could be better embedded. This chapter has added to the existing literature by noting just some of the actions which could be taken to improve the protection of human rights in a unique combination of key areas of elder law, particularly in relation to the cross-cutting issues of ageism, capacity and elder abuse.

A human rights-based approach to elder law offers the potential to reconceptualise our responsibilities towards older generations and reposition older persons as active rights-holders, not merely passive beneficiaries of care or a 'burden on society'. This shift in thinking represents a major change in the way that aged care and elder law have

---

[21] Mitchell (1994); Millar (1998).

traditionally been approached within our society. A human rights-based approach has the potential not only to introduce minimum standards within the various substantive areas of elder law, but, if approached systematically and supported by an appropriate legal architecture, could positively influence multidisciplinary approaches to ageing and lead to genuine, practical improvement. It may also lead to an attitudinal shift within societies more broadly and help to combat the ageism which has become all too pervasive.

As noted above, a human rights-based approach to elder law is best achieved within a framework of genuine commitment to human rights across the whole life course. Many, though clearly not all, of the reasons that older people are vulnerable to human rights violations can be linked to factors which emerged earlier in their lives, such as financial insecurity, lower levels of education, social marginalisation or untreated health issues. While the particular areas of elder law justifiably warrant attention in line with the recommendations above, achieving a society which truly respects and promotes the human rights of older people must inevitably involve respecting the rights and dignity of people of all ages.

References

8 80 Cities (n.d.) Creating Cities for All. https://www.880cities.org/. Accessed 17 February 2020

Adams S (2009) No place like home? Housing inequality in later life. In: Cann P, Dean M (eds) Unequal ageing: the untold story of exclusion in old age. Policy Press, University Press Scholarship Online, chapter 4

Adelman RD, Greene MG, Ory MG (2000) Communication between older patients and their physicians. Clin Geriatr Med 16(1):1–24

Barbosa Neves B et al (2019) 'It's the worst bloody feeling in the world': experiences of loneliness and social isolation among older people living in care homes. J Aging Stud 49:74–84

Cockburn T, Hamilton B (2009) Assessment of capacity: disciplinary issues and potential liability. Proctor (March): 15

Hamilton B, Cockburn T (2008) Capacity to make a will and enduring power of attorney: issues new and old. Queensland law Soc J (December): 14

Harding B (2006) Age equality in housing: a guide to tackling age discrimination for housing providers, commissioners, planners and builders. Housing and older people development group. https://hoop.eac.org.uk/downloads/kbase/Age-Equality-LeafletOld.pdf. Accessed 10 Feb 2020

Ibrahim JE, Davis M (2013) impediments to applying the 'dignity of risk' principle in residential aged care services. Australas J Ageing 32(3):188–193

Lonie J, Purser K (2017) Assessing testamentary capacity from the medical perspective. Aust Bar Rev 44(3):297–317

Mégret F (2011) The human rights of older persons: a growing challenge. Hum Rights Law Rev 11(1):33–66

Millar B (1998) The dignity of risk and the right to failure: a lesson for patient-focused care. J Clin Nurs 7(4):295–296

Mitchell G (1994) The dignity of risk and the right to failure: one profile of patient-focused care. Perspectives 18(3):10

Morgan S (2004) Positive risk-taking: an idea whose time has come. Health Care Risk Rep 10(10):18–19

Nay R (2002) The dignity of risk. Aust Nurs J 9(9):33

Open-ended Working Group on Ageing (2017) Report of the open-ended working group on ageing on its eighth working session. UN Doc A/AC.278/2017/2
Open-ended Working Group on Ageing (2018) Report of the open-ended working group on ageing on its ninth working session. UN Doc A/AC.278/2018/2
Open-ended Working Group on Ageing (2019a) Report of the open-ended working group on ageing on its 10th working session. UN Doc A/AC.278/2019/2
Open-ended Working Group on Ageing (2019b) overview of 11th session. https://social.un.org/ageing-working-group/eleventhsession.shtml. Accessed 15 Jan 2020
Power E (2016) Older women and home. transforming cities for care (13 Sept 2016) https://dremmapower.com/older-women-and-home/. Accessed 17 Feb 2020
Purser K, Cockburn T, Ulrick E (2020) Examining access to formal justice mechanisms for vulnerable older people in the context of enduring powers of attorney. Elder Law Review 12(2)
Purser K, Rosenfeld T (2014) Evaluation of legal capacity by doctors and lawyers: the need for collaborative assessment. Med J Aust 201(8):483–485
Ries NM (2018) Elder abuse and lawyers' ethical responsibilities: incorporating screening into practice. Legal Ethics 21(1):23–45
Ries NM, Mansfield E (2018) Elder abuse: the role of general practitioners in community-based screening and multidisciplinary action. Aust J Gen Pract 47(4):235–238
Stewart E, Stein L (2019) One whole pension goes on helen and peter's rent. this is australia's rental crisis. abc news (28 August 2019). https://www.abc.net.au/news/2019-08-28/what-can-be-done-about-rental-stress/11452486. Accessed 19 Sept 2019
Wyman MF, Shiovitz-Ezra S, Bengel J (2018) Ageism in the health care system: providers, patients, and systems. Contemporary perspectives on ageism. In: Ayalon L, Tesch-Römer C (eds) International perspectives on aging. Springer International Publishing, Cham, p 202

International Instruments

International Covenant on Civil and Political Rights (1966) Opened for signature 16 December 1966, entered into force 23 March 1976. 999 UNTS 171
International Covenant on Economic, Social and Cultural Rights (1966) Opened for signature 16 December 1966, entered into force 3 January 1976. 993 UNTS 3
Madrid International Plan of Action on Ageing (2002) Adopted by Second World Conference on Ageing, Madrid, 8–12 April 2002
United Nations General Assembly (2013) *Towards a Comprehensive and Integral International Legal Instrument to Promote and Protect the Rights and Dignity of Older Persons*. Resolution 67/139. UN Doc A/RES/67/139 (13 February 2013)

The manufacturer's authorised representative in the EU is Springer Nature Customer Service Centre GmbH, Europaplatz 3, 69115 Heidelberg, Germany. If you have any concerns regarding our products, please contact ProductSafety@springernature.com

Printed and bound by CPI Group (UK) Ltd, Croydon, CR0 4YY
30/09/2026
02225559-0016